METHODS IN MOLECULAR BIOLOGY

Series Editor
John M. Walker
School of Life and Medical Sciences
University of Hertfordshire
Hatfield, Hertfordshire, AL10 9AB, UK

For further volumes:
http://www.springer.com/series/7651

Cell Microencapsulation

Methods and Protocols

Edited by

Emmanuel C. Opara

Wake Forest Institute for Regenerative Medicine, Wake Forest School of Medicine, Winston-Salem, NC, USA; Virginia Tech-Wake Forest School of Biomedical, Engineering & Sciences (SBES), Wake Forest School of Medicine, Winston-Salem, NC, USA

Humana Press

Editor
Emmanuel C. Opara
Wake Forest Institute for Regenerative Medicine
Wake Forest School of Medicine
Winston-Salem, NC, USA

Virginia Tech-Wake Forest School of Biomedical
Engineering & Sciences (SBES)
Wake Forest School of Medicine
Winston-Salem, NC, USA

ISSN 1064-3745 ISSN 1940-6029 (electronic)
Methods in Molecular Biology
ISBN 978-1-4939-6362-1 ISBN 978-1-4939-6364-5 (eBook)
DOI 10.1007/978-1-4939-6364-5

Library of Congress Control Number: 2016951446

Printed on acid-free paper

This Humana Press imprint is published by Springer Nature
The registered company is Springer Science+Business Media LLC New York

Preface

The technique of microencapsulation was introduced approximately eight decades ago, and its potential application in cell therapy was described almost 50 years later. The technique offers a wide range of applications in both industry and medicine. Cell microencapsulation comes under the broader field of bioencapsulation and in most cases involves the immobilization of therapeutic cells in polymeric scaffolds or semipermeable hydrogel capsules that provide the cells with a favorable protective environment allowing the exchange of nutrients and oxygen and protecting them from the host's immune system by blocking the entry of antibodies and cytotoxic immune cells. Thus, the technology has been widely used in various efforts to develop transplantable bioartificial organs without the need to immunosuppress transplant recipients, to expand donor cell sources, and in cancer cell therapy. More recently, this technique has received attention for applications in stem cell therapy, and the encapsulation of bacterial cells and other cell types for drug delivery. In stem cell therapy, cell microencapsulation has the potential to be used for the restriction of migratory mesenchymal stromal cells within a defective target tissue/organ to induce tissue repair.

The purpose of this book is to provide a unique forum to review the promise of cell microencapsulation in a broad sense that encompasses various cell types that have been encapsulated for different purposes, the different approaches/devices used for microencapsulation, the biomaterials used in cell microencapsulation, the challenges to the technology, and the current status of its application in different clinical situations. To achieve this purpose, the book is divided into five parts. Part I is the introductory part of the book that discusses historical developments of the technology and the current challenges facing it, as well as the various applications of cell microencapsulation. Part II discusses the main approaches and devices currently used in cell microencapsulation while Part III presents an overview of the various polymeric materials currently in use for cell microencapsulation, procedures for assessment of the quality of microcapsules used for cell encapsulation, and the enabling technologies to either monitor or enhance encapsulated cell function. In Part IV, specific examples of the methods used to encapsulate various cell types are discussed while Part V provides an overview of the different clinical situations in which cell microencapsulation has been applied.

Cell Microencapsulation is intended to be a reference handbook for researchers, engineers, clinicians, and other health-care professionals, as well as food technologists, who will find in the book detailed descriptions of methods for the microencapsulation of specific cell types and their current or potential clinical and industrial applications. The book also includes detailed information about the design and manufacture of different devices including large-scale production devices for use in cell microencapsulation. Furthermore, individuals or family members of individuals afflicted with certain diseases for which cell microencapsulation has potential application will find the book to be a good source of information as they research new therapeutic options in cell therapy for those disease states. In addition, entrepreneurs/investors and the scientists that work with them in Biotechnology will find this book to be a great resource for assessment of the promise of the cell microencapsulation technology to enable them make informed decisions about where to make rewarding investments.

I would like to express my sincere gratitude to certain individuals with whom I have interacted personally since I have been working on the microencapsulation technology. First, I would like to thank Dr. Gerrit H. J. Wolters, formerly of the University of Groningen, Groningen, The Netherlands, who provided me with a copy of his published article on the instrumentation for the air-syringe-pump droplet generator that enabled my research team at the Duke University Medical Center to build our first apparatus for cell microencapsulation in 1995. The key members of that research team who worked with the gifted engineers and technicians at the Duke Instrument Shop to build that first device were Marc R. Garfinkel, M.D., and Robert C. Harland, M.D., to whom I am greatly indebted. The contributions of some of my former students and fellows at Duke, most notably, William F. Kendall Jr., M.D., and Marcus D. Darrabie, M.D., to the development and perfection of the technique in my lab at Duke deserve my eternal gratitude. I am also indebted to my long-term collaborator, Dr. M.K. Ramasubramanian, the D.W. Reynolds Distinguished Professor and Department Chair, Department of Mechanical Engineering at Clemson University in South Carolina, USA, and two former students, Dr. Sameer Tendulkar and Dr. John Patrick McQuilling, who worked with us on the first generation of our microfluidic devices for cell microencapsulation. In addition, I am deeply grateful to Dr. Joel Stitzel, Professor and Chair of Biomedical Engineering at the Wake Forest School of Medicine, for offering me the opportunity to serve as the Graduate Program Director at the Wake Forest campus of the joint Virginia Tech-Wake Forest School of Biomedical Engineering & Sciences (SBES), and for providing me what has turned out to be the best working relationship experience of my entire academic career. In addition, I appreciate immensely the moral support of some friends, including Engr. and Mrs. Ben Akah, Dr. and. Mrs. Nnaemeka J. Ojukwu, Dr. and. Mrs. Obiora Ogbuawa, and Rev. Dr. Donatus N. Nwachukwu, as it has been a continuous source of inspiration in my career.

Finally, I would like to acknowledge the great patience, personal sacrifice, and unqualified support of my lovely wife, Clarice, and our children, Ogechi, Chiedu, Chucky, and Ike, in my career. I am grateful to each and every one of them for giving me the luxury of long periods of absence from home as I vigorously pursued my research efforts. For their support, I would like to dedicate this book to all my family and my lifelong mentor, Vay Liang W. Go, M.D., currently Distinguished University Professor of Medicine and Director, UCLA Center for Excellence in Pancreatic Diseases, who gave me the opportunity to work with him at the National Institutes of Health, Bethesda, Maryland, where I started my research in islet biology that set the stage for the development of my interest and expertise in cell microencapsulation.

Winston-Salem, NC *Emmanuel C. Opara*

Contents

Part IV Isolation of Cells for Microencapsulation

Part V Cell Microencapsulation in Clinical Applications

Contributors

A. AIJAZ • *Department of Biomedical Engineering, Rutgers University, Piscataway, NJ, USA*
BRAD P. BARNETT • *Wilmer Ophthalmological Institute, Johns Hopkins Hospital, Baltimore, MD, USA*
GIUSEPPE BASTA • *Laboratory for Endocrine Cell Transplants and Biohybrid Organs, Department of Medicine, Section of Cardiovascular, Endocrine and Metabolic Clinical Physiology, University of Perugia, Perugia, Italy*
A. BIDORET • *Oniris, UMR CNRS 6144 GEPEA, Nantes, France*
B.E. BINTZ • *Cytosolv, Inc., Providence, RI, USA; Brown University, Providence, RI, USA*
BARBARA D. BOYAN • *School of Engineering, Virginia Commonwealth University, Richmond, VA, USA; Wallace H. Coulter Department of Biomedical Engineering, Georgia Institute of Technology, Georgia Tech and Emory University, Atlanta, GA, USA*
ERIC M. BREY • *Department of Biomedical Engineering, Illinois Institute of Technology, Chicago, IL, USA; Research Service, Hines Veterans Administration Hospital, Hines, IL, USA*
JEFF W.M. BULTE • *Division of MR Research, Russell H. Morgan Department of Radiology and Radiological Science, The Johns Hopkins University School of Medicine, Baltimore, MD, USA; Cellular Imaging Section and Vascular Biology Program, Institute for Cell Engineering, The John Hopkins University School of Medicine, Baltimore, MD, USA; Department of Biomedical Engineering, The Johns Hopkins University School of Medicine, Baltimore, MD, USA; Department of Chemical and Biomolecular Engineering, The Johns Hopkins Whiting School of Engineering, Baltimore, MD, USA; Department of Oncology, The Johns Hopkins University School of Medicine, Baltimore, MD, USA*
RICCARDO CALAFIORE • *Laboratory for Endocrine Cell Transplants and Biohybrid Organs, Department of Medicine, Section of Cardiovascular, Endocrine and metabolic Clinical Physiology, University of Perugia, Perugia, Italy; Lions International Diabetes Research Center, Terni, Italy*
J. CIRIZA • *NanoBioCel Group, Laboratory of Pharmaceutics, School of Pharmacy, University of the Basque Country (UPV/EHU), Vitoria-Gasteiz, Spain; Networking Center in Bioengineering, Biomaterials and Nanomedicine (CIBER-BBN), Vitoria-Gasteiz, Spain*
CLANCY J. CLARK • *Department of Surgery, Wake Forest School of Medicine, Winston-Salem, NC, USA*
PABLO CAVIEDES • *Molecular & Clinical Pharmacology Program, ICBM, Faculty of Medicine, University of Chile, Independencia, Santiago, Chile*
MARCUS D. DARABBIE • *Department of Surgery, Duke University Medical Center, Durham, NC, USA*
BART DE HAAN • *Division of Immuno-Endocrinology, Departments of Pathology and Laboratory Medicine, University of Groningen, Groningen, The Netherlands*
PAUL DE VOS • *Division of Immuno-Endocrinology, Departments of Pathology and Laboratory Medicine, University of Groningen, Groningen, The Netherlands*

D.F. EMERICH • *NS Gene, Providence, RI, USA*
KEVIN ENCK • *Wake Forest Institute for Regenerative Medicine, Wake Forest School for Medicine, Winston-Salem, NC, USA; Virginia Tech-Wake Forest School of Biomedical Engineering and Science (SBES), Wake Forest School of Medicine, Winston-Salem, NC, USA*
ALAN C. FARNEY • *Institute for Regenerative Medicine, Wake Forest School of Medicine, Winston-Salem, NC, USA; Department of Surgery, Wake Forest School of Medicine, NC, USA*
CLARENCE E. FOSTER III • *Department of Surgery, University of California Irvine, Orange, CA, USA; Department of Transplantation, University of California Irvine, Orange, CA, USA*
SHIN-HUEI FU • *Department and Graduate Institute of Microbiology and Immunology, National Defense Medical Center, Taipei, Taiwan*
SUBROTA HATI • *Dairy Microbiology Department, Anand Agricultural University, Anand, Gujarat, India*
R.M. HERNÁNDEZ • *NanoBioCel Group, Laboratory of Pharmaceutics, School of Pharmacy, University of the Basque Country (UPV/EHU), Vitoria-Gasteiz, Spain; Networking Center in Bioengineering, Biomaterials and Nanomedicine (CIBER-BBN), Vitoria-Gasteiz, Spain*
BREND RAY-SEA HSU • *Division of Endocrinology and Metabolism, Department of Internal Medicine, and School of Traditional Chinese Medicine, College of Medicine, Chang-Gung Memorial Hospital, Chang Gung University, Taoyuan, Taiwan*
MICHAEL HUNCKLER • *Institute for Regenerative Medicine, Wake Forest School of Medicine, Winston-Salem, NC, USA*
KALYAN C. KATURI • *Lawrence Berkeley National Laboratory, Berkeley, CA, USA*
WILLIAM F. KENDALL Jr. • *Department of Surgery, Sanford Health, Sioux Falls, SD, USA*
OMADITYA KHANNA • *Chicago Medical School at Rosalind Franklin, University of Medicine and Science, North Chicago, IL, USA*
RAMSEY C. KINNEY • *Wallace H. Coulter Department of Biomedical Engineering, Georgia Institute of Technology, Georgia Tech and Emory University, Atlanta, GA, USA*
DAVID KO • *Department of Surgery, University of California Irvine, Orange, CA, USA*
RAHUL KRISHNAN • *Department of Surgery, University of California Irvine, Orange, CA, USA*
JONATHAN R.T. LAKEY • *Department of Surgery, University of California Irvine, Orange, CA, USA; Department of Transplantation, University of California Irvine, Irvine, CA, USA; Department of Biomedical Engineering, University of California Irvine, Irvine, CA, USA*
D.M. LAVIN • *Cytosolv, Inc., Providence, RI, USA; Brown University, Providence, RI, USA*
SHIRAE K. LESLIE • *School of Engineering, Virginia Commonwealth University, Richmond, VA, USA*
LANJUAN LI • *State Key Laboratory for Diagnosis and Treatment of Infectious Diseases, Collaborative Innovation Center for Diagnosis and Treatment of Infectious Diseases, The First Affiliated Hospital, College of Medicine, Zhejiang University, Hangzhou, China*
WENDY LIU • *Department of Biomedical Engineering, University of California Irvine, Irvine, CA, USA*
SURAJIT MANDAL • *Dairy Microbiology Division, ICAR-National Dairy Research Institute (Deemed University), Karnal, Haryana, India*
E. MARTINS • *Oniris, UMR CNRS 6144 GEPEA, Nantes, France*

SHINICHI MATSUMOTO • *National Center for Global Health and Medicine, Otsuka Pharmaceutical Factory Inc., Tokushima, Japan*
JOHN PATRICK MCQUILLING • *Wake Forest Institute for Regenerative Medicine, Wake Forest School of Medicine, Winston-Salem, NC, USA; Virginia Tech-Wake Forest University School of Biomedical Engineering and Sciences, Winston-Salem, NC, USA*
PIA MONTANUCCI • *Laboratory for Endocrine Cell Transplants and Biohybrid Organs, Department of Medicine, Section of Cardiovascular, Endocrine and metabolic Clinical Physiology, University of Perugia, Perugia, Italy*
RONKE M. OLABISI • *Department of Biomedical Engineering, Rutgers University, Piscataway, NJ, USA*
EMMANUEL C. OPARA • *Wake Forest Institute for Regenerative Medicine, Wake Forest School of Medicine, Winston-Salem, NC, USA; Virginia Tech-Wake Forest School of Biomedical, Engineering & Sciences (SBES), Wake Forest School of Medicine, Winston-Salem, NC, USA*
G. ORIVE • *NanoBioCel Group, Laboratory of Pharmaceutics, School of Pharmacy, University of the Basque Country (UPV/EHU), Vitoria-Gasteiz, Spain; Networking Center in Bioengineering, Biomaterials and Nanomedicine (CIBER-BBN), Vitoria-Gasteiz, Spain*
GIUSEPPE ORLANDO • *Wake Forest Institute for Regenerative Medicine, Wake Forest School for Medicine, Winston-Salem, NC, USA; Department of Surgery, Section of Transplantation, Wake Forest School of Medicine, Winston-Salem, NC, USA*
XIAOPING PAN • *State Key Laboratory for Diagnosis and Treatment of Infectious Diseases, Collaborative Innovation Center for Diagnosis and Treatment of Infectious Diseases, The First Affiliated Hospital, College of Medicine, Zhejiang University, Hangzhou, China*
GENARO A. PAREDES-JUAREZ • *Division of MR Research, Russell H. Morgan Department of Radiology and Radiological Science, The Johns Hopkins University School of Medicine, Baltimore, MD, USA; Cellular Imaging Section and Vascular Biology Program, Institute for Cell Engineering, The Johns Hopkins University School of Medicine, Baltimore, MD, USA*
RAJESH PARETA • *Institute for Regenerative Medicine, Wake Forest School of Medicine, Winston-Salem, NC, USA*
J.L. PEDRAZ • *NanoBioCel Group, Laboratory of Pharmaceutics, School of Pharmacy, University of the Basque Country (UPV/EHU), Vitoria-Gasteiz, Spain; Networking Center in Bioengineering, Biomaterials and Nanomedicine (CIBER-BBN), Vitoria-Gasteiz, Spain*
SAMUEL PENDERGRAFT • *Institute for Regenerative Medicine, Wake Forest School of Medicine, Winston-Salem, NC, USA*
D. PERERA • *Department of Biomedical Engineering, Rutgers University, Piscataway, NJ, USA*
D. PONCELET • *Oniris, UMR CNRS 6144 GEPEA, Nantes, France*
B. PONCELET DE SMET • *Impascience, Sucé sur Erdre, France*
MELUR K. RAMASUBRAMANIAN • *Department of Mechanical Engineering, Clemson University, Clemson, SC, USA*
RICARDO L. ROSSI • *Department of Surgery, Clínica Alemana de Santiago, Independencia, Santiago, Chile*
L. SAENZ DEL BURGO • *NanoBioCel Group, Laboratory of Pharmaceutics, School of Pharmacy, University of the Basque Country (UPV/EHU), Vitoria-Gasteiz, Spain; Networking Center in Bioengineering, Biomaterials and Nanomedicine (CIBER-BBN), Vitoria-Gasteiz, Spain*

M.A. Sandrof • *Cytosolv, Inc., Providence, RI, USA*
Justin M. Saul • *Department of Chemical, Paper, and Biomedical Engineering, Miami University, Oxford, OK, USA*
Zvi Schwartz • *School of Engineering, Virginia Commonwealth University, Richmond, VA, USA*
Varna Sharma • *Department of Mechanical Engineering, Clemson University, Clemson, SC, USA*
Masayuki Shimoda • *National Center for Global Health and Medicine, Tokyo, Japan*
Sivanandane Sittadjody • *Wake Forest Institute for Regenerative Medicine, Wake Forest School for Medicine, Winston-Salem, NC, USA*
Sittadjody Sivanandane • *Wake Forest Institute for Regenerative Medicine, Wake Forest School for Medicine, Winston-Salem, NC, USA*
A.M. Smink • *Division of Immuno-Endocrinology, Departments of Pathology and Laboratory Medicine, University of Groni, Groningen, The Netherlands*
Sami I. Somo • *Department of Biomedical Engineering, Illinois Institute of Technology, Chicago, IL, USA; Research Service, Hines Veterans Administration Hospital, Hines, IL, USA*
Riccardo Tamburrini • *Wake Forest Institute for Regenerative Medicine, Wake Forest School for Medicine, Winston-Salem, NC, USA; Department of Surgery, Section of Transplantation, Wake Forest School of Medicine, Winston-Salem, NC, USA*
Chris G. Thanos • *Cytosolv, Inc., Providence, RI, USA*
Patricio Cabané Toledo • *Head and Neck Surgeon, University of Chile Clinical Hospital, Independencia, Santiago, Chile*
Yanhong Zhang • *State Key Laboratory for Diagnosis and Treatment of Infectious Diseases, Collaborative Innovation Center for Diagnosis and Treatment of Infectious Diseases, The First Affiliated Hospital, College of Medicine, Zhejiang University, Hangzhou, China*

Part I

Introduction

Chapter 1

Historical Perspectives and Current Challenges in Cell Microencapsulation

Paul de Vos

Abstract

The principle of immunoisolation of cells is based on encapsulation of cells in immunoprotective but semipermeable membranes that protect cells from hazardous effects of the host immune system but allows ingress of nutrients and outgress of therapeutic molecules. The technology was introduced in 1933 but has only received its deserved attention for its therapeutic application for three decades now.

In the past decade important advances have been made in creating capsules that provoke minimal or no inflammatory responses. There are however new emerging challenges. These challenges relate to optimal nutrition and oxygen supply as well as standardization and documentation of capsule properties.

It is concluded that the proof of principle of applicability of encapsulated grafts for treatment of human disease has been demonstrated and merits optimism about its clinical potential. Further innovation requires a much more systematic approach in identifying crucial properties of capsules and cellular grafts to allow sound interpretations of the results.

Key words Encapsulation, Natural polymers, Synthetic polymers, Insulin, Alginate, Biocompatibility, Biotolerability

1 Introduction

Encapsulation involves the protection of living cells from hazardous effects of the host immune system by enveloping the cells in an immunoprotective membrane. The technology has received much attention by the scientific community in the past three decades but its introduction dates back to as far as 1933. At that time Bisceglie et al. [1] studied the effect of absence of vascularization on tumor cells by encapsulating the cells in immunoisolating membranes and transplanted them in the abdominal cavity of pigs. As membrane Bisceglie applied amnion tissue-sheets and demonstrated prolonged cell survival of the enveloped cells in the immunoprotective membranes [1]. Unfortunately, Bisceglie did not recognize the potential of his findings for the treatment of disease. It took until 1950, when Algire et al. [2] introduced the concept of immunoisolation by creating diffusion chambers for implantation purposes to cure

Emmanuel C. Opara (ed.), *Cell Microencapsulation: Methods and Protocols*, Methods in Molecular Biology, vol. 1479,
DOI 10.1007/978-1-4939-6364-5_1,

disease by therapeutic cells. Algire also recognized as the first the importance of application of biocompatible materials and the need for constant, predictable properties of those materials for therapeutic application [2]. Since that time encapsulation devices have been produced in different conformations [3], with application of many different types of biomaterials [4] and has been applied for the treatment of many diseases that require a minute-to-minute regulation of metabolites such as in hemophilia B [5], anemia [6], dwarfism [7], kidney [8] and liver failure [9], pituitary disorders [10], central nervous system insufficiency [11], and diabetes mellitus [12].

2 Macrocapsules and Microcapsules

In the past three decades encapsulation of living cells has been applied in two families of geometries, i.e., macrocapsules and microcapsules. In macrocapsules, cells are packed in relatively large diffusion chambers. These chambers have semipermeable properties. The semipermeable macrocapsules have been produced in the form of hollow fibers, flat sheets, and disks [13]. Macrocapsules can be implanted as intravascular or extravascular devices [14]. When applied as intravascular device, a connection is made with the blood circulation. Usually the cells are seeded in polymeric capillaries and connected by anastomosis to the circulation. The advantage of this approach is the fast exchange of metabolites, nutrients, and oxygen [15]. A major disadvantage of this system is that thrombosis may occur after implantation. Up to now lifelong anticoagulation therapy has been a prerequisite with intravascular devices. This risk of thrombosis makes most intravascular approaches an unacceptable alternative for conventional treatment [16]. The side effects of implantation of the devices are simply too severe.

Because of the need of anticoagulation therapy most group nowadays focus on extravascular devices. In this approach therapeutic cells are encapsulated in semipermeable membranes and implanted without direct vascular connection. Exchange of therapeutic molecules, metabolites, and oxygen between the enveloped cells and the surrounding tissue depends on free diffusion over the membranes. An advantage of the system is that it does not need major surgery and can, in most cases, be easily replaced in case of failure of the graft. The numerous reports demonstrating successful application of extravascular devices in experimental animals and humans [17–20] illustrate the principle applicability of the approach.

Two types of extravascular devices are distinguished, i.e., the aforementioned macrocapsules and microcapsules. Conceptionally, encapsulation in macrocapsules is the simplest approach. Groups of cells are mixed within an immobilizing matrix and brought into one, large membrane and are subsequently implanted. Successful application has been reported [21–28] but the system has a major drawback. All macrocapsules have compared to smaller spherical microcapsules a relatively large surface-to-volume ratio. This implies that high amounts of therapeutic molecules, nutrients, and oxygen are required to build an effective diffusion gradient for ingress and outgress of molecules. This interferes with an adequate response to changes in metabolites in the recipients and also delays supply of nutrients to the cells. Another issue is that the cell density in macrocapsules cannot be high. The cells compete for nutrients such as oxygen and will stop functioning or will become even necrotic when the density is too high [29–31]. As a consequence the seeding density should be quite low to guarantee adequate nutrition [14]. Within most applications, the cell density should not exceed 5–10% of the volume fraction [15]. This suggests that if large numbers of cells are required to cure disease [15], either an artificial supply of nutrients should be incorporated in the design or several or very large devices must be implanted. In the past 5 years much attention has been given by the scientific community to means to enhance nutrition [23, 31–40] in the hope to be able to create smaller devices.

3 Insufficient Supply of Nutrients is Achilles Heel of Encapsulation

A transplantation area in which immunoisolation by encapsulation has a pertinent position is in research towards transplantation of pancreatic islets for the treatment of diabetes without immunosuppression. In this area most of the work on improving nutrition has been performed. Pancreatic islets are obtained from rare cadaveric donors and are also very sensitive for low oxygen tensions [41, 42]. Devices with optimal supply of nutrients are therefore an absolute requirement for application of pancreatic islets.

Research efforts to improve nutrition of immunoisolated islets started more than two decades ago by including different fenestrated membranes and angiogenic compound [43] in devices to enhance vascularization of the surface [32–34, 44, 45]. This however could never prevent the development of necrotic zones and loss of functional capacity of the grafts [44, 45]. Pancreatic islets need relatively high oxygen tensions for producing insulin and for regulation of glucose levels. The extravascular oxygen tensions around the macrocapsules in vivo of not more than 40 mmHg or lower are not sufficient to maintain optimal function and viability of the grafts [4, 46]. This has inspired a number of researchers to develop means to enhance the oxygen tensions in vivo. This has been done by inclusion of oxygen-generating chemicals in the capsules [47, 48], by inclusion of oxygen generating organisms [39], or by external supplementation of oxygen by means of an oxygen pump [31, 38, 40, 49].

The latter, the supply of oxygen by means of a pump, requires some further consideration as a recent series of experiments has shown that islets can function for a long time and keep on producing insulin in several animal models and even in a human in the absence of application of immunosuppression when oxygen is supplied in sufficient amounts [40, 49]. In this concept islets are surrounded by a polymeric shell and fed with oxygen by means of a manually operated oxygen pump. Supraphysiological amounts of oxygen are being pumped into the device inducing a gradient of oxygen in the islet graft [40, 49]. Islets proofed to perform quite well in this concept. They produce c-peptide in response to a glucose load in a human recipient [49] and histological examination confirmed the viability and absence of significant necrosis in the islets. This illustrates that the relative unfavorable surface to volume ratio of macrocapsules can be overcome by enhancing supply of nutrients.

The other promising concept that has been subject of intensive studies in the past three decades is the application of microencapsulation. Microcapsules are not associated with surface-to-volume ratio issues and are therefore preferred by some groups [4, 50]. Microcapsules have a size smaller than 700 μm and envelop individual clusters of cells or islets. The spherical geometry allows fast exchange of therapeutic molecules and nutrients between the surroundings and the encapsulated cells. Pancreatic islets in capsules have been shown to closely mimic the release of insulin and glucose of free, unencapsulated pancreatic islets [51, 52]. Many issues related to biocompatibility have been studied with application of microcapsules and are discussed in the following sections. Also, many new polymers, immunoprotective membranes, and strategies to enhance longevity of grafts have been designed in the context of microcapsules.

4 Biocompatibility and Biotolerability

A subject that after three decades of research is still subject of intensive debate is the biocompatibility of encapsulation devices. Biocompatibility is a complex, difficult concept with an even more complicated definition. Biocompatibility is usually defined as "the ability of a biomaterial to perform with an appropriate host response in a specific application." This definition was formulated at a time of emerging application of fully artificial organs, such as artificial hips and knees [53]. These fully artificial constructs induce innate immune responses resulting in fibrosis around the devices and integration of the artificial materials in the surrounding tissue. The immune responses and integration of the fully artificial devices in tissue is desired and therefore defined as an "appropriate host response". For bioartificial organs, such as encapsulated cells,

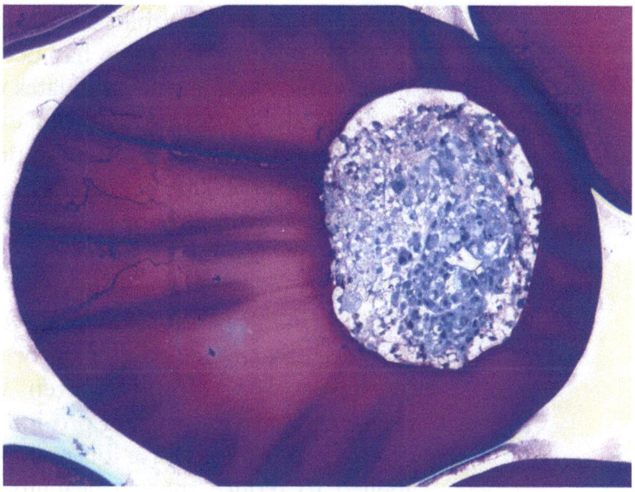

Fig. 1 Preferably no or minimal inflammatory responses should be elicited by encapsulate cells. This requires full control of critical capsules parameters. The proof of principle: an encapsulated pig pancreatic islet 4 weeks after xenotransplantation in a streptozotocin diabetic C57/b6 mice. GMA-embedded capsules. Original magnification ×40. Note absence of cellular adhesion

defining the appropriate host response is more complicated as inflammatory responses are associated with diffusion of harmful cytokines into the capsules with death of the cells as a consequence [50]. Preferably no or at least minimal cellular overgrowth should be provoked by the capsules to ensure free diffusion of nutrients and oxygen and to guarantee exchange of therapeutic molecules (Fig. 1). Because of the complexity of the biocompatibility issues the field has adopted a new term and definition for describing tissue responses against capsules. This term is "biotolerability" [50] and defined as "the ability of a material to reside in the body for long periods of time with only low degrees of inflammatory reactions" [54]. This definition also covers another important requirement which involves the "friendliness" or compatibility of the biomaterial with the encapsulated cells. Cells should grow and function in the polymer network as adequately as in their natural environment [55, 56].

Biotolerability of immunoisolating devices is defined by a complex combination of factors. A significant amount of research efforts has been spend in the past two decades on creating pure polymers that contain no or minimal amounts of endotoxin. It has been shown by many that even minimal contaminations with endotoxins can induce severe inflammatory responses with fibrosis of the capsules as a consequence [57–61]. Alginate is the best-studied encapsulation material and still the most commonly applied polymer in encapsulation research because of the friendly circumstances by which is can be applied to envelope cells and because of its relative high biotolerability. Almost 20 years ago the essence of applying purified alginates has been reported [62]. In the years after, commercial purified alginates were being brought to the market. The alginates were without exception low in the endotoxin lipopolysaccharide (LPS) but still provoked responses in the hands of some groups. This has stimulated research on the identification of factors determining the presence or absence of immune responses against capsules. It was found that capsules with a too high surface roughness provoke strong innate responses [28, 63, 64] with cell adhesion in mice and rats as a consequence. For that reason we usually keep the surface roughness below 10 nM to avoid these type responses [55]. However, even with low surface roughnesses responses may occur when pro-inflammatory chemical groups are exposed. Notorious are for example polyamide groups that are applied to reduce the permeability of the membranes to provide immunoprotection. Poly-L-lysine and poly-L-ornithine are often applied for these purposes but are highly proinflammatory when not in the correct conformation [65–69]. Poly-L-lysine has to be forced into a superhelical core with alginate and beta-sheets to prevent responses [65, 68]. Seemingly minor changes in the encapsulation procedure such as changes in temperature can have a profound effect on the surface chemistry and induce strong inflammatory responses.

Besides poor control of the aforementioned physicochemical variations, also simple items in the production process can influence biotolerability. In the past years there is an emerging trend of introducing new technologies to produce encapsulated islets. Sophisticated emerging technologies such as microfluids and electrospinning [70] are being proposed as alternative for the more conventional droplet formation technologies [71]. What has been overlooked and is still insufficiently recognized in the field is that the new technologies produce capsules with unique and technology dependent variations in capsule surface properties. Even with application with exactly the same polymer, our group has observed differences in surface roughness and chemistry. This has an impact on biotolerability. Also, physicochemical variations between capsules were observed and even inhomogeneity on surface characteristics on a single capsule. This is in our opinion an important item in the reported variations in success of encapsulated islets.

Recently it was reported that also the size of the capsules matters [72]. It was reported that smaller capsules (500 μm) provoke a stronger response than larger capsules (1800 μm). Remarkably, this difference in response was never observed by others [73–76]. In fact we never see any responses against planar alginate beads as applied by the authors [77]. Variations in the production process might be held responsible for the finding that smaller and larger capsules provoke different responses in some groups [73]. The most likely explanation is however the following. The group [72] does not apply purified alginates but a crude alginate that based on an in vivo skin study was selected for provoking lesser tissue responses than other unpurified alginates [78, 79]. When unpurified alginates are applied, endotoxins will be exposed on the surface of the capsules. Due to the higher surface to volume ratio of smaller capsules more endotoxins will be present on the smaller capsules compared to the larger capsules with differences in responses as a consequence [46, 73, 80]. This phenomenon is known for many years and emphasizes the importance of controlling and documenting all known factors responsible for induction of inflammatory responses against capsules. Side-by-side comparison of data sets is only possible when all physicochemical variations, including endotoxin content, are known [50]. In our hands and that of others size does not matter, at least not in the size ranges that are conventionally applied.

5 Impurities Revisit

Although most groups nowadays apply homemade or commercially available purified alginates, unrecognized contaminations might still contribute to the enormous variations in reported immune responses against capsules [46, 73, 80].

LPS is the best-recognized endotoxin present in encapsulation polymers. LPS is an endotoxin and also referred to as pathogen associated molecular pattern (PAMP). LPS as PAMP binds to Toll-like receptor 4 and induces proinflammatory responses in a wide variety of immune cells [81–83]. There are however other PAMPs that are having at least the same immune stimulatory capacity as LPS. In a recent study our group found some of these strong immune stimuli in alginates and some other polymers applied for immunoisolating devices. In unpurified alginates lipoteichoic acids (LTA) and proteoglycans (PG) was found [80]. These PAMPs stimulate TLR-2 and were responsible for proinflammatory responses in vitro and in vivo [46, 73, 80, 84]. Removal of these contaminations was found to be cumbersome as all purification procedures up to now were focused on removal of LPS. Novel methodologies for removal had to be developed. Also it was observed that PAMPs could be reintroduced during purification procedures or during storage [80] by for example unsterile conditions or use of equipment or disposables that contained PAMPs.

A technology platform was developed to screen encapsulation polymers for PAMPs [80] (Fig. 2). This platform involves a cell-based series of experiments combined with ELISA approaches to identify the contaminant present in the polymers [80]. This platform has been applied to screen and determine the purity of commercially produced purified alginates that were in some cases recommended for human application. Without exception, all the commercially available alginates tested up to now contained significant amounts

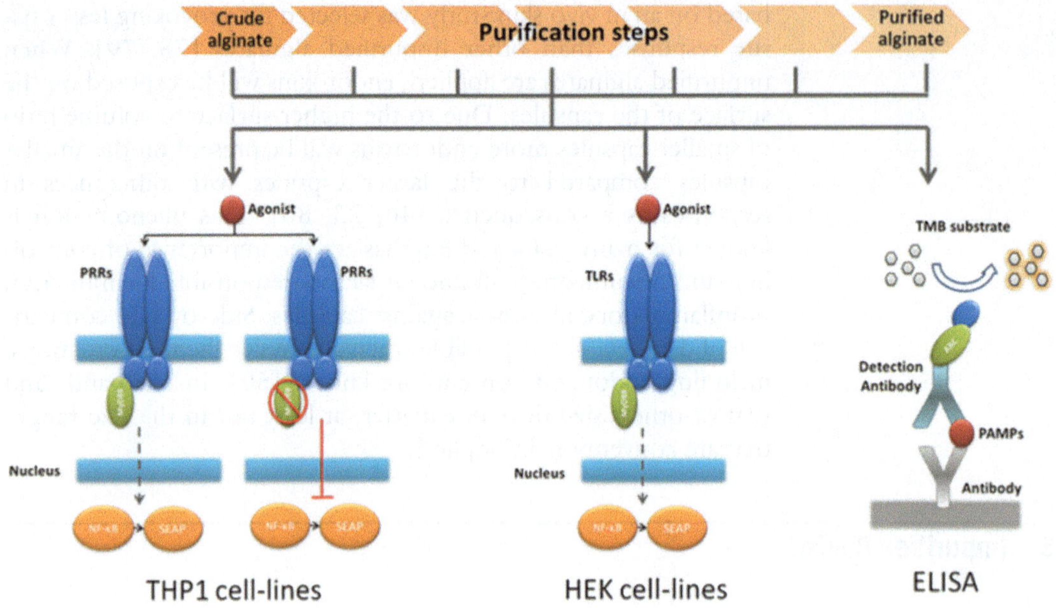

Fig. 2 Schematic presentation of the technology platform to identify the presence of pathogen associated molecular patterns (PAMPS) in polymers applicable for cell encapsulation as published in ref. [80]

of PAMPs and induced immune responses in immune cells. This is a serious issue impeding application and is not solved yet by the companies involved.

6 Strategies to Improve Tolerability

As outlined above it is far from simple to produce biotolerable capsules. After three decades of intensive research most factors interfering with biotolerability are identified but only a few groups have the skills and technology available to control, measure, and document the critical parameters for capsule formation [75]. Producing biotolerable capsules requires the application of a combination of physical and chemical approaches and an extreme control of the production process. It is laborious and requires experienced and trained technicians. Unfortunately the skills for producing capsules are underestimated. This is interfering with side-by-side comparison of data and is frustrating progress in the field [50]. Strategies to improve tolerability should therefore meet more prerequisites than just reduction of tissue responses. It should be simple and readily applicable in laboratories that lack physicochemical methodologies.

Many have been the efforts to change the surface properties of capsules by adding chemical groups to reduce the surface properties or to change coagulation and adhesion of proteins [64, 85–101]. Improvement of performance by reducing tissue responses was reported but often the coating was not permanent or interfered with diffusion of essential molecules such as insulin. A relative new emerging approach is the building of the so-called polymer brushes on surfaces of capsules. Normally molecules that bind to capsule surfaces bind in the so-called mushroom formats [102–107]. By increasing the grafting density more molecules will be bound on an identical surface area impeding with the formation of mushrooms. When a critical grafting density is accomplished the molecules will stretch and form the so-called polymer brush [106]. These brushes prevent adhesion of proteins, do not interfere in the desired length ranges with permeability [106, 107] and can bring improvement of biotolerability. This can even be accomplished in systems that are not produced according to the international standards [107].

Polymer brushes need to be created from relative long molecules for optimal efficacy. They can be di-block polymers with a group that readily binds to the core capsule materials and an outside that is known to provide beneficial properties such as biotolerability. In our group we have selected a series of candidate molecules [108] applicable for encapsulation of cells and for improvement of biotolerability by relative simple and disturbance free procedures that can be performed in labs with minimal equipment. The hope is that these types of approaches will make encapsulation a more straightforward approach and facilitate biotolerability and application.

7 Biotolerability from the Inside Out

Much attention has been focused in the past decades on creating capsules that provoke minimal or no tissue responses but even if this has been accomplished the grafts are having a limited survival time [4, 50, 109]. The longevity of a graft is not only influenced by the events from the outside but also by the intracapsular environment that should support long-term survival of tissue. A too rigid intracapsular environment does contribute to cellular rearrangements with either formation of nonfunctional multinucleated giant cells or death cells as a consequence [55]. The process responsible for these rearrangements is called mechanotransduction [55, 110–115]. During mechanotransduction mechanical forces on cells are transformed into biochemical signals [55, 110–115]. Cells respond by adjustments in cellular and extracellular structure. This results in modulation of cellular functions such as proliferation, differentiation, migration and apoptosis, and is harmful for cellular homeostasis. The molecular processes and sensors by which mechanotransduction in cells occurs are largely unknown, but recent studies suggest that integrins play a key role in mechanotransduction. With application of very rigid alginate matrixes to enforce the mechanical resistance of capsules, we recently encountered mechanotransduction as a homeostasis interfering process [55]. Different types of cells lines were encapsulated in alginates of different composition. Specific alginates were found to interfere with cell survival in capsules while other types supported cell survival [55]. Alginates are composed of mannuronic acid (M) and guluronic acids (G). The guluronic acid components, that link constitutive molecules in an egg-box model, provide rigidity to the inner matrix and are therefore preferred by many for application in sites where high shear forces are to be expected such as in the brain [56]. Despite a long term history of application of high-G alginates, we found that this matrix induced mechanotransduction associated cell death in a few days after encapsulation. This mechanotransduction induced cell death was not observed in less rigid matrices or in inner capsules with an alginate of a different type. More research efforts are required to determine the circumstances under which mechanotransduction does or does not occur after encapsulation.

The presence of an extracellular matrix (ECM) may also be a requirement for long-term survival of encapsulated cells. The majority of polymers currently applied do not support interaction with the many integrins and other cell-regulatory anchoring receptors on cells [56]. Especially in those applications where cadaveric cells are being applied to restore organ function, an ECM may be a requirement. ECM provides a matrix for cells to grow on [116–119], it serves as depot or binding site for many growth factors, or is a scavenger for many deleterious molecules. A few groups have applied RGD in their concept of encapsulation [120, 121] and have reported long periods of grafts survival in small as well as large animal models. It remains subject of debate but application of ECM might be a mandatory step when long-term function of grafts is envisioned such as in transplantation of pancreatic islets.

The consequences of a nonoptimal intracapsular environment for longevity of an encapsulated cellular graft may be larger than considered up to now. Currently, loss of cells is assumed to be an issue that can easily be overcome by transplanting a larger cellular mass. This however might be an underestimation of the consequences of dying cells as they may contribute to inflammatory responses by releasing specific alarm molecules and induce immune responses leading to graft failure [122]. Loss of cells in capsules occurs via three processes as recently reported [122]. It involves autophagy, necrosis, necroptosis, and apoptosis [122]. All these cell processes are associated with release of intracellular components such as DNA, RNA, and HMGB1 that bind to specific receptors on immune cells. These specific receptors are the so-called pattern recognition receptors (PRRs). Examples of PRRs are Toll-like receptors (TLRs), NOD receptors, and C-type lectins [123–127]. The DNA, RNA, and HMGB1 are also referred to as

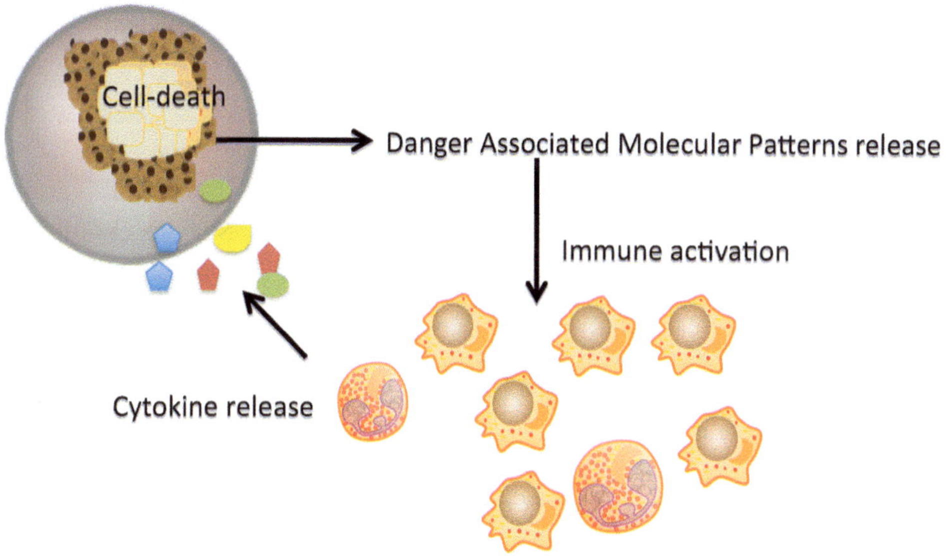

Fig. 3 Cell death does not only induce loss of the functional mass of cells in an encapsulated graft but also contributes to inflammatory responses. Dying cells release danger-associated molecular pattern molecules (DAMPs) which are potent activators of immune cells found in the transplantation site. After immune activation, cytokines are released that are small enough to pass the capsule membrane and induce more cell death

alarmins or Danger Associated Molecular Patterns (DAMPs) and bind or activate the PRRs and are very potent stimulators for immune cells. In recent studies it has been shown that encapsulated pancreatic islets do release DAMPs and activate immune cells in vitro [122] (Fig. 3). By applying a membrane that does not allow for entry of molecules smaller that 100 kDa some DAMPs can be retained but diffusion of DAMPs and associated immune activation cannot be completely blocked by the current generation of immunoprotective membranes [122]. We have indications that the quality of islet grafts and associated release of molecules such as DAMPs has a profound influence on the functional survival of islet grafts [122]. DAMP release can only be prevented by supplementation pharmaceuticals that stop cell death processes such as necroptosis [122] or by making the intracapsular environment a friendly environment that supports survival of cells.

8 Concluding Remarks and Future Considerations

Important advances have been made in the past decades with encapsulated cell therapy. Many new biomaterials and concepts have been introduced and many factors have been identified that determine the presence or absence of a tissue response against capsules. Clinical trials have started with some degree of success [49, 128].

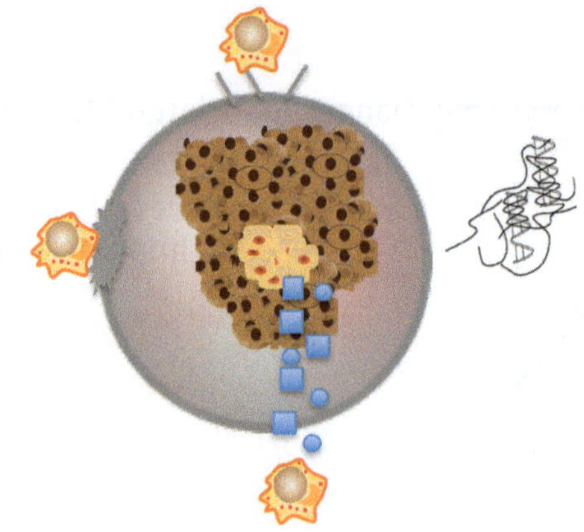

Fig. 4 Identified capsule properties that have a profound influence on the biotolerability of the encapsulated cellular graft. Minor changes in surface roughness, chemistry, presence of proinflammatory molecules such as pathogen associated molecular patterns may induce inflammatory cell adhesion and tissue responses against the capsules. Unfortunately only rarely these essential properties are reported on in transplantation studies

A major issue that is influencing the general opinion about applicability of the technology is the enormous lab-to-lab variation in success of the encapsulated cellular grafts [4, 50]. It is well known that the cause of this lab-to-lab variation is the poorly controlled physicochemical circumstances under which capsules are produced (Fig. 4). In a European consortium a stepwise analysis of capsule properties was performed to identify the minimum set of characteristics that has to be documented in order to reproduce the encapsulation procedure [75]. These were simple parameters such as polymer molecular weight and composition, permeability, mechanical strength, surface properties, and characteristics of the production process [75]. Unfortunately, until now it are only the former members of the European consortium that document and report on these essential parameters [50]. It is essential however for understanding the factors responsible for success and failure that the aforementioned critical parameters will be documented by the scientific community as a whole. Without it is currently impossible to do side-by-side comparisons of data-sets from different research groups.

A current regulatory trend in the USA is the suggestion that nonhuman primates are the ultimate models for human application of encapsulated islets grafts. As recently reviewed this might be a suggestion that interferes with progress [46]. Nonhuman primates have innate and adaptive immune responses that divert on essential parts from humans. Especially innate immune pathways that are involved in responses against capsules are different in nonhuman primates. Responses that will be seen in nonhuman primates will be very different from that in humans [46].

It has repeatedly been shown that capsules do work in small species but that scaling up to larger animals and humans in cumbersome. In our experience changing species of encapsulated islet graft recipients means many adaptations in capsule properties such as mechanics, osmolarity, and permeability. Up to now these adaptations have not been documented and it is not known what the requirements are that capsules have to meet to work in large animals or humans [4, 50]. A systematic approach to identify the values is lacking up to now. A major challenge for the near future is to set the essential parameters for application in humans and to make the procedures more reproducible. This is not only a scientific challenge but also a regulatory challenge. Bioartificial organs are a relative new area of application and many regulatory issues are coming from other areas such as organ transplantation. Safety and functionality are associated with different items in the bioartificial organ field; for example, regulations about presence of endotoxins are set too high for the cell encapsulation field [75]. Also there is no regulation for types of endotoxins present in the capsular grafts. The aforementioned LTA and PG are highly inflammatory molecules and present in most polymers currently applied for encapsulation. Not only does their presence negatively impact biotolerability, it also may influence human health. Regulation might therefore facilitate progress.

The historical and more recent review of challenges hopefully illustrates two important critical recommendations. The first is that the proof-of-principle studies with encapsulated grafts demonstrate the principal applicability of the technology for treatment of human disease. The second is that the field needs a much more systematic approach in characterizing properties of capsules and cellular grafts to allow for sound interpretations of the results and further innovation.

Acknowledgements

PdV is grateful for financial support from the Dutch Diabetes Foundation and Juvenile Diabetes Research Foundation.

References

1. Bisceglie VV (1933) Uber die antineoplastische Immunitat. Krebsforsch 40:141–158
2. Algire GH, Weaver JM, Prehn RT (1954) Growth of cells in vivo in diffusion chambers. I Survival of homografts in mice. J Natl Cancer Inst 15:493–507
3. Vaithilingam V, Tuch BE (2011) Islet transplantation and encapsulation: an update on recent developments. Rev Diabet Stud 8(1): 51–67
4. de Vos P, Lazarjani HA, Poncelet D, Faas MM (2014) Polymers in cell encapsulation

from an enveloped cell perspective. Adv Drug Deliv Rev 67–68:15–34. doi:10.1016/j.addr.2013.11.005
5. Liu HW, Ofosu FA, Chang PL (1993) Expression of human factor IX by microencapsulated recombinant fibroblasts. Hum Gene Ther 4:291–301
6. Koo J, Chang TSM (1993) Secretion of erythropoietin from microencapsulated rat kidney cells. Int J Artif Organs 16:557–560
7. Chang PL, Shen N, Westcott AJ (1993) Delivery of recombinant gene products with microencapsulated cells in vivo. Hum Gene Ther 4:433–440
8. Cieslinski DA, David Humes H (1994) Tissue engineering of a bioartificial kidney. Biotechnol Bioeng 43(7):678–681. doi:10.1002/bit.260430718
9. Uludag H, Sefton MV (1993) Microencapsulated human hepatoma (HepG2) cells: in vitro growth and protein release. J Biomed Mater Res 27(10):1213–1224
10. Colton CK (1995) Implantable biohybrid artificial organs. Cell Transplant 4:415–436
11. Aebischer P, Goddard M, Signore AP, Timpson RL (1994) Functional recovery in hemiparkinsonian primates transplanted with polymer-encapsulated PC12 cells. Exp Neurol 126:151–158
12. Lim F, Sun AM (1980) Microencapsulated islets as bioartificial endocrine pancreas. Science 210:908–910
13. Opara EC, Kendall WF Jr (2002) Immunoisolation techniques for islet cell transplantation. Expert Opin Biol Ther 2(5):503–511
14. Uludag H, De Vos P, Tresco PA (2000) Technology of mammalian cell encapsulation. Adv Drug Deliv Rev 42(1-2):29–64
15. De Vos P, Marchetti P (2002) Encapsulation of pancreatic islets for transplantation in diabetes: the untouchable islets. Trends Mol Med 8(8):363–366
16. Wilson JT, Chaikof EL (2008) Challenges and emerging technologies in the immunoisolation of cells and tissues. Adv Drug Deliv Rev 60(2):124–145. doi:10.1016/j.addr.2007.08.034
17. Aebischer P, Pochon NAM, Heyd B, DÇglon N, Joseph JM, Zurn AD, Baetge EE, Hammang JP, Goddard M, Lysaght M, Kaplan F, Kato AC, Schluep M, Hirt L, Regli F, Porchet F, De Tribolet N (1996) Gene therapy for amyotrophic lateral sclerosis (ALS) using a polymer encapsulated xenogenic cell line engineered to secrete hCNTF. Hum Gene Ther 7:851–860
18. Haisch A, Groger A, Radke C, Ebmeyer J, Sudhoff H, Grasnick G, Jahnke V, Burmester GR, Sittinger M (2000) Macroencapsulation of human cartilage implants: pilot study with polyelectrolyte complex membrane encapsulation. Biomaterials 21(15):1561–1566
19. Risbud MV, Bhargava S, Bhonde RR (2003) In vivo biocompatibility evaluation of cellulose macrocapsules for islet immunoisolation: Implications of low molecular weight cut-off. J Biomed Mater Res A 66(1):86–92. doi:10.1002/jbm.a.10522
20. Prochorov AV, Tretjak SI, Goranov VA, Glinnik AA, Goltsev MV (2008) Treatment of insulin dependent diabetes mellitus with intravascular transplantation of pancreatic islet cells without immunosuppressive therapy. Adv Med Sci 53(2):240–244
21. McQuilling JP, Arenas-Herrera J, Childers C, Pareta RA, Khanna O, Jiang B, Brey EM, Farney AC, Opara EC (2011) New alginate microcapsule system for angiogenic protein delivery and immunoisolation of islets for transplantation in the rat omentum pouch. Transplant Proc 43(9):3262–3264. doi:10.1016/j.transproceed.2011.10.030
22. Opara EC, McQuilling JP, Farney AC (2013) Microencapsulation of pancreatic islets for use in a bioartificial pancreas. Methods Mol Biol 1001:261–266. doi:10.1007/978-1-62703-363-3_21
23. Pareta R, McQuilling JP, Sittadjody S, Jenkins R, Bowden S, Orlando G, Farney AC, Brey EM, Opara EC (2014) Long-term function of islets encapsulated in a redesigned alginate microcapsule construct in omentum pouches of immune-competent diabetic rats. Pancreas 43(4):605–613. doi:10.1097/mpa.0000000000000107
24. Bilensoy E, Hincal AA (2009) Recent advances and future directions in amphiphilic cyclodextrin nanoparticles. Expert Opin Drug Deliv 6(11):1161–1173
25. Cheung CY, Anseth KS (2009) Immunoisolative encapsulation system. CO, US Patent US 20090028945
26. Cornolti R, Figliuzzi M, Remuzzi A (2009) Effect of micro- and macroencapsulation on oxygen consumption by pancreatic islets. Cell Transplant 18(2):195–201
27. Pattani A, Patravale VB, Panicker L, Potdar PD (2009) Immunological effects and membrane interactions of chitosan nanoparticles. Mol Pharm 6(2):345–352
28. Tam SK, Dusseault J, Bilodeau S, Langlois G, Halle JP, Yahia L (2011) Factors influencing alginate gel biocompatibility. J Biomed Mater Res A 98(1):40–52

29. Avgoustiniatos ES, Colton CK (1997) Effect of external oxygen mass transfer resistances on viability of immunoisolated tissue. Ann NY Acad Sci 31(831):145–167
30. Dionne KE, Colton CK, Yarmush ML (1993) Effect of hypoxia on insulin secretion by isolated rat and canine islets of Langerhans. Diabetes 42(1):12–21
31. Ludwig B, Rotem A, Schmid J, Weir GC, Colton CK, Brendel MD, Neufeld T, Block NL, Yavriyants K, Steffen A, Ludwig S, Chavakis T, Reichel A, Azarov D, Zimermann B, Maimon S, Balyura M, Rozenshtein T, Shabtay N, Vardi P, Bloch K, de Vos P, Schally AV, Bornstein SR, Barkai U (2012) Improvement of islet function in a bioartificial pancreas by enhanced oxygen supply and growth hormone releasing hormone agonist. Proc Natl Acad Sci U S A 109(13):5022–5027. doi:10.1073/pnas.1201868109
32. Gandhi JK, Opara EC, Brey EM (2013) Alginate-based strategies for therapeutic vascularization. Ther Deliv 4(3):327–341. doi:10.4155/tde.12.163
33. Khanna O, Moya ML, Greisler HP, Opara EC, Brey EM (2010) Multilayered microcapsules for the sustained-release of angiogenic proteins from encapsulated cells. Am J Surg 200(5):655–658. doi:10.1016/j.amjsurg.2010.08.001
34. Khanna O, Moya ML, Opara EC, Brey EM (2010) Synthesis of multilayered alginate microcapsules for the sustained release of fibroblast growth factor-1. J Biomed Mater Res A 95(2):632–640. doi:10.1002/jbm.a.32883
35. Sakai S, Ono T, Ijima H, Kawakami K (2000) Control of molecular weight cut-off for immunoisolation by multilayering glycol chitosan-alginate polyion complex on alginate-based microcapsules. J Microencapsul 17(6):691–699
36. Chia SM, Wan AC, Quek CH, Mao HQ, Xu X, Shen L, Ng ML, Leong KW, Yu H (2002) Multi-layered microcapsules for cell encapsulation. Biomaterials 23(3):849–856
37. Khattak SF, Chin KS, Bhatia SR, Roberts SC (2007) Enhancing oxygen tension and cellular function in alginate cell encapsulation devices through the use of perfluorocarbons. Biotechnol Bioeng 96(1):156–166. doi:10.1002/bit.21151
38. Barkai U, Weir GC, Colton CK, Ludwig B, Bornstein SR, Brendel MD, Neufeld T, Bremer C, Leon A, Evron Y, Yavriyants K, Azarov D, Zimermann B, Maimon S, Shabtay N, Balyura M, Rozenshtein T, Vardi P, Bloch K, de Vos P, Rotem A (2013) Enhanced oxygen supply improves islet viability in a new bioartificial pancreas. Cell Transplant 22(8):1463–1476. doi:10.3727/096368912x657341
39. Evron Y, Zimermann B, Ludwig B, Barkai U, Colton CK, Weir GC, Arieli B, Maimon S, Shalev N, Yavriyants K, Goldman T, Gendler Z, Eizen L, Vardi P, Bloch K, Barthel A, Bornstein SR, Rotem A (2015) Oxygen supply by photosynthesis to an implantable islet cell device. Horm Metab Res 47(1):24–30. doi:10.1055/s-0034-1394375
40. Neufeld T, Ludwig B, Barkai U, Weir GC, Colton CK, Evron Y, Balyura M, Yavriyants K, Zimermann B, Azarov D, Maimon S, Shabtay N, Rozenshtein T, Lorber D, Steffen A, Willenz U, Bloch K, Vardi P, Taube R, de Vos P, Lewis EC, Bornstein SR, Rotem A (2013) The efficacy of an immunoisolating membrane system for islet xenotransplantation in minipigs. PLoS One 8(8), e70150. doi:10.1371/journal.pone.0070150
41. Papas KK, Bellin MD, Sutherland DE, Suszynski TM, Kitzmann JP, Avgoustiniatos ES, Gruessner AC, Mueller KR, Beilman GJ, Balamurugan AN, Loganathan G, Colton CK, Koulmanda M, Weir GC, Wilhelm JJ, Qian D, Niland JC, Hering BJ (2015) Islet oxygen consumption rate (OCR) dose predicts insulin independence in clinical islet autotransplantation. PLoS One 10(8), e0134428. doi:10.1371/journal.pone.0134428
42. Papas KK, Colton CK, Nelson RA, Rozak PR, Avgoustiniatos ES, Scott WE 3rd, Wildey GM, Pisania A, Weir GC, Hering BJ (2007) Human islet oxygen consumption rate and DNA measurements predict diabetes reversal in nude mice. Am J Transplant 7(3):707–713. doi:10.1111/j.1600-6143.2006.01655.x
43. Kamba T, Tam BY, Hashizume H, Haskell A, Sennino B, Mancuso MR, Norberg SM, O'Brien SM, Davis RB, Gowen LC, Anderson KD, Thurston G, Joho S, Springer ML, Kuo CJ, McDonald DM (2006) VEGF-dependent plasticity of fenestrated capillaries in the normal adult microvasculature. Am J Physiol Heart Circ Physiol 290(2):H560–H576
44. Trivedi N, Steil GM, Colton CK, Bonner-Weir S, Weir GC (2000) Improved vascularization of planar membrane diffusion devices following continuous infusion of vascular endothelial growth factor. Cell Transplant 9(1):115–124
45. Isayeva IS, Kasibhatla BT, Rosenthal KS, Kennedy JP (2003) Characterization and performance of membranes designed for macroencapsulation/implantation of pancreatic islet cells. Biomaterials 24(20):3483–3491

46. Paredes Juarez GA, Spasojevic M, Faas MM, de Vos P (2014) Immunological and technical considerations in application of alginate-based microencapsulation systems. Front Bioeng Biotechnol 2:26. doi:10.3389/fbioe.2014.00026
47. Coronel MM, Stabler CL (2013) Engineering a local microenvironment for pancreatic islet replacement. Curr Opin Biotechnol 24(5):900–908. doi:10.1016/j.copbio.2013.05.004
48. Pedraza E, Coronel MM, Fraker CA, Ricordi C, Stabler CL (2012) Preventing hypoxia-induced cell death in beta cells and islets via hydrolytically activated, oxygen-generating biomaterials. Proc Natl Acad Sci U S A 109(11):4245–4250. doi:10.1073/pnas.1113560109
49. Ludwig B, Reichel A, Steffen A, Zimerman B, Schally AV, Block NL, Colton CK, Ludwig S, Kersting S, Bonifacio E, Solimena M, Gendler Z, Rotem A, Barkai U, Bornstein SR (2013) Transplantation of human islets without immunosuppression. Proc Natl Acad Sci U S A 110(47):19054–19058. doi:10.1073/pnas.1317561110
50. Rokstad AM, Lacik I, de Vos P, Strand BL (2014) Advances in biocompatibility and physico-chemical characterization of microspheres for cell encapsulation. Adv Drug Deliv Rev 67–68:111–130. doi:10.1016/j.addr.2013.07.010
51. De Haan BJ, Faas MM, De Vos P (2003) Factors influencing insulin secretion from encapsulated islets. Cell Transplant 12(6):617–625
52. De Vos P, Smedema I, van Goor H, Moes H, van Zanten J, Netters S, de Leij LF, de Haan A, De Haan BJ (2003) Association between macrophage activation and function of micro-encapsulated rat islets. Diabetologia 46(5):666–673
53. Williams DF (2008) On the mechanisms of biocompatibility. Biomaterials 29(20):2941–2953. doi:10.1016/j.biomaterials.2008.04.023
54. Hilborn J, Bjursten LM (2007) A new and evolving paradigm for biocompatibility. J Tissue Eng Regen Med 1(2):110–119. doi:10.1002/term.4
55. Bhujbal SV, de Haan B, Niclou SP, de Vos P (2014) A novel multilayer immunoisolating encapsulation system overcoming protrusion of cells. Sci Rep 4:6856. doi:10.1038/srep06856
56. Bhujbal SV, Paredes-Juarez GA, Niclou SP, de Vos P (2014) Factors influencing the mechanical stability of alginate beads applicable for immunoisolation of mammalian cells. J Mech Behav Biomed Mater 37:196–208. doi:10.1016/j.jmbbm.2014.05.020
57. Zimmermann U, Thurmer F, Jork A, Weber M, Mimietz S, Hillgartner M, Brunnenmeier F, Zimmermann H, Westphal I, Fuhr G, Noth U, Haase A, Steinert A, Hendrich C (2001) A novel class of amitogenic alginate microcapsules for long-term immunoisolated transplantation. Ann NY Acad Sci 944:199–215
58. Hasse C, Zielke A, Klock G, Schlosser A, Barth P, Zimmermann U, Sitter H, Lorenz W, Rothmund M (1998) Amitogenic alginates: key to first clinical application of microencapsulation technology. World J Surg 22:659–665
59. De Vos P, Vegter D, Strubbe JH, De Haan BJ, Van Schilfgaarde R (1997) Impaired glucose tolerance in recipients of an intraperitoneally implanted microencapsulated islet allograft is caused by the slow diffusion of insulin through the peritoneal membrane. Transplant Proc 29:756–757
60. Klôck G, Pfeffermann A, Ryser C, Grîhn P, Kuttler B, Hahn HJ, Zimmermann U (1997) Biocompatibility of mannuronic acid-rich alginates. Biomaterials 18:707–713
61. De Vos P, Van Hoogmoed CG, Busscher HJ (2002) Chemistry and biocompatibility of alginate-PLL capsules for immunoprotection of mammalian cells. J Biomed Mater Res 60:252–259
62. De Vos P, De Haan BJ, Wolters GHJ, Strubbe JH, Van Schilfgaarde R (1997) Improved biocompatibility but limited graft survival after purification of alginate for microencapsulation of pancreatic islets. Diabetologia 40:262–270
63. Tam SK, Bilodeau S, Dusseault J, Langlois G, Halle JP, Yahia LH (2011) Biocompatibility and physicochemical characteristics of alginate-polycation microcapsules. Acta Biomater 7(4):1683–1692. doi:10.1016/j.actbio.2010.12.006
64. Bunger CM, Gerlach C, Freier T, Schmitz KP, Pilz M, Werner C, Jonas L, Schareck W, Hopt UT, De Vos P (2003) Biocompatibility and surface structure of chemically modified immunoisolating alginate-PLL capsules. J Biomed Mater Res 67A(4):1219–1227
65. Van Hoogmoed CG, Busscher HJ, De Vos P (2003) Fourier transform infrared spectroscopy studies of alginate-PLL capsules with varying compositions. J Biomed Mater Res 67A(1):172–178
66. Tam KT, Dusseault J, Polizu S, MÇnard M, HallÇ JP, L'Hocine Y (2005) Physicochemical

model of alginate-poly-l-lysine microcapsules defined at the micrometric/nanometric scale using ATR-FTIR, XPS, and ToF-SIMS. Biomaterials 34:6950–6961

67. De Vos P, De Haan BJ, Kamps JA, Faas MM, Kitano T (2007) Zeta-potentials of alginate-PLL capsules: a predictive measure for biocompatibility? J Biomed Mater Res A 80(4):813–819
68. De Vos P, Van Hoogmoed CG, van Zanten J, Netter S, Strubbe JH, Busscher HJ (2003) Long-term biocompatibility, chemistry, and function of microencapsulated pancreatic islets. Biomaterials 24(2):305–312
69. Ponce S, Orive G, Hernandez R, Gascon AR, Pedraz JL, De Haan BJ, Faas MM, Mathieu HJ, De Vos P (2006) Chemistry and the biological response against immunoisolating alginate-polycation capsules of different composition. Biomaterials 27(28):4831–4839
70. Poncelet D, de Vos P, Suter N, Jayasinghe SN (2012) Bio-electrospraying and cell electrospinning: progress and opportunities for basic biology and clinical sciences. Adv Healthc Mater 1(1):27–34. doi:10.1002/adhm.201100001
71. De Vos P, De Haan BJ, Van Schilfgaarde R (1997) Upscaling the production of encapsulated islets. Biomaterials 18:1085–1090
72. Veiseh O, Doloff JC, Ma M, Vegas AJ, Tam HH, Bader AR, Li J, Langan E, Wyckoff J, Loo WS, Jhunjhunwala S, Chiu A, Siebert S, Tang K, Hollister-Lock J, Aresta-Dasilva S, Bochenek M, Mendoza-Elias J, Wang Y, Qi M, Lavin DM, Chen M, Dholakia N, Thakrar R, Lacik I, Weir GC, Oberholzer J, Greiner DL, Langer R, Anderson DG (2015) Size- and shape-dependent foreign body immune response to materials implanted in rodents and non-human primates. Nat Mater 14(6):643–651. doi:10.1038/nmat4290
73. Paredes-Juarez GA, de Haan BJ, Faas MM, de Vos P (2013) The role of pathogen-associated molecular patterns in inflammatory responses against alginate based microcapsules. J Control Release 172(3):983–992. doi:10.1016/j.jconrel.2013.09.009
74. Hall KK, Gattas-Asfura KM, Stabler CL (2011) Microencapsulation of islets within alginate/poly(ethylene glycol) gels cross-linked via Staudinger ligation. Acta Biomater 7(2):614–624. doi:10.1016/j.actbio.2010.07.016
75. de Vos P, Bucko M, Gemeiner P, Navratil M, Svitel J, Faas M, Strand BL, Skjak-Braek G, Morch YA, Vikartovska A, Lacik I, Kollarikova G, Orive G, Poncelet D, Pedraz JL, Ansorge-Schumacher MB (2009) Multiscale requirements for bioencapsulation in medicine and biotechnology. Biomaterials 30(13):2559–2570. doi:10.1016/j.biomaterials.2009.01.014
76. Orive G, Tam SK, Pedraz JL, Halle JP (2006) Biocompatibility of alginate-poly-L-lysine microcapsules for cell therapy. Biomaterials 27(20):3691–3700. doi:10.1016/j.biomaterials.2006.02.048
77. De Vos P, De Haan B, Van Schilfgaarde R (1997) Effect of the alginate composition on the biocompatibility of alginate-polylysine microcapsules. Biomaterials 18:273–278
78. Dang TT, Thai AV, Cohen J, Slosberg JE, Siniakowicz K, Doloff JC, Ma M, Hollister-Lock J, Tang KM, Gu Z, Cheng H, Weir GC, Langer R, Anderson DG (2013) Enhanced function of immuno-isolated islets in diabetes therapy by co-encapsulation with an anti-inflammatory drug. Biomaterials 34(23):5792–5801. doi:10.1016/j.biomaterials.2013.04.016
79. Liu WF, Ma M, Bratlie KM, Dang TT, Langer R, Anderson DG (2011) Real-time in vivo detection of biomaterial-induced reactive oxygen species. Biomaterials 32(7):1796–1801. doi:10.1016/j.biomaterials.2010.11.029
80. Paredes Juarez GA, De Haan BJ, Faas MM, De Vos P (2014) A technology platform to test the efficacy of purification of alginate. Materials 7:2087–2103. doi:10.3390/ma7032087
81. Tang D, Kang R, Coyne CB, Zeh HJ, Lotze MT (2012) PAMPs and DAMPs: signal 0s that spur autophagy and immunity. Immunol Rev 249(1):158–175. doi:10.1111/j.1600-065X.2012.01146.x
82. Kawai T, Akira S (2010) The role of pattern-recognition receptors in innate immunity: update on Toll-like receptors. Nat Immunol 11(5):373–384. doi:10.1038/ni.1863
83. Bauer S, Muller T, Hamm S (2009) Pattern recognition by Toll-like receptors. Adv Exp Med Biol 653:15–34
84. Zeuthen LH, Fink LN, Frokiaer H (2008) Toll-like receptor 2 and nucleotide-binding oligomerization domain-2 play divergent roles in the recognition of gut-derived lactobacilli and bifidobacteria in dendritic cells. Immunology 124(4):489–502. doi:10.1111/j.1365-2567.2007.02800.x
85. Mahou R, Tran NM, Dufresne M, Legallais C, Wandrey C (2012) Encapsulation of Huh-7 cells within alginate-poly(ethylene glycol) hybrid microspheres. J Mater Sci Mater Med 23(1):171–179. doi:10.1007/s10856-011-4512-3

86. Yang D, Guo S, Qiao J, Nie J (2011) Investigation on the preparation and application of chitosan/alginate microcapsules. J Control Release 152(Suppl 1):e71–e72. doi:10.1016/j.jconrel.2011.08.130
87. Rokstad AM, Brekke OL, Steinkjer B, Ryan L, Kollarikova G, Strand BL, Skjak-Braek G, Lacik I, Espevik T, Mollnes TE (2011) Alginate microbeads are complement compatible, in contrast to polycation containing microcapsules, as revealed in a human whole blood model. Acta Biomater 7(6):2566–2578. doi:10.1016/j.actbio.2011.03.011
88. Mazzitelli S, Luca G, Mancuso F, Calvitti M, Calafiore R, Nastruzzi C, Johnson S, Badylak SF (2011) Production and characterization of engineered alginate-based microparticles containing ECM powder for cell/tissue engineering applications. Acta Biomater 7(3):1050–1062. doi:10.1016/j.actbio.2010.10.005
89. Gardner CM, Burke NA, Chu T, Shen F, Potter MA, Stover HD (2011) Poly(methyl vinyl ether-alt-maleic acid) polymers for cell encapsulation. J Biomater Sci Polym Ed 22(16):2127–2145. doi:10.1163/092050610x535149
90. De Haan BJ, Rossi A, Faas MM, Smelt MJ, Sonvico F, Colombo P, de Vos P (2011) Structural surface changes and inflammatory responses against alginate-based microcapsules after exposure to human peritoneal fluid. J Biomed Mater Res A 98(3):394–403
91. Santos E, Zarate J, Orive G, Hernandez RM, Pedraz JL (2010) Biomaterials in cell microencapsulation. Adv Exp Med Biol 670:5–21
92. Tam SK, de Haan BJ, Faas MM, Halle JP, Yahia L, de Vos P (2009) Adsorption of human immunoglobulin to implantable alginate-poly-L-lysine microcapsules: effect of microcapsule composition. J Biomed Mater Res A 89(3):609–615. doi:10.1002/jbm.a.32002
93. Gattas-Asfura KM, Stabler CL (2009) Chemoselective cross-linking and functionalization of alginate via Staudinger ligation. Biomacromolecules 10(11):3122–3129. doi:10.1021/bm900789a
94. De Castro M, Orive G, Hernandez RM, Bartkowiak A, Brylak W, Pedraz JL (2009) Biocompatibility and in vivo evaluation of oligochitosans as cationic modifiers of alginate/Ca microcapsules. J Biomed Mater Res A 91(4):1119–1130. doi:10.1002/jbm.a.32270
95. Qi M, Strand BL, Morch Y, Lacik I, Wang Y, Salehi P, Barbaro B, Gangemi A, Kuechle J, Romagnoli T, Hansen MA, Rodriguez LA, Benedetti E, Hunkeler D, Skjak-Braek G, Oberholzer J (2008) Encapsulation of human islets in novel inhomogeneous alginate-Ca^{2+}/Ba^{2+} microbeads: in vitro and in vivo function. Artif Cells Blood Substit Immobil Biotechnol 36(5):403–420. doi:10.1080/10731190802369755
96. Marsich E, Borgogna M, Donati I, Mozetic P, Strand BL, Salvador SG, Vittur F, Paoletti S (2008) Alginate/lactose-modified chitosan hydrogels: a bioactive biomaterial for chondrocyte encapsulation. J Biomed Mater Res A 84(2):364–376
97. Thanos CG, Calafiore R, Basta G, Bintz BE, Bell WJ, Hudak J, Vasconcellos A, Schneider P, Skinner SJ, Geaney M, Tan P, Elliot RB, Tatnell M, Escobar L, Qian H, Mathiowitz E, Emerich DF (2007) Formulating the alginate-polyornithine biocapsule for prolonged stability: evaluation of composition and manufacturing technique. J Biomed Mater ResA 83(1):216–224
98. Baruch L, Machluf M (2006) Alginate-chitosan complex coacervation for cell encapsulation: effect on mechanical properties and on long-term viability. Biopolymers 82(6):570–579. doi:10.1002/bip.20509
99. Shen F, Li AA, Cornelius RM, Cirone P, Childs RF, Brash JL, Chang PL (2005) Biological properties of photocrosslinked alginate microcapsules. J Biomed Mater Res 75B(2):425–434
100. Orive G, Carcaboso AM, Hernandez RM, Gascon AR, Pedraz JL (2005) Biocompatibility evaluation of different alginates and alginate-based microcapsules. Biomacromolecules 6(2):927–931
101. Orive G, Bartkowiak A, Lisiecki S, De CM, Hernandez RM, Gascon AR, Pedraz JL (2005) Biocompatible oligochitosans as cationic modifiers of alginate/Ca microcapsules. J Biomed Mater Res B Appl Biomater 74(1):429–439
102. Bernards M, He Y (2014) Polyampholyte polymers as a versatile zwitterionic biomaterial platform. J Biomater Sci Polym Ed 25(14-15):1479–1488. doi:10.1080/09205063.2014.938976
103. Jiang H, Xu FJ (2013) Biomolecule-functionalized polymer brushes. Chem Soc Rev 42(8):3394–3426. doi:10.1039/c2cs35453e
104. Liao J, Wang C, Wang Y, Luo F, Qian Z (2012) Recent advances in formation, properties, and applications of polymersomes. Curr Pharm Des 18(23):3432–3441
105. Moroni L, Klein Gunnewiek M, Benetti EM (2014) Polymer brush coatings regulating cell behavior: passive interfaces turn into active. Acta Biomater 10(6):2367–2378. doi:10.1016/j.actbio.2014.02.048
106. Spasojevic M, Bhujbal S, Paredes G, de Haan BJ, Schouten AJ, de Vos P (2013)

Considerations in binding diblock copolymers on hydrophilic alginate beads for providing an immunoprotective membrane. J Biomed Mater Res A. doi:10.1002/jbm.a.34863

107. Spasojevic M, Paredes-Juarez GA, Vorenkamp J, de Haan BJ, Schouten AJ, de Vos P (2014) Reduction of the inflammatory responses against alginate-poly-L-lysine microcapsules by anti-biofouling surfaces of PEG-b-PLL diblock copolymers. PLoS One 9(10), e109837. doi:10.1371/journal.pone.0109837
108. Spasojevic M, Vorenkamp J, Jansen M, de Vos P, Schouten AJ (2014) Synthesis and Phase Behavior of Poly(N-isopropylacrylamide)-b-Poly(L-Lysine Hydrochloride) and Poly(N-Isopropylacrylamide-co-Acrylamide)-b-Poly(L-Lysine Hydrochloride). Materials 7(7):5305–5326. doi:10.3390/ma7075305
109. Orive G, Emerich D, De Vos P (2014) Encapsulate this: the do's and don'ts. Nat Med 20(3):233. doi:10.1038/nm.3486
110. Gasparski AN, Beningo KA (2015) Mechanoreception at the cell membrane: More than the integrins. Arch Biochem Biophys. doi:10.1016/j.abb.2015.07.017
111. Ivanovska IL, Shin JW, Swift J, Discher DE (2015) Stem cell mechanobiology: diverse lessons from bone marrow. Trends Cell Biol 25(9):523–532. doi:10.1016/j.tcb.2015.04.003
112. Jansen KA, Donato DM, Balcioglu HE, Schmidt T, Danen EH, Koenderink GH (2015) A guide to mechanobiology: Where biology and physics meet. Biochim Biophys Acta. doi:10.1016/j.bbamcr.2015.05.007
113. Schaefer A, Hordijk PL (2015) Cell-stiffness-induced mechanosignaling - a key driver of leukocyte transendothelial migration. J Cell Sci 128(13):2221–2230. doi:10.1242/jcs.163055
114. Tsimbouri PM (2015) Adult stem cell responses to nanostimuli. J Funct Biomater 6(3):598–622. doi:10.3390/jfb6030598
115. Walsh CM, Bautista DM, Lumpkin EA (2015) Mammalian touch catches up. Curr Opin Neurobiol 34:133–139. doi:10.1016/j.conb.2015.05.003
116. Bi Y, Hubbard C, Purushotham P, Zimmer J (2015) Insights into the structure and function of membrane-integrated processive glycosyltransferases. Curr Opin Struct Biol 34:78–86. doi:10.1016/j.sbi.2015.07.008
117. Iozzo RV, Schaefer L (2015) Proteoglycan form and function: a comprehensive nomenclature of proteoglycans. Matrix Biol 42:11–55. doi:10.1016/j.matbio.2015.02.003
118. Kuehn C, Vermette P, Fulop T (2014) Cross talk between the extracellular matrix and the immune system in the context of endocrine pancreatic islet transplantation. A review article. Pathol Biol 62(2):67–78. doi:10.1016/j.patbio.2014.01.001
119. Teschler JK, Zamorano-Sanchez D, Utada AS, Warner CJ, Wong GC, Linington RG, Yildiz FH (2015) Living in the matrix: assembly and control of Vibrio cholerae biofilms. Nat Rev Microbiol 13(5):255–268. doi:10.1038/nrmicro3433
120. Dufrane D, Steenberghe M, Goebbels RM, Saliez A, Guiot Y, Gianello P (2006) The influence of implantation site on the biocompatibility and survival of alginate encapsulated pig islets in rats. Biomaterials 27(17):3201–3208
121. Dufrane D, Goebbels RM, Saliez A, Guiot Y, Gianello P (2006) Six-month survival of microencapsulated pig islets and alginate biocompatibility in primates: proof of concept. Transplantation 81(9):1345–1353
122. Paredes-Juarez GA, Sahasrabudhe NM, Tjoelker RS, de Haan BJ, Faas MM, de Vos P (2015) Danger-associated molecular patterns production by human pancreatic islets under low oxygen and nutrient conditions in the presence and absence of an immunoisolating capsule and necrostatin-1. Sci Rep 5:14623
123. Dempsey A, Bowie AG (2015) Innate immune recognition of DNA: a recent history. Virology 479–480:146–152. doi:10.1016/j.virol.2015.03.013
124. Freitag J, Castro CN, Berod L, Lochner M, Sparwasser T (2015) Microbe-associated immunomodulatory metabolites: influence on T cell fate and function. Mol Immunol. doi:10.1016/j.molimm.2015.07.025
125. Iwasaki A, Medzhitov R (2015) Control of adaptive immunity by the innate immune system. Nat Immunol 16(4):343–353. doi:10.1038/ni.3123
126. Jaeger M, Stappers MH, Joosten LA, Gyssens IC, Netea MG (2015) Genetic variation in pattern recognition receptors: functional consequences and susceptibility to infectious disease. Future Microbiol 10(6):989–1008. doi:10.2217/fmb.15.37
127. Sellge G, Kufer TA (2015) PRR-signaling pathways: learning from microbial tactics. Semin Immunol 27(2):75–84. doi:10.1016/j.smim.2015.03.009
128. Basta G, Montanucci P, Luca G, Boselli C, Noya G, Barbaro B, Qi M, Kinzer KP, Oberholzer J, Calafiore R (2011) Long-term metabolic and immunological follow-up of nonimmunosuppressed patients with type 1 diabetes treated with microencapsulated islet allografts: four cases. Diabetes Care 34(11):2406–2409. doi:10.2337/dc11-0731

Chapter 2

Applications of Cell Microencapsulation

Emmanuel C. Opara

Abstract

The goal of this chapter is to provide an overview of the different purposes for which the cell microencapsulation technology can be used. These include immunoisolation of non-autologous cells used for cell therapy; immobilization of cells for localized (targeted) delivery of therapeutic products to ablate, repair, or regenerate tissue; simultaneous delivery of multiple therapeutic agents in cell therapy; spatial compartmentalization of cells in complex tissue engineering; expansion of cells in culture; and production of different probiotics and metabolites for industrial applications. For each of these applications, specific examples are provided to illustrate how the microencapsulation technology can be utilized to achieve the purpose. However, successful use of the cell microencapsulation technology for whatever purpose will ultimately depend upon careful consideration for the choice of the encapsulating polymers, the method of fabrication (cross-linking) of the microbeads, which affects the permselectivity, the biocompatibility and the mechanical strength of the microbeads as well as environmental parameters such as temperature, humidity, osmotic pressure, and storage solutions.

The various applications discussed in this chapter are illustrated in the different chapters of this book and where appropriate relevant images of the microencapsulation products are provided. It is hoped that this outline of the different applications of cell microencapsulation would provide a good platform for tissue engineers, scientists, and clinicians to design novel tissue constructs and products for therapeutic and industrial applications.

Key words Microencapsulation, Applications, Goals, Cell therapy, Immunoisolation, Tissue engineering, Drug delivery, Cancer treatment, Probiotics

1 Introduction

Cell microencapsulation is an aspect of the broader field of bioencapsulation which involves the immobilization of therapeutic cells using polymer scaffolds or semi permeable hydrogel capsules that provide the cells with a favorable protective environment for a variety of purposes [1–4], which is the focus of this chapter. Historically, an experiment designed for a different purpose provided a model platform for the takeoff of this technology that dates back to 1933 when Bisceglie enclosed tumor cells in a permselective polymeric membrane and transplanted them in the abdominal cavity of a pig to determine the effect of loss of vasculature on the survival of

Emmanuel C. Opara (ed.), *Cell Microencapsulation: Methods and Protocols*, Methods in Molecular Biology, vol. 1479,
DOI 10.1007/978-1-4939-6364-5_2, © Springer Science+Business Media New York 2017

implanted tissues [5]. The results from this early study showed that the cells survived long enough for it to be concluded that they were shielded from destruction by the immune system. However, as recently proposed by David Scharp and Piero Marchetti the credit for the development of extravascular diffusion device should go to Algire and his colleagues who developed the technology in order to study both cellular mechanisms of tissue rejection and tumor growth in a series of publications that spanned over a decade [6].

For some applications of cell microencapsulation, the required properties of the microbead matrix may be quite different. For example applications directed towards the repair or regeneration of tissue, biodegradability of the matrix may be desirable. When the matrix degrades, the entrapped cells may proliferate and create their own extracellular matrix in place of the artificial one used to entrap them [7]. Since there are many different applications of cell encapsulation technology, the specific requirements to achieve the desired goal in each case will include the choice of the biomaterials used as the entrapping matrix. It has been suggested that suitable materials for cell encapsulation should mimic the extracellular matrix and should be processed under conditions compatible with the presence of cells [7]. A classic example of cell encapsulation application utilizing injectable and degradable hydrogel formulations is cartilage and bone tissue engineering, where it has been shown that the swelling and degradation properties of the hydrogels influence the chondrogenic and osteogenic differentiation of encapsulated bone-marrow derived mesenchymal stromal cells [8, 9]. In addition, it has been shown that the implantation of biodegradable oligo(poly(ethylene glycol) fumarate) hydrogel microbeads encapsulating the pigment epithelial cells (PECs) of the dorsal iris in lentectomized newts resulted in a regenerated lens 30 days after explantation, thus demonstrating that the degrading hydrogel did not adversely affect regeneration [10].

2 Cell Immunoisolation

Following the pioneering work of Algire and his colleagues, Chang subsequently introduced the idea of using encapsulation for the immunoisolation (immunoprotection) of transplanted cells and coined the term 'artificial cells' in describing this concept [11], which was successfully put into practice when it was used to immobilize non-autologous islet cells in a construct termed a bioartificial endocrine pancreas for glycemic control in a diabetic rat model [12]. Generally, this construct shown in Fig. 1 is now simply referred to as a bioartificial pancreas [4, 13]. The principle of immunoisolation entails the enclosure of cells in semi permeable hydrogel microcapsules that provide the cells with an appropriate environment that allows the exchange of nutrients and oxygen

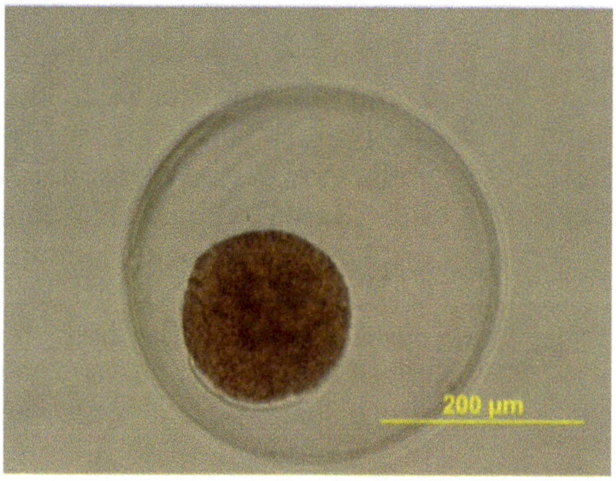

Fig. 1 Picture of a rat islet in alginate–PLO–alginate microcapsule

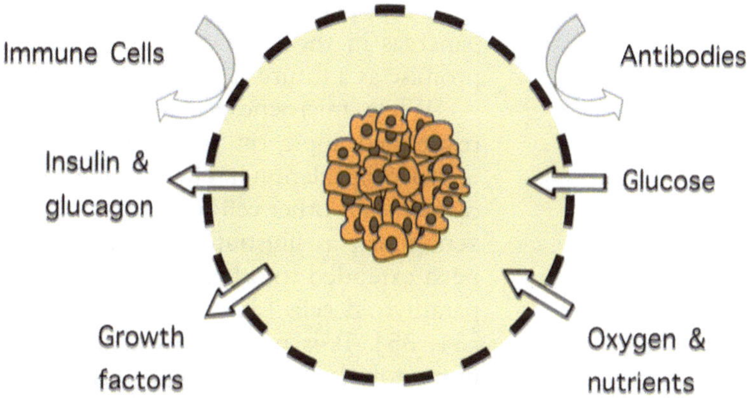

Fig. 2 Illustration of the principle of immunoisolation of islets encapsulated in a semipermeable microcapsule that allows entry and exit of molecules with permissible sizes

while protecting the cells from the host's immune system by blocking the entry of cytotoxic antibodies and cells, as illustrated in Fig. 2.

The report by Lim and Sum on the successful use of microencapsulation for the immunoisolation of transplanted islet cells generated enormous enthusiasm and provided the impetus for the belief that microencapsulation has significant potential to solve two major barriers to islet cell transplantation, namely the severe shortage of human pancreas, and the need to use immunosuppression to prevent transplant rejection [4, 6, 13, 14]. Eventually, this concept has been tested successfully with allografts in numerous studies performed in diabetic rodents [6, 12, 15], but has only had limited success in human studies [16–20]. There are numerous reasons for the limited success of microencapsulated islet cells in human studies [4, 6, 14]. However, there is room for improvement, which is why there is still significant enthusiasm for further development of this technology [4, 6, 10] that many believe has the promise of a cure for Type 1 diabetes. It is noteworthy that the immunoisolation of xenogeneic islet cells by microencapsulation has been shown to be successful in both small and large animal studies [21–24]. Interestingly, de Vos et al. have shown that encapsulated islet xenografts in humans may not be prone to cytotoxic cytokine destruction, in contrast to islet allografts, perhaps at least in part because xenogeneic islet cells are resistant to the binding of human cytokines [25], which may explain why encapsulated porcine islet xenografts have been shown to survive for nearly a decade in a diabetic patient [26]. Another study with microencapsulated porcine islets has also reported long-term function >6 years after transplantation in many Type 1 diabetic patients [27]. In addition, a recent report has also shown significant efficacy of microencapsulated neonatal porcine islets in Type 1 diabetic patients many years after transplantation [28], thus suggesting that the bioartificial pancreas in the form of xenografts may indeed hold significant promise as a future treatment option for patients.

While microencapsulation of islet cells has been the most researched topic on the concept of cell immunoisolation, cell microencapsulation has been applied to a wide variety of endocrine diseases and other cell replacement therapies either as allografts or xenografts, as illustrated in this book. Thus, its application has been extended to other cell types including hepatocytes [29, 30]; parathyroid cells [31], myoblasts [32, 33], stem cells and others [34, 35]. It is noteworthy that although a few studies on cell immunoisolation by microencapsulation have been performed with other polymeric materials, the overwhelming majority of studies in the literature on this subject have utilized the complex polysaccharide, alginate, as the encapsulating material [36]. Uncoated non-permselective alginate microbeads have been reported to have high permeability for molecules with molecular sizes >600 kD. Uptake studies with IgG (150 kD) and thyroglobulin (669 kD) have suggested these molecules are able to get into these uncoated microbeads. Similarly, uncoated alginate microbeads implanted in the peritoneum were positive for both IgG and C3 component after only 1 week [37]. Therefore, molecules with a range of sizes from macrophages and T-cells to smaller cytokine molecules such as IL-1β, TNF-α, and IFN-γ can easily penetrate into the alginate microcapsules and cause damage or destruction of the encapsulated islets [38]. To provide immunoisolation for the microcapsules, it is essential to apply a permeability barrier between the encapsulated islets and host immune system. Coating the alginate microcapsules with a polyamino acid layer, followed by an additional outer coating with alginate, typically creates this barrier. The positively charged polyamino acid molecules will readily bind to the negatively charged alginate molecules forming a complex membrane [39], which significantly reduces the pore size of the microcapsule, and prevents immune cells from entering the microcapsule. In order to prevent interactions of non-bound polyamines to host tissue, a thin second layer of alginate is added [4, 14, 40]. The polyamino barrier acts as a shell, providing mechanical stability to the microcapsule, allowing for the liquefaction of the inner alginate [41], if desired. The thickness of this barrier can be varied through manipulations of incubation time and polymer concentration. The most routinely used perm-selective biomaterial is poly-L-lysine (PLL), which was the first material used to create this barrier [12]; however more recent research has shown that poly-L-ornithine (PLO) coating provides stronger mechanical support to the microcapsules with markedly reduced immune response [42, 43].

3 Immobilization of Cells for Localized Delivery of Therapeutic Products

The second area in which cell microencapsulation plays a role is the immobilization of cells for therapeutic applications [44]. As is the case in cell immunoisolation, in this application the rationale is to protect the cells from destruction by the immune system while they deliver products for different therapeutic purposes. One prominent application of encapsulated cells is in blood vessel engineering. Properly structured microvascular networks are essential for normal tissue function. Physiologically, blood vessels are required for transport of nutrients and oxygen to cells and tissues. When blood perfusion of tissue is altered, cells undergo starvation, which can lead to a pathologic state. Examples include some of the leading causes of death in the USA such as chronic heart failure, myocardial infarction, and stroke. In addition, altered vascular supply plays a role in peripheral vascular disease, can contribute to poor outcomes in reconstructive and transplant surgery, and limits the broad clinical application of engineered tissues [44].

Neovascularization, the formation of new blood vessels, can occur via either vasculogenesis or angiogenesis. While vasculogenesis is the assembly of precursor cells into vascular networks angiogenesis is the formation of new blood vessels sprouting from existing vessels and is the primary mechanism of neovascularization in adults [44]. While vascular networks are relatively stable, angiogenesis occurs during embryogenesis, tissue healing and regeneration, the menstrual cycle, and in some pathological situations. It is a highly complex process that involves a complex series of steps: including endothelial cell (EC) activation, degradation of the extracellular matrix (ECM), EC migration, alignment, proliferation, lumen formation, branching, and anastomosis. Vessel stabilization is then achieved, in part, by mural cell recruitment and proliferation followed by basement membrane production [44].

In angiogenesis as a therapy for pathological conditions a typical approach is to investigate the delivery of cells (mature, progenitor, etc.) and/or growth factors (in the form of proteins or genes) to the target tissue. Therapeutic stimulation of vessel growth to improve tissue perfusion is an area of significant interest in many regenerative medicine and tissue engineering applications, and the delivery of multiple factors may improve outcomes. Cell delivery approaches focus on stimulating vascularization either via cell release of soluble factors, cell proliferation and incorporation into new vessels or prevascularization of tissue construct prior to implantation. Our group has successfully used the approach of co-encapsulating islet cells with angiogenic protein in alginate microcapsules to enhance tissue vascularization after implantation [4, 15, 45]. Another group has reported that the co-encapsulation of pancreatic islets with bioengineered IGF-II-producing cells promotes islet cell survival [46]. Other groups have also shown that mesenchymal stromal cells (MSCs) can be co-encapsulated with islets to improve microencapsulated islet cell function in vitro and in vivo [47, 48].

Another illustration of the utilization of microencapsulation to immobilize cells for delivery of therapeutic molecules is the microencapsulation of stem cells. For instance, mesenchymal stromal cell (MSC) therapy has emerged as a potential treatment option for a variety of medical conditions [49–58]. While MSC's innate migratory ability is invaluable for treating diseases involving multiple tissues, it has been shown that systemic infusion of MSC results in lung entrapment and consequent decreased numbers at the injury site [59]. Unfortunately this migratory property also results in transient effect of MSC therapy. In addition, despite possessing immunomodulatory properties, it has been shown that MSC are not immuno-inert, and allogeneic MSC are likely to be recognized by the recipient's immune system in a non-myeloablative setting [60–64]. Moreover, even in an autologous setting, once MSC arrive at sites of injury/inflammation within the body, they encounter a relatively hostile environment, which is often pro-apoptotic, resulting in rapid loss of the transplanted MSC. Preliminary studies in our laboratory indicate that MSC microencapsulation in alginate microbeads results in long-term viability when maintained in in vitro cultures. One can hypothesize then that MSC microencapsulation prior to transplantation will restrict cells to the site of injury, avoid their immune destruction when delivered as allografts, as well as protect MSC from apoptotic signals, thus maximizing MSC's therapeutic effect. Indeed, it has been shown that encapsulated GLP-1-producing MSC have a beneficial effect on failing pig hearts [65].

4 Compartmentalization of Cells in Multilayered Tissue Engineering

In 2010, our group described a multilayer alginate microencapsulation system suitable for coencapsulation of islet cells with angiogenic proteins [4, 66]. We have subsequently used this multilayer microbead approach to fabricate an ovarian tissue construct (Fig. 3) for hormone replacement [67]. This tissue construct is a bioinspired mimic of the follicular architecture of theca cells in an intermediate layer that surrounds granulosa cells in an inner layer of alginate microbeads. The two layers are separated by a semipermeable membrane made of poly-L-ornithine (PLO) for the immunoisolation of the encapsulated granulosa cells while acting as a basement membrane for the theca cells on top. The theca cells in the intermediate alginate layer are also immunoisolated by a PLO membrane that precedes a final coating with alginate.

To test the hypothesis that this natural arrangement of the follicular cells in our multilayer microcapsules was necessary for optimal construct function, we examined two other encapsulated cell culture schemes as controls. In these two other culture schemes, theca cells and granulosa cells were either encapsulated in separate

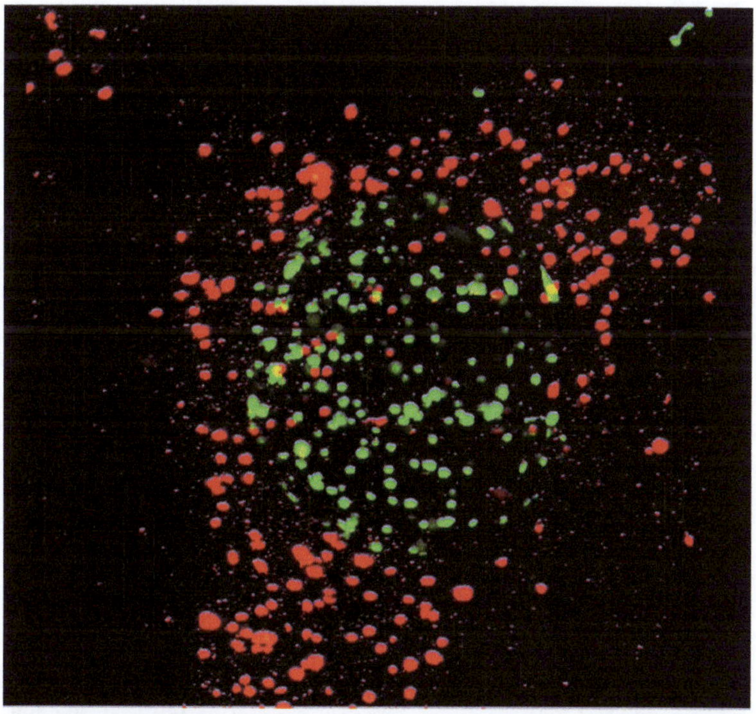

Fig. 3 Confocal Image (Composite of Z-stack) of compartmentalized ovarian cells in a multilayer alginate microbead (*red external layer* = theca cells; *green inner layer* = granulosa cells)

microcapsules and cocultured or encapsulated randomly together (non-layered) in the same microbeads and maintained in culture for 30 days. We found that encapsulated cells showed sustained viability during long-term in vitro culture with those encapsulated in the multilayered microcapsules secreting significantly higher and sustained concentrations of 17 β-estradiol (E_2) than the two other encapsulation schemes in response to follicle-stimulating hormone (FSH) and luteinizing hormone (LH). In addition, the cells in the multilayer microcapsules also secreted activin and inhibin in vitro. In contrast, when granulosa and theca cells were cultured in 2D culture, progesterone (P_4) secretion increased while E_2 secretion decreased over a 30-day period, thus demonstrating the importance of the 3D culture provided by the microcapsules in the maintenance of cell phenotype [67]. Furthermore, this study represents an elegant example of the application of cell encapsulation technique for spatial compartmentalization of cells in the engineering of complex tissues.

5 Expansion of Cells in Culture

Induced pluripotent stem cells (iPSCs) represent an emerging cell source for biomedical applications, and such applications usually require a large number of cells. Suspension culture of iPSC aggregates can augment cell yields but may result in excess aggregation or cell death by shear stress. Hydrogel-based microencapsulation may solve such problems observed in cell suspension culture, albeit, this remains to be determined [68]. In general, for cell therapies, donor primary cells are often difficult to obtain and expand to appropriate numbers, rendering stem cells an attractive alternative because of their capacities for self-renewal, differentiation, and trophic factor secretion. Microencapsulation of stem cells offers several benefits, namely the creation of a defined microenvironment which can be designed to modulate stem cell phenotype, protection from hydrodynamic forces and prevention of agglomeration during expansion in suspension bioreactors, and a means to transplant cells behind a semipermeable barrier, allowing for molecular secretion while avoiding immune reaction [69].

Prior to the application of microencapsulation, successful methods for culturing human hematopoietic cells employed some form of perfused bioreactor system, which does not permit the clonal outgrowth of single progenitor cells. Levee et al. successfully used alginate–poly-L-lysine-alginate microencapsulation of human bone marrow, combined with rapid medium exchange, as a mechanism to achieve the clonal outgrowth of single progenitor cells for the purpose of studying the kinetics of progenitor cell growth [70]. They reported that a 12 to 24-fold multilineage expansion of adult human bone marrow cells was achieved in about 16–19 days with this system and that visually identifiable colonies within the capsules were responsible for the increase in cell number. The colonies that represented the majority of cell growth originated from cells that appeared to be present in a frequency of about 1 in 4000 in the encapsulated cell population. These colonies were predominantly granulocytic and contained greater than 40,000 cells each. Large erythroid colonies were also present in the capsules, and they often contained over 10,000 cells each. Time profiles of the erythroid progenitor cell density over time were obtained [70].

Microencapsulation is also a very useful tool for therapeutic and industrial technologies that require the production of sufficient numbers of well-characterized cells and their efficient long-term storage. In order to establish a scalable bioprocess, human embryonic stem cells (hESC)-microcapsules were cultured in stirred tank bioreactors. The combination of microencapsulation and microcarrier technology resulted in a highly efficient protocol for the production and storage of pluripotent hESCs. This strategy ensured high expansion ratios (an approximately 20-fold increase in cell concentration) and high cell recovery yields (>70%) after cryopreservation. When compared with non-encapsulated cells, cell survival post-thawing demonstrated a threefold improvement without compromising hESC characteristics. Microencapsulation also improved the culture of hESC aggregates by protecting cells from hydrodynamic shear stress, controlling aggregate size and maintaining cell pluripotency for 2 weeks [71].

6 Cell Microencapsulation in Cancer Treatment

In cancer treatment, cell microencapsulation has been used as a technique to deliver therapeutic molecules to destroy (ablate) cancer cells. In this application, the cell microencapsulation approach is used as a strategy to protect the cells delivering the therapeutic products to cancer cells. An example of such therapeutic molecules is immunostimulatory monoclonal antibodies directed toward surface proteins of immune cells where they enhance immune response against cancer. Exogenous administration of the recombinant humanized immunoglobulins is being tested in clinical trials using this approach [72]. To illustrate a role for microencapsulation in this treatment modality, encapsulated antibody-producing hybridoma cells have been tested and compared with systemic administration of monoclonal antibodies. Hybridomas producing anti-CD137 and anti-OX40 mAb were encapsulated in alginate to generate microcapsules containing viable cells that secrete antibody. Immobilized cells in vitro were able to release the rat immunoglobulin produced by the hybridomas into the incubation medium. When the hybridoma-loaded microcapsules were implanted by injection into the subcutaneous tissue of mice, they provided a platform for viable secreting cells, which lasted for more than 1 week [73].

Also, it is known that tumors may develop resistance to specific angiogenic inhibitors via activation of alternative pathways. Therefore, multiple angiogenic pathways may be targeted to achieve significant angiogenic blockade [74, 75]. In a study investigating the effects of a combined delivery of the angiogenic inhibitors, endostatin and tumstatin, in a model of human glioblastoma multiforme, microencapsulated transfected porcine aortic endothelial (PAE) cells producing these inhibitors were applied as localized therapy in a subcutaneous glioblastoma model. When endostatin (ES) or tumstatin (Tum) were delivered separately, in vivo tumor growth was inhibited by 58% and 50%, respectively. However, the combined application of ES + Tum resulted in a significantly more pronounced inhibition of tumor growth (83%). cDNA microarrays of tumors treated with ES + Tum showed an upregulation of prolactin receptor (PRLR). ES + Tum-induced upregulation of PRLR in glioma cells was also observed in vitro [76].

Microencapsulated cells have also been studied for the treatment of bone cancer pain. Cancer-induced bone pain (CIP) is the most common cause of cancer pain [77, 78], and approximately 70% of patients with terminal breast or prostate cancer have evidence of bone metastases [79]. It has been suggested that human pheochromocytoma cells, which are known to contain and release met-enkephalin and norepinephrine, may be a promising resource for cell therapy in cancer-induced intractable pain [80]. In a recent study by Li et al.; microencapsulated human pheochromocytoma cells (micro-HPC) were intrathecally implanted to determine if this was a viable procedure to reduce pain in the rat model of bone cancer. The results showed that human pheochromocytoma cell implant-induced antinociception, which was mediated by met-enkephalin and norepinephrine secreted from the cell implants and acting on spinal receptors [80].

7 Microencapsulation of Bacteria Cells for Probiotic Applications

There is also an emerging interest in microencapsulation to deliver probiotic bacteria in controlled or targeted release in the gastrointestinal tract. With the explosion of probiotic health-based products, many reports indicate that there is poor survival of probiotic bacteria in these products. Further, the survival of these bacteria in the human gastrointestinal system is presently unknown. Providing probiotic living cells with a physical barrier against adverse environmental conditions is therefore an approach currently receiving considerable attention [81, 82]. In this application, microencapsulation of probiotics is associated with the protection of the probiotic cells in food products [83]. Functional foods provide a health benefit that goes beyond general nutritional content [84, 85]. In particular, foods containing probiotics are a natural way of enhancing the functionality of food products [86]. The viability of probiotic cells must be maintained throughout a food product's shelf-life and under gastrointestinal conditions after consumption. Maintenance of their survival until they reach the gastrointestinal tract is one of the key requirements for health benefit. The International Dairy Federation (IDF) has suggested that a minimum of 10^7 CFU probiotic bacterial cells should be alive at the time of consumption per gram of the product [85].

In a recent study microencapsulation of *Bifidobacterium longum* (B. longum) with *Eleutherine americana* extract, oligosaccharides extract, and commercial fructo-oligosaccharides was assessed for bacterial survival after sequential exposure to simulated gastric and intestinal juices, and refrigeration storage. Microencapsulated B. longum with the extract and oligosaccharides extract in the food products showed better survival than free cells under adverse conditions. Sensory analysis demonstrated that the products containing co-encapsulated bacterial cells were more acceptable by consumers than free cells. In particular, pineapple juice prepared with co-encapsulated cells had lower values for over acidification, compared with the juice with free cells added. These observations suggested that microencapsulated *B. longum* with *E. americana* could enhance functional properties of fresh milk tofu and pineapple juice [85]. It has also been recently shown that incorporation of maize starch (2 %) in alginate microbeads followed by coating of the beads with stearic acid (2 %) led to better protection and the complete release of entrapped lactobacilli in simulated colonic pH solution. Thus, the resulting encapsulated probiotics can be exploited in the development of probiotic functional foods with better survival of sensitive probiotic organisms [87].

8 Other Industrial Applications

Microencapsulation has also been applied to other industrial purposes as a means to improve the efficiency of production of different metabolites [3]. For instance, in fermentation it has been applied to enhancing cell density, aroma, and capacity of systems [3, 88, 89]. It has also been used to prevent the washout of biological catalysts from fermentation reactors [3]. Recently, microencapsulation using an emulsification method to immobilize *Clostridium acetobutylicum* ATCC 824 spores has been utilized for biobutanol production. The encapsulated spores were revived using heat shock treatment and the fermentation efficiency of the resultant encapsulated cells was compared with that of the free (non-encapsulated) cells. The microspheres were easily recovered from the fermentation medium by filtration and reused up to five cycles of fermentation. In contrast, the free (non-encapsulated) cells could be reused for two cycles only. The microspheres remained intact throughout repeated use demonstrating their capability as microbial cell nurseries in fermentation [90]. In an earlier study it had been shown that the production of lactic acid from glucose was enhanced by microencapsulation of *Lactococcus lactis* IO-1 cells [91].

9 Summary

It is apparent from the above review that there are many different applications of the cell microencapsulation technology that range from biomedicine to bio-industry. In biomedicine, cell microencapsulation has been mostly applied in the creation of bioartificial organs that specifically target a variety of endocrine disorders as well as kidney and liver failures [3]. However, it has also been applied in industry for a variety of purposes ranging from increasing cell manufacturing efficiency to enhancement of food aroma.

Each application may require peculiar specifications for the polymeric material used for encapsulation. In particular, adequate chemical characterization of the material to be used is necessary as it is critically important for the quality and function of the resultant product. One peculiarity required of polymers used for immunoisolation of encapsulated cells is adequate permselectivity for optimal protection of cells from immune destruction when such cells are implanted for therapeutic purposes in non-autologous setting. Currently, it does not appear that a naturally occurring biopolymer with adequate endogenous permselectivity for perfect immunoisolation is available. As discussed earlier, alginate is currently the most widely used material for encapsulation of cells designated for cell therapy, and we have noted that this polymer does not have appreciable permselectivity for immunoisolation. Hence polyamino acid polymeric (PLL or PLO) coatings of alginate microbeads have generally been utilized to enhance the pore-size exclusion of toxic immune molecules. Another polymer that has been utilized to enhance the permselectivity of alginate microcapsules is chitosan (β-[1–4]-linked-D-glucosamine) obtained by the deacetylation of chitin (β-[1–4]-linked *N*-acetyl-D-glucosamine), which is one of the most abundant naturally occurring polysaccharides [92]. Chitosan coating of alginate microbeads has been performed with or without modifications utilizing other materials and cross-linking agents to improve the mechanical stability and immunoisolation of the microbeads [93].

Obviously, for sustained performance of the encapsulated cells, it is necessary to fabricate microcapsules that possess long-term stability during storage or after implantation. There are various ways to manipulate the mechanical strength of microcapsules without adversely affecting the desired characteristics of other capsule parameters including permeability properties. It has been suggested that successful achievement of this goal will depend on the following factors: the type of biomaterials used for fabricating the capsule matrix and membrane [42, 94, 95], the type of gelling ion [96, 97], the type of cell, and the selected encapsulation procedure [3].

Another critically important consideration for efficient use of microencapsulated cells is the storage of microcapsules, a requirement that appears to be presently underappreciated [3]. Optimal storage conditions for microcapsules are necessary for transportation between manufacturing and implantation sites as well as during the period of product manufacture. Invariably the method of fabrication (cross-linking) of the microbeads comes into play in this matter. Several studies have shown that microcapsule characteristics and performance are often very sensitive to environmental parameters such as temperature, humidity, osmotic pressure, storage solution, or solvent [3, 98–101]. Successful use of the cell microencapsulation technology for whatever purpose will ultimately depend upon careful consideration of these parameters.

Acknowledgement

The author would like to thank Michael Hunckler for help with the illustration shown in Fig. 2.

References

1. Orive G, Hernández RM, Gascón AR et al (2004) History, challenges and perspectives of cell microencapsulation. Trends Biotechnol 22(2):87–92
2. Lim GJ, Zare S, Dyke MV, Atala A (2010) Cell microencapsulation. In: Pedraz JL, Orive G (eds) Therapeutic applications of cell microencapsulation. Springer, New York, pp 126–136
3. De Vos P, Bucko M, Gemeiner P et al (2009) Multiscale requirements for bioencapsulation in medicine and biotechnology. Biomaterials 30:2559–2570
4. Opara EC, Mirmalek-Sani S-H, Khanna O et al (2010) Design of a bioartificial pancreas. J Invest Med 58(7):831–837
5. Bisceglie V (1933) Uber die antineoplastische immunitat; heterologe Einpflnzung von Tumoren in Huhner-embryonen. Ztschr Krebsforsch 40:122–140
6. Scharp DW, Marchetti P (2014) Encapsulated islets for diabetes therapy: History, current progress, and critical issues requiring solution. Adv Drug Deliv 67–68:35–73
7. Gasperini L, Mano JF, Reis RL (2014) Natural polymers for the microencapsulation of cells. J R Soc Interface 11(100):20140817
8. Park H, Guo X, Temenoff JS et al (2009) Effect of swelling ratio of injectable hydrogel composites on chondrogenic differentiation of encapsulated rabbit marrow mesenchymal stem cells in vitro. Biomacromolecules 10:541–546
9. Temenoff JS, Park H, Jabbari E et al (2004) In vitro osteogenic differentiation of marrow stromal cells encapsulated in biodegradable hydrogels. J Biomed Mater Res Part A 70:235–244
10. Zhang MW, Park H, Guo X et al (2010) Adapting biodegradable oligo(poly(ethylene glycol) fumarate) hydrogels for pigment epithelial cell encapsulation and lens regeneration. Tissue Eng Part C Methods 16(2):261–267
11. Chang TMS (1964) Semipermeable microcapsules. Science 146(3643):524–525
12. Lim F, Sun M (1980) Microencapsulated islets as bioartificial endocrine pancreas. Science 210(4472):908–910

13. Kizilel S, Garfinkel M, Opara E (2005) The bioartificial pancreas: progress and challenges. Diabetes Technol Ther 7:968–985
14. O'Sullivan ES, Vegas A, Anderson DG, Weir GC (2011) Islets transplanted in immunoisolation devices: a review of the progress and the challenges that remain. Endocr Rev 32(6):827–844
15. Pareta R, McQuilling JP, Sivanandane S et al (2014) Long-term function of islets encapsulated in a re-designed alginate microcapsule construct in omentum pouches of immune-competent diabetic rats. Pancreas 43(4):605–613
16. Calafiore R (1992) Transplantation of microencapsulated pancreatic human islets for therapy of diabetes mellitus. A preliminary report. ASAIO J 38:34–37
17. Soon-Shiong P, Heintz RE, Merideth N et al (1994) Insulin independence in a type 1 diabetic patient after encapsulated islet transplantation. Lancet 343:950–951
18. Calafiore R, Basta G, Luca G et al (2006) Microencapsulated pancreatic islet allograft into non-immunosuppressed patients with Type 1 diabetes. Diabetes Care 29:137–138
19. Tuch BE, Keogh GW, Williams LJ et al (2009) Safety and viability of microencapsulated human islets transplanted into diabetic humans. Diabetes Care 32:1887–1889
20. Basta G, Montanucci P, Luca G et al (2011) Long-term metabolic and immunological follow-up of nonimmunosuppressed patients with type 1diabetes treated with microencapsulated islet allografts: four cases. Diabetes Care 34:2406–2409
21. Lanza RP, Jackson R, Sullivan A, Ringeling J et al (1999) Xenotransplantation of cells using biodegradable microcapsules. Transplantation 67(8):1105–1111
22. Meyer T, Höcht B, Ulrichs K (2008) Xenogeneic islet transplantation of microencapsulated porcine islets for therapy of type I diabetes: long-term normoglycemia in STZ-diabetic rats without immunosuppression. Pediatr Surg Int 24(12):1375–1378
23. Sun Y, Ma X, Zhou D et al (1996) Normalization of diabetes in spontaneously diabetic cynomolgus monkeys by xenografts of microencapsulated porcine islets without immunosuppression. J Clin Invest 98:1417–1422
24. Dufrane D, Goebbels RM, Saliez A, Guiot Y, Gianello P (2006) Six-month survival of microencapsulated pig islets and alginate biocompatibility in primates: proof of concept. Transplantation 81(9):1345–1353
25. de Vos P, Marchetti P (2002) Encapsulation of pancreatic islets for transplantation in diabetes: the untouchable islets. Trends Mol Med 8:363–366
26. Elliott RB, Escobar L, Tan PL, Muzina M, Zwain S, Buchanan C (2007) Live encapsulated porcine islets from a type 1 diabetic patient 9.5 yr after xenotransplantation. Xenotransplantation 14:157–161
27. Valdes-Gonzalez R, Rodriguez-Ventura AL, White DJ et al (2010) Long-term follow-up of patients with type 1 diabetes transplanted with neonatal pig islets. Clin Exp Immunol 162(3):537–542
28. Matsumoto S, Tan P, Baker J et al (2014) Clinical porcine islet xenotransplantation under comprehensive regulation. Transplant Proc 46(6):1992–1995
29. Chen Y, Yu C, Lv G et al (2014) Rapid large-scale culturing of microencapsulated hepatocytes: a promising approach for cell-based hepatic support. Transplant Proc 46(5):1649–1657
30. Jitraruch S, Dhawan A, Hughes RD et al (2014) Alginate microencapsulated hepatocytes optimised for transplantation in acute liver failure. PLoS One 9(12), e113609
31. Cabané P, Gac P, Amat J et al (2009) Allotransplant of microencapsulated parathyroid tissue in severe postsurgical hypoparathyroidism: a case report. Transplant Proc 41(9):3879–3883
32. Murua A, Orive G, Hernández RM, Pedraz JL (2009) Cryopreservation based on freezing protocols for the long-term storage of microencapsulated myoblasts. Biomaterials 30(20):3495–3501
33. Ahmad HF, Sambanis A (2013) Cryopreservation effects on recombinant myoblasts encapsulated in adhesive alginate hydrogels. Acta Biomater 9(6):6814–6822
34. Gurruchaga H, Ciriza J, Saenz Del Burgo L et al (2015) Cryopreservation of microencapsulated murine mesenchymal stem cells genetically engineered to secrete erythropoietin. Int J Pharm 485(1-2):15–24
35. Acarregui A, Orive G, Pedraz JL, Hernández RM (2013) Therapeutic applications of encapsulated cells. Methods Mol Biol 1051:349–364
36. Farney AC, Sutherland DER, Opara EC (2015) Evolution of islet transplantation for the last 30 years. Pancreas 45(1):8–20
37. Lanza RP, Kuhtreiber WM, Ecker D et al (1995) Xenotransplantatton of porcine and bovine islets without immunosuppression using uncoated alginate microspheres. Transplantation 59:1377–1384

38. van Schilfgaarde R, de Vos P (1999) Factors influencing the properties and performance of microcapsules for immunoprotection of pancreatic islets. J Mol Med (Berl) 77:199–205
39. Uludag H, de Vos P, Tresco PA (2000) Technology of mammalian cell encapsulation. Adv Drug Deliv Rev 42:29–64
40. Khanna O, Lawson JC, Moya ML, Opara EC, Brey EM (2012) Generation of alginate microspheres for biomedical applications. J Vis Exp 66, e3388
41. Darrabie M, Freeman BK, Kendall WF et al (2001) Durability of polylysine-alginate microcapsules. J Biomed Mater Res 54:396–399
42. Darrabie MD, Kendall WF, Opara EC (2005) Characteristics of poly-L-ornithine-coated alginate microcapsules. Biomaterials 26/34:6846–6852
43. Tam SK, Bilodeau S, Dusseault J et al (2011) Biocompatibility and physicochemical characteristics of alginate-polycation microcapsules. Acta Biomater 7:1683–1692
44. Ghandi JK, Opara EC, Brey EM (2013) Alginate-based strategies for therapeutic vascularization. Ther Deliv 4(3):327–341
45. McQuilling JP, Arenas-Herrera J, Childers C et al (2011) New alginate microcapsule system for angiogenic protein delivery and immunoisolation for transplantation in the rat omentum pouch. Transplant Proc 43:3262–3264
46. Jourdan G, Dusseault J, Benhamou PY, Rosenberg L, Hallé JP (2011) Co-encapsulation of bioengineered IGF-II-producing cells and pancreatic islets: effect on beta-cell survival. Gene Ther 18(6):539–545
47. Davis NE, Beenken-Rothkopf LN, Mirsoian A et al (2012) Enhanced function of pancreatic islets co-encapsulated with ECM proteins and mesenchymal stromal cells in a silk hydrogel. Biomaterials 33(28):6691–6697
48. Vériter S, Gianello P, Igarashi Y et al (2014) Improvement of subcutaneous bioartificial pancreas vascularization and function by coencapsulation of pig islets and mesenchymal stem cells in primates. Cell Transplant 23(11):1349–1364
49. Amado LC, Saliaris AP, Schuleri KH et al (2005) Cardiac repair with intramyocardial injection of allogeneic mesenchymal stem cells after myocardial infarction. Proc Natl Acad Sci U S A 102(32):11474–11479
50. Wu Y, Chen L, Scott PG, Tredget EE (2007) Mesenchymal stem cells enhance wound healing through differentiation and angiogenesis. Stem Cells 25(10):2648–2659
51. Atsma DE, Fibbe WE, Rabelink TJ (2007) Opportunities and challenges for mesenchymal stem cell-mediated heart repair. Curr Opin Lipidol 18(6):645–649
52. Firth AL, Yao W, Ogawa A et al (2010) Multipotent mesenchymal progenitor cells are present in endarterectomized tissues from patients with chronic thromboembolic pulmonary hypertension. Am J Physiol Cell Physiol 298(5):C1217–C1225
53. Aguayo-Mazzucato C, Bonner-Weir S (2010) Stem cell therapy for type 1 diabetes mellitus. Nat Rev Endocrinol 6:139–148
54. Fortino C, Ricordi C, Lauriola V, Alejandro R, Pileggi A (2010) Bone marrow-derived stem cell transplantation for the treatment of insulin-dependent diabetes. Rev Diabet Stud 7:144–157
55. Volarevic V, Arsennievic N, Lukic ML, Stojkovic M (2011) Concise review: mesenchymal stem cell treatment of the complications of diabetes mellitus. Stem Cells 29(1):5–10
56. Tolar J, Villeneuve P, Keating A (2011) Mesenchymal stromal cells for graft-versus-host disease. Human Gene Ther 22:1–6
57. Figliuzzi M, Bonanrini B, Silvani S, Remuzzi A (2014) Mesenchymal stem cells help pancreatic islet transplantation to control type 1 diabetes. World J Stem Cells 6(2):163–172
58. Volarevic V, Nurkovic J, Arsennievic N, Stojkovic M (2014) Concise review: therapeutic potential of mesenchymal stem cells for the treatment of acute liver failure and cirrhosis. Stem Cells 32(11):2813–2823
59. Caplan AI, Dennis JE (2006) Mesenchymal stem cells as trophic mediators. J Cell Biochem 98(5):1076–1084
60. Porada CD, Almeida-Porada G (2010) Mesenchymal stem cells as therapeutics and vehicles for gene and drug delivery. Adv Drug Deliv Rev 62(12):1156–1166
61. Porada CD, Zanjani ED, Almeida-Porad G (2006) Adult mesenchymal stem cells: a pluripotent population with multiple applications. Curr Stem Cell Res Ther 1(3):365–369
62. Nauta AJ, Westerhuis G, Kruisselbrink AB et al (2006) Donor-derived mesenchymal stem cells are immunogenic in an allogeneic host and stimulate donor graft rejection in a nonmyeloablative setting. Blood 108(6):2114–2120
63. Lim JH, Kim JS, Yoon IH et al (2010) Immunomodulation of delayed-type hypersensitivity responses by mesenchymal stem cells is associated with bystander T cell

apoptosis in the draining lymph node. J Immunol 185(7):4022–4029
64. Le Blanc K, Tammik L, Sundberg B, Haynesworth SE, Ringdén O (2003) Mesenchymal stem cells inhibit and stimulate mixed lymphocyte cultures and mitogenic responses independently of the major histocompatibility complex. Scand J Immunol 57(1):11–20
65. Wright EJ, Farrell KA, Malik N et al (2012) Encapsulated glucagon-like peptide-1-producing mesenchymal stem cells have a beneficial effect on failing pig hearts. Stem Cells Transl Med 1(10):759–769
66. Khanna O, Moya ML, Opara EC, Brey EM (2010) Synthesis of multi-layered alginate microcapsules for the sustained release of fibroblast growth factor-1. J Biomed Mater Res A 95(2):632–640
67. Sittadjody S, Saul SM, Joo S, Yoo JJ, Atala A, Opara EC (2013) Engineered multilayer ovarian tissue that secretes sex steroids and peptide hormones in response to gonadotropins. Biomaterials 34(10):2412–2420
68. Horiguchi I, Chowdhury MM, Sakai Y, Tabata Y (2014) Proliferation, morphology, and pluripotency of mouse induced pluripotent stem cells in three different types of alginate beads for mass production. Biotechnol Prog 30(4):896–904
69. Wilson JL, Najia MA, Saeed R, McDevitt TC (2014) Alginate encapsulation parameters influence the differentiation of microencapsulated embryonic stem cell aggregates. Biotechnol Bioeng 111(3):618–631
70. Levee MG, Lee GM, Paek SH, Palsson BO (1994) Microencapsulated human bone marrow cultures: a potential culture system for the clonal outgrowth of hematopoietic progenitor cells. Biotechnol Bioeng 43(8): 734–739
71. Serra M, Correia C, Malpique R et al (2011) Microencapsulation technology: a powerful tool for integrating expansion and cryopreservation of human embryonic stem cells. PLoS One 6(8), e23212
72. Melero I, Hervas-Stubbs S, Glennie M, Pardoll DM, Chen L (2007) Immunostimulatory monoclonal antibodies for cancer therapy. Nat Rev Cancer 7(2):95–106
73. Dubrot J, Portero A, Orive G et al (2010) Delivery of immunostimulatory monoclonal antibodies by encapsulated hybridoma cells. Cancer Immunol Immunother 59(11): 1621–1631
74. Jain RK, di Tomaso E, Duda DG, Loeffler JS, Sorensen AG, Batchelor TT (2007) Angiogenesis in brain tumours. Nat Rev Neurosci 8(8):610–622
75. Batchelor TT, Sorensen AG, di Tomaso E et al (2007) AZD2171, a pan-VEGF receptor tyrosine kinase inhibitor, normalizes tumor vasculature and alleviates edema in glioblastoma patients. Cancer Cell 11(1):83–95
76. Oliveira-Ferrer L, Wellbrock J, Bartsch U et al (2013) Combination therapy targeting integrins reduces glioblastoma tumor growth through antiangiogenic and direct antitumor activity and leads to activation of the pro-proliferative prolactin pathway. Mol Cancer 12(1):144
77. Banning A, Sjogren P, Henriksen H (1991) Pain causes in 200 patients referred to a multidisciplinary cancer pain clinic. Pain 45:45–48
78. Coleman RE (1997) Skeletal complications of malignancy. Cancer 80:1588–1594
79. Coleman RE (2006) Clinical features of metastatic bone disease and risk of skeletal morbidity. Clin Cancer Res 12:6243–6249
80. Li X, Li G, Wu S et al (2014) Antinociceptive effect of intrathecal microencapsulated human pheochromocytoma cell in a rat model of bone cancer pain. Int J Mol Sci 15(7): 12135–12148
81. Kailasapathy K (2002) Microencapsulation of probiotic bacteria: technology and potential applications. Curr Issues Intest Microbiol 3(2):39–48
82. Riaz QU, Masud T (2013) Recent trends and applications of encapsulating materials for probiotic stability. Crit Rev Food Sci Nutr 53(3):231–244
83. Heidebach T, Forst P, Kulozik U (2012) Microencapsulation of probiotic cells for food applications. Crit Rev Food Sci Nutr 52(4):291–311
84. Lacroix C, Yildirim S (2007) Fermentation technologies for the production of probiotics with high viability and functionality. Curr Opin Biotechnol 18(2):176–183
85. Phoem AN, Chanthachum S, Voravuthikunchai SP (2015) Applications of microencapsulated Bifidobacterium longum with Eleutherine Americana in fresh milk tofu and pineapple juice. Nutrients 7(4):2469–2484
86. Muthukumarasamy P, Holley RA (2006) Microbiological and sensory quality of dry fermented sausages containing alginate-microencapsulated Lactobacillus reuteri. Int J Food Microbiol 111(2):164–169
87. Mandal S, Hati S, Puniya AK, Khamrui K, Singh K (2014) Enhancement of survival of alginate-encapsulated Lactobacillus casei

NCDC 298. J Sci Food Agric 94(10): 1994–2001
88. Shah NP (2000) Probiotic bacteria: selective enumeration and survival in dairy foods. J Dairy Sci 83(4):894–907
89. Lofty SN, Fadel HH, El-Ghorab AH, Shaheen MS (2015) Stability of encapsulated beef-like flavorings prepared for enzymatically hydrolysed mushroom proteins with other precursors under conventional and microwave heating. Food Chem 187:7–13
90. Rathore S, Wan Sia Heng P, Chan LW (2015) Microencapsulation of Clostridium acetobutylicum ATCC 824 spores in gellan gum microspheres for the production of biobutanol. J Microencapsul 12:1–10
91. Sirisansaneeyakul S, Luangpipat T, Vanichsriratana W et al (2007) Optimization of lactic acid production by immobilized Lactococcus lactis IO-1. J Ind Microbiol Biotechnol 34(5):381–391
92. Abdel-Rahman RM, Hrdina R, Abdel-Mohsen AM et al (2015) Chitin and chitosan from Brazilian Atlantic coast: isolation, characterization and antibacterial activity. Int J Biol Macromol 80:107–120
93. Chandy T, Mooradian DL, Rao GH (1999) Evaluation of modified alginate-chitosan-polyethylene glycol microcapsules for cell encapsulation. Artif Organs 23(10):894–903
94. Chia SM, Wan AC, Quek CH et al (2002) Multi-layered microcapsules for cell encapsulation. Biomaterials 23:849–856
95. Orive G, Hernandez RM, Gascon AR, Igartua M, Pedraz JL (2003) Development and optimisation of alginate-PMCG-alginate microcapsules for cell immobilisation. Int J Pharm 259:57–68
96. Morch YA, Donati I, Strand BL, Skjåk-Bræk G (2006) Effect of Ca, Ba, and Sr on alginate microbeads. Biomacromolecules 7: 1471–1480
97. Darrabie MD, Kendall WF, Opara EC (2006) Effect of alginate composition and gelling cation on microbead swelling. J Microencapsul 23(1):29–37
98. Gibbs BF, Kermasha S, Alli I, Muligan C, Anon A (2002) Constant air quality humidity and temperature control during soft capsule manufacturing. Chem Plants Process 1: 22–23
99. Vinogradova OI (2004) Mechanical properties of polyelectrolyte multilayer microcapsules. J Physiol 16:32–36
100. Gupta V, Verma S, Nanda A, Nanda S (2005) Osmotically controlled drug delivery. Drug Del Technol 5:68–76
101. Pirvu C (2005) Chemical characterization and applications of microcapsules. Farmacia 53:61–68

Part II

Approaches to Cell Microencapsulation

Chapter 3

Cell Microencapsulation: Dripping Methods

A. Bidoret, E. Martins, B. Poncelet De Smet, and D. Poncelet

Abstract

Microencapsulation processes may be divided into three steps, namely: incorporation of the bioactive substance in the matrix, dispersion of the matrix in droplets, and conversion in microcapsules. This contribution is focused on the second step and more specifically using the dripping approach to form droplets by extrusion of liquid through a nozzle. Different technologies of dripping are described, using as an example the production of alginate beads.

Key words Dripping, Droplet size, Production

1 Introduction

As described in Part 1 of this book, there are many reasons and methods for immobilizing cells. Encapsulation is probably an easier, faster, efficient, reproducible, and cost-effective way to mimic the natural environment and the organization of immobilized cells. Encapsulation may be described in three steps:

1. Mixing the cells with the future capsule core, generally a liquid,
2. Dispersing the resulting suspension as droplets,
3. Converting the droplets in microcapsules.

This contribution will focus on the second step and more specifically on dripping technologies, i.e., forming droplets by extrusion through a nozzle. This approach is still the main method for cell encapsulation [1]. As they are formed, droplets may be converted in beads or capsules by ionic or thermal gelation, coacervation, polymerization, or interfacial cross-linking [2]. The matrix of the particles may be composed of a large range of materials [3] but for simplicity, we will refer in this chapter to the formation of hydrogel beads by extruding drop by drop an alginate solution in a calcium solution.

Emmanuel C. Opara (ed.), *Cell Microencapsulation: Methods and Protocols*, Methods in Molecular Biology, vol. 1479, DOI 10.1007/978-1-4939-6364-5_3, © Springer Science+Business Media New York 2017

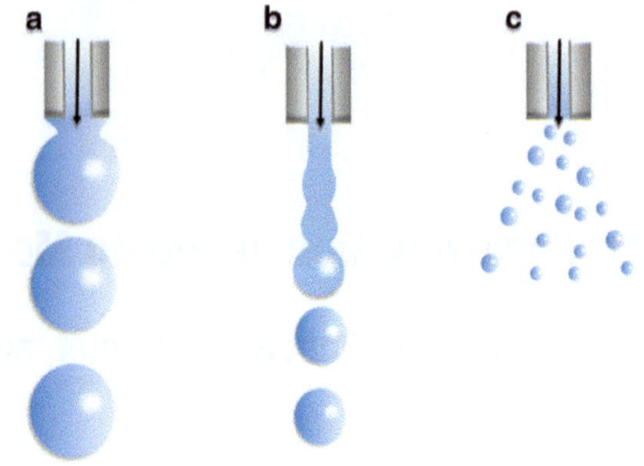

Fig. 1 Dripping regimes: (**a**) Dropwise, (**b**) Jet-breaking, (**c**) Spray regime (E. Martins ©)

Formation of droplets by extrusion of a liquid through an orifice is largely function of the velocity of the extruded liquid. Three regimes may be considered (Fig. 1):

1. For low velocities, droplets are formed at the tip of the orifice and break off one by one: dropwise regime.
2. For intermediate velocities, the liquid exits the orifice as a jet and may break-up as droplets: jet-breaking regime.
3. For large velocities, the jet becomes unstable and breaks in multiple small droplets. This is the spray regime. In the frame of the actual review, this regime will be not treated but it is the basis of spray-drying and spray-cooling technology consisting in spraying the liquid in a large chamber and either drying droplets by warm air or solidifying melt liquid droplets by cool air.

The transition from dropwise to jet-breaking regime is defined by the minimum jet velocity, u, $_{\min}j$ [4]

$$u_{j,\min} = 2\sqrt{\frac{\gamma}{\rho d_j}} \tag{1}$$

Where γ is the liquid surface tension, ρ the liquid density, and dj the internal diameter of the nozzle. The maximum flow rate for the dropwise regime, $F_{\max,\ \text{dropwise}}$, may be obtained by multiplying the minimum jet velocity, u_j, by the section of the jet (assuming that the jet diameter is equal to the internal nozzle diameter):

$$F_{\max,\text{drop wise}} = \frac{\pi d_i^2}{2}\sqrt{\frac{\gamma}{\rho d_i}} \tag{2}$$

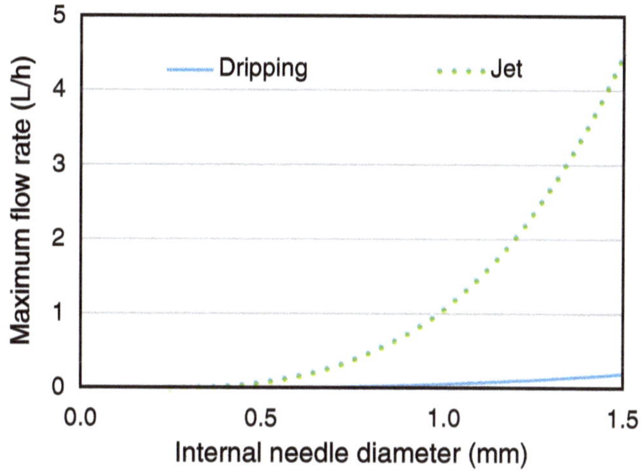

Fig. 2 Influence of the droplet diameter on the maximum productivity

The jet velocity must be limited to the terminal drop falling velocity, *ut*, to avoid collision between the jet and the droplet (reference):

$$u_t = \sqrt{\frac{4gd_d}{3\rho C_D}} \quad (3)$$

where d_d is the droplet diameter, *CD* is the drag coefficient equal to 0,44 [5]. The maximum flow rate for jet-breaking regime, $F_{max,\ jet\text{-}breaking}$ is again found by multiplying by the jet section:

$$F_{max,\ jet\ breaking} = \frac{\pi d_i^2}{4}\sqrt{\frac{4gd_d}{3\rho C_D}} \quad (4)$$

Figure 2 shows the flow rate or productivity in function of the internal diameter diameter. In jet-breaking regime, the droplet diameter is approximately twice the internal jet diameter (see later),

It is clear that the dropwise regime will be suitable only for small productions (50–250 mL/h), while the jet-breaking, especially using multi-nozzle systems, may produce quite large amounts (up to hundreds L/h for large droplets). On another side, Fig. 2 shows that the maximum productivity for both regimes is mainly proportional to the square of the droplet diameter. Reducing by two the droplet/capsule diameter thus decreases by four the productivity.

1.1 General Equipment

Figure 3 presents two assemblings for producing alginate beads by extrusion/dripping.

1. Figure 3a presents an installation based on a syringe pump, which allows easy fix precise flow rate. However, this setting is mainly adequate for dropwise regime, at low production (30–60 mL) and is limited to low viscosity (200 mPa.s).

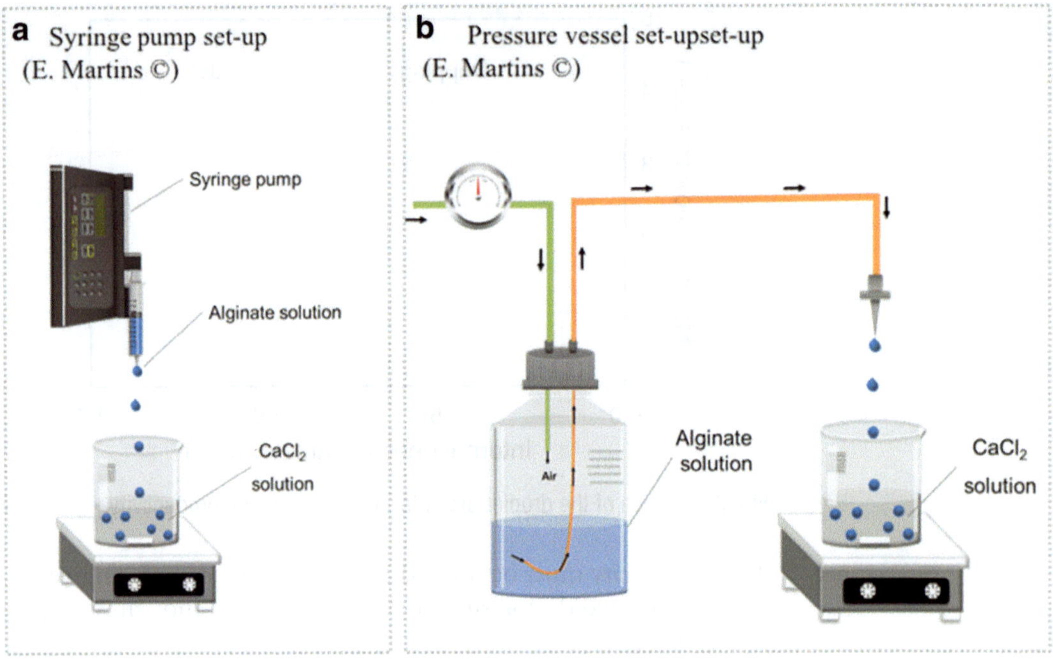

Fig. 3 Assembling for extrusion/dripping production of alginate beads

2. The system presented in Fig. 3b is based on pressure vessel. Normally, the flow rate is proportional to the applied pressure. However, any variation in the liquid viscosity may modify the flow rate making this system more difficult to control. This solution is mainly required for break-jet regime.

In dropwise regime, needles are generally used as nozzles (Fig. 4a). Smaller is the internal diameter and longer is the needle, higher are the pressure needed to reach a certain flow rate and the risk of needle blockage. In practice, it is advisable to select needles and cones with an internal diameter larger than 0.4 mm. Using cones (Fig. 4b) in place of needles reducing the pressure by a factor 5 to 10. Some models are compatible with Luer-Lock connections (Fig. 4c), allowing to alternatively use needles or cones on the same setup.

In jet-breaking regime, liquid must form a linear and vertical jet, avoiding turbulence. Most commercial equipments have a specific design and nozzles manufactured by high technologies such as laser. For in-house setup, one would prefer nozzles specifically developed to produce high-quality jet (Fig. 4d) in place of needles or plastic cones.

1.2 Protocol

The following generic protocol will be applied for the following sections, with some specific remarks linked to the methods.

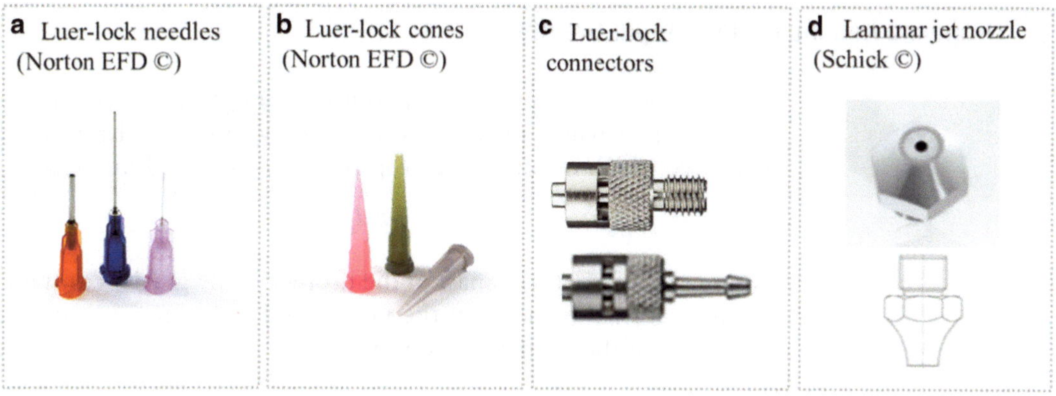

Fig. 4 Nozzles and connectors

1. An alginate solution (18 g/L) is produced by adding slowly 3, 6 g sodium alginate powder on 200 mL of distilled water under gentle agitation. Let stand for 1 h until complete dissolution.
2. Either
   (a) Fill with the alginate solution a syringe equipped with a needle or a cone, and place it on the syringe pump, on top of the collecting bath.
   (b) Fill the pressure vessel with the alginate solution, connect the vessel to the nozzle support.
3. Prepare the collecting bath by filling a beaker with 200 mL of calcium chloride solution (50 mM), place the beaker on a magnetic stirrer and fix the agitation to get a low vortex; droplets must penetrate the liquid at half distance between the beaker wall and center.
4. Start the syringe pump or fix the vessel pressure to get adequate flow rate.
5. After alginate solution extrusion, let stand the beads 15 min in the collecting bath.
6. Filter the beads on a 50–100 μm mesh, rinse the beads, and resuspend them in appropriate media for future application.

Alginate droplets contract during gelation; the resulting beads have a diameter of around 80 % of the droplet size.

2 Dropwise Regime Under Simple Gravity

In the dropwise regime, the liquid flow through the nozzle forming a droplet which grows until the gravity forces are larger than the force which maintain the droplet on the device (Tate's law):

$$mg = \pi d_e \gamma \tag{5}$$

where m represents the mass of the droplet, g the gravity constant, d_e the external diameter of the nozzle and γ the surface tension. The mass of the droplet is given by:

$$m = \frac{\pi}{6} d^3 \rho \tag{6}$$

where ρ is the density of the extruded solution. Combining Eqs. 1 and 2 allows to define the droplet diameter, d_d

$$d = \sqrt[3]{\frac{6 d_e \gamma}{g \rho}} \tag{7}$$

Figure 5 shows that the size of the droplets will be ranging between 3 to 4 mm, relatively independent to the nozzle diameter.

2.1 Equipment

Developing in-house a module for maintaining the needle/cone is not too complex. As an alternative, equipment may be purchased from Nisco Swiss company, either with a single or multi-nozzle setup (http://nisco.ch/) (Fig. 6). The company provides also a single nozzle with coaxial liquid system allowing to make core/shell capsules

2.2 Protocol

The protocol is identical to the generic one presented above. The flow rate will simply be adjusted to a value inferior to the minimum jet flow rate (100–200 mL, mainly depending of the liquid viscosity).

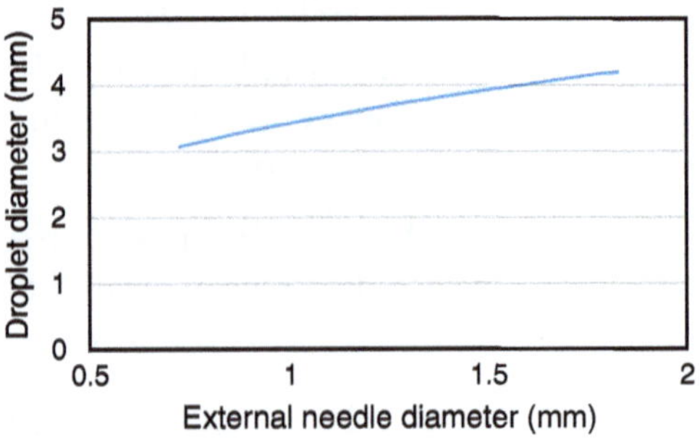

Fig. 5 Size of the droplet, d_p, versus the external needle diameter, d_e (surface tension of the extruded solution 73 mN/m at 20 °C)

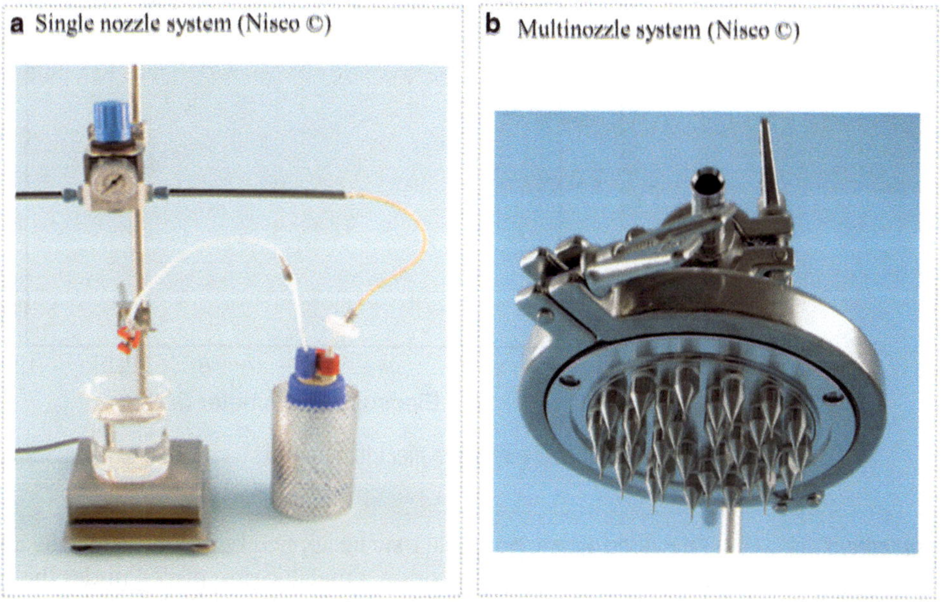

Fig. 6 Assembling for dripping under dropwise and simple gravity regime

3 Dropwise Regime Under Electrostatic Potential: Electro-dripping

Under simple gravity, only large droplets/beads can be produced. By applying an electrostatic potential to the needle, ions migrate to the surface of the droplets [6]. This results in some repulsion at the surface of the droplet, which interferes and reduces the surface tension [6]:

$$\gamma = \gamma_o \left(1 - \frac{U^2}{U_c^2}\right), \tag{8}$$

where γo is the surface tension without applied electrostatic potential, U the applied electrostatic potential, Uc, the critical electrostatic potential. As a result of combining Eqs. 7 and 8, the Fig. 7 gives a typical evolution of the droplet size in function of the electrostatic potential applied to produce alginate beads. As the electrostatic potential U increases, the droplet size decreases until a minimum value, reached while the electrostatic potential is equal to the critical electrostatic potential, Uc. At higher electrostatic potential the liquid will exit the nozzle as jet. This jet may separate into droplets whose size is independent of the applied voltage.

Different factors may affect the critical electrostatic potential Uc as well as the minimum droplet size, including the diameter of the nozzle, sign of the electrostatic potential, composition of the solution, flow rate [4].

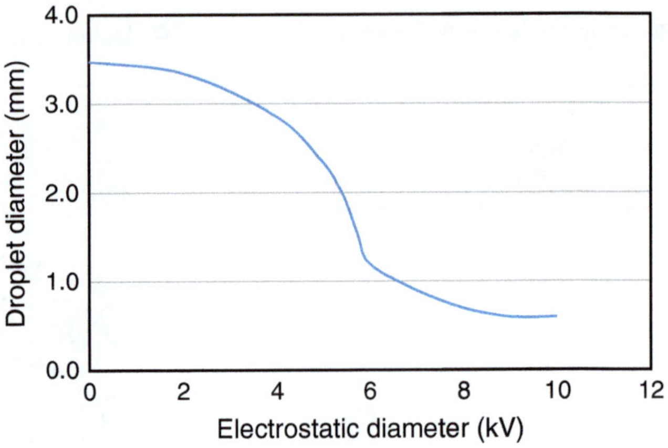

Fig. 7 Droplet size versus the electric potential

3.1 Equipment

The electrostatic potential may be applied between the needle and the collecting bath (Fig. 7a) or on a metallic ring placed under the needle (Fig. 7b). The electrostatic potential must be applied using a high-voltage power supply delivering a very low current intensity (Bertan, USA). The collecting vessel or the ring will be preferably connected to the ground. Distance between the needle and either the collecting solution or the ring should be at least one centimeter per 10 kV applied to avoid electric arch. The risk of electrocution is very limited but the electric discharge may be painful. Always turn off the power supply and wait a few seconds before to manipulate the system.

It exists at least one commercial electro-dripping device developed by the Nisco Swiss company (http://nisco.ch/, Fig. 8c).

3.2 Protocol

Follow the generic protocol. As the liquid is extruded as droplets, turn on the power supply and increase the electric potential until you reach the expected droplet size. When the solution is extruded, decrease the electric potential, turn off the power supply, and wait a few seconds before manipulating the system.

4 Dropwise Regime with Coaxial Air Flow

To decrease the size of droplet, air flow can be applied coaxially around the needle. The air drag force will support the gravity force. Establishing equations to predict the size of the droplets in function of air flow rate is challenging. However, depending on the design and the air flow, size as low as a few micrometers could be obtained. As expected, the productivity of the system decreases quickly as the droplet size decreases.

They are two configurations for the coaxial air flow:

1. The needle tip extends outside of the air flow tube (Fig. 9a). The air simply applies on the forming droplet a drag force leading to

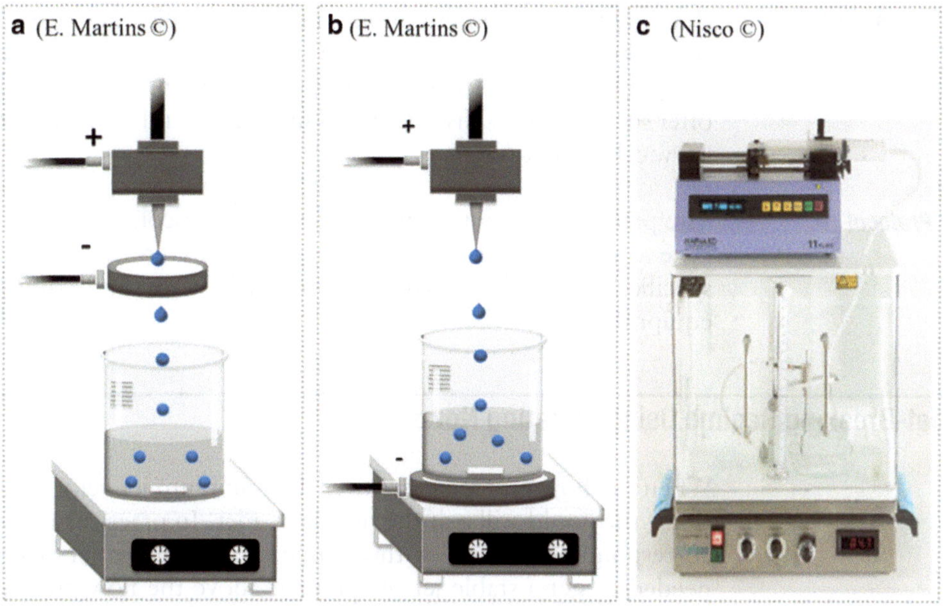

Fig. 8 Assembling for production of alginate beads by electro-dripping

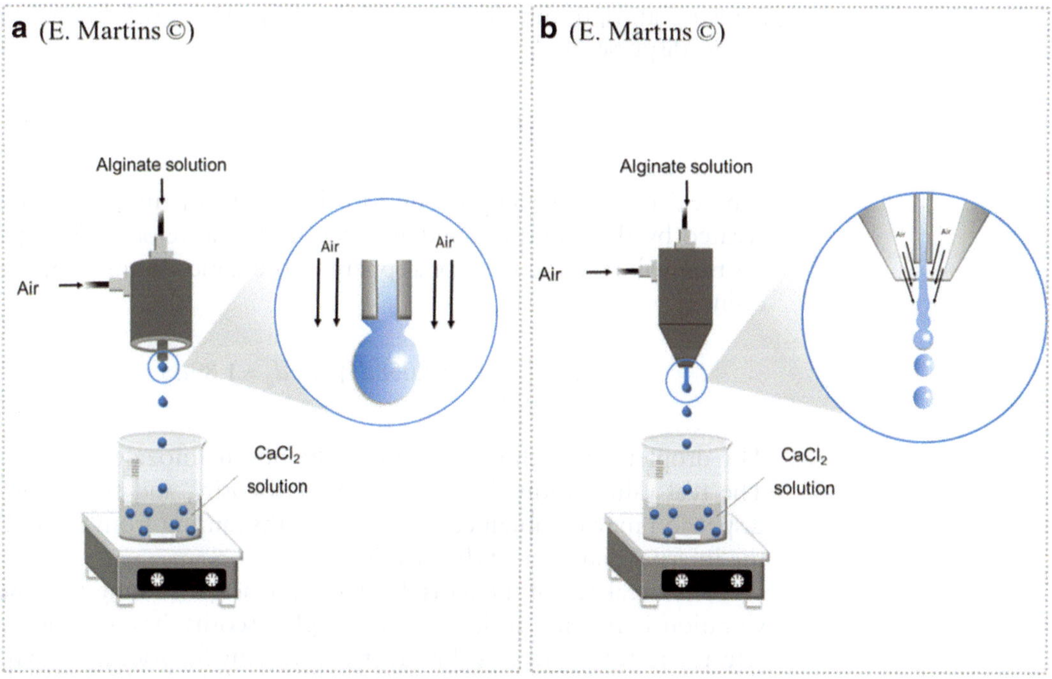

Fig. 9 Assembling for alginate bead production using coaxial air flow

smaller size. The droplet diameter generally may be reduced to 800 μm without forming a spray.

2. The needle tip is inside the air flow tube (Fig. 9b). The air compresses the liquid, generally leading to a jet, which breaks into droplets down to a few micrometers.

4.1 Equipment

Building the first configuration is not too complex but again Nisco company proposes a ready-to-use unit. For the second configuration, it is best to buy a commercial equipment. Two companies offer such a system: Nisco (http://nisco.ch/) and Cellena (http://www.cellena.net/en/).

4.2 Protocol

The protocol is identical to the generic protocol, except that you have to set up the air flow rate to get the desired droplet size. Some specific instructions may be provided with the commercial equipment.

5 Jet-Breaking Regime Using Vibrating Nozzle

When the flow rate increases over the minimum jet velocity conditions, the liquid will exit the nozzle as a jet. In practice, a flow rate corresponding to two or three time the minimum jet velocity is needed to get a stable jet. As pointed above the flow rate may be limited to avoid collision between the jet and the droplets but also to remain in laminar flow conditions. The jet will break into droplets due to Rayleigh instability. By avoiding any vibration but applying an optimum frequency f, the droplet size will be mainly mono-dispersed [7]:

$$f = \frac{u_j}{\lambda} \text{ and } \lambda = 4.0458\mathrm{d_j} \tag{9}$$

Where λ is the wavelength but also the length of the jet portion created by the vibration, and u_j is the jet linear velocity. This jet portion will take the form of a sphere of the same volume that the cylinder:

$$\frac{\pi}{6}d_d^3 = \left(\frac{\pi}{4}d_j^2\right)\left(4.058d_j\right) \text{ or } d_d = 1.89d_j \tag{10}$$

The droplet size is thus about twice the internal nozzle diameter. The real value is function of the liquid viscosity, and the mono-dispersion may be obtained for wave lengths ranting from 3.5 to 7 the internal diameter of the nozzle.

The breakage of the jet is due to resonance, assuming that the vibration is moving inside the jet. High viscosity has a damping effect, and this process will correctly work only for solution of viscosity lower than 200 mPa.s

5.1 Equipment

This technology is relatively difficult to develop in-house. At our knowledge, three companies propose such equipment: Buchi (http://www.buchi.com/), Nisco (http://nisco.ch/) and Brace (http://www.brace.de/), the last one provides also equipment for

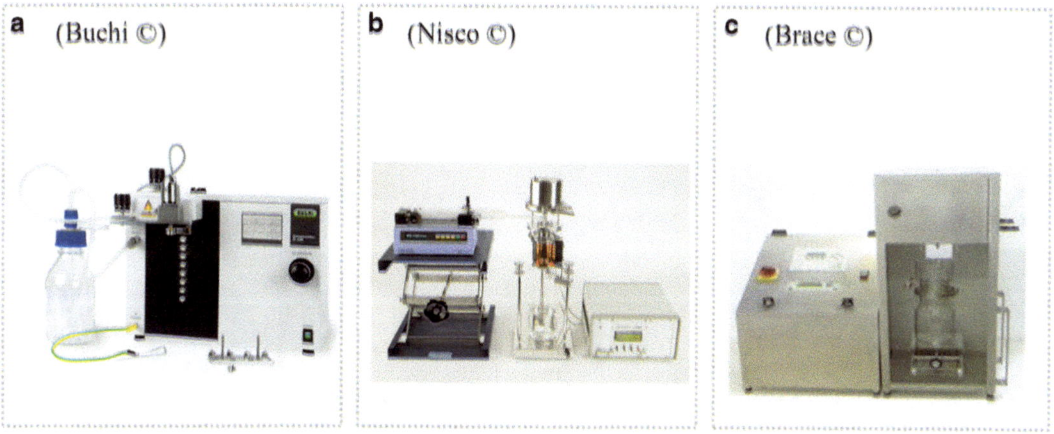

Fig. 10 Commercial equipment for producing alginate beads by vibrated nozzle jet breaking

industrial production. Nisco and Brace offer multi-nozzle equipment. All of them also propose a coaxial system (Fig. 10) allowing production of core/shell capsule.

These equipments are supplied with LEDs flashing at the same frequency that the vibration applied on the liquid. When the conditions are optimum, the jet breaks into mono-dispersed droplets and light flashes happen at the exact stage of droplet detachment. The droplets seem immobile. If the droplets seem to move, the flow rate or the frequency has to be modified until droplets become apparently immobile.

5.2 Protocol

General protocol may be applied here, adjusting the parameters as follow:

1. Select a nozzle with an internal diameter half of the requested droplet diameter.
2. Set the frequency and the flow rate, according either to the Eqs. 9 and 11, or to supplier recommendations.
3. If the flow of droplets seems to move up or down, modify either the frequency or the liquid flow rate until droplets seem immobile. This step is critical and often not understood by the operator. Supplier recommendations are based on a specific solution and must be adjusted in function of the physical properties of the real used liquid.

6 Jet-Breaking Regime Using a Rotating Wheel

If the solution viscosity is too high, the jet may be broken using a wheel equipped with very fine wires: jet-cutter (Fig. 11). This system was developed by Vorlop's group in Germany [8] and exploited

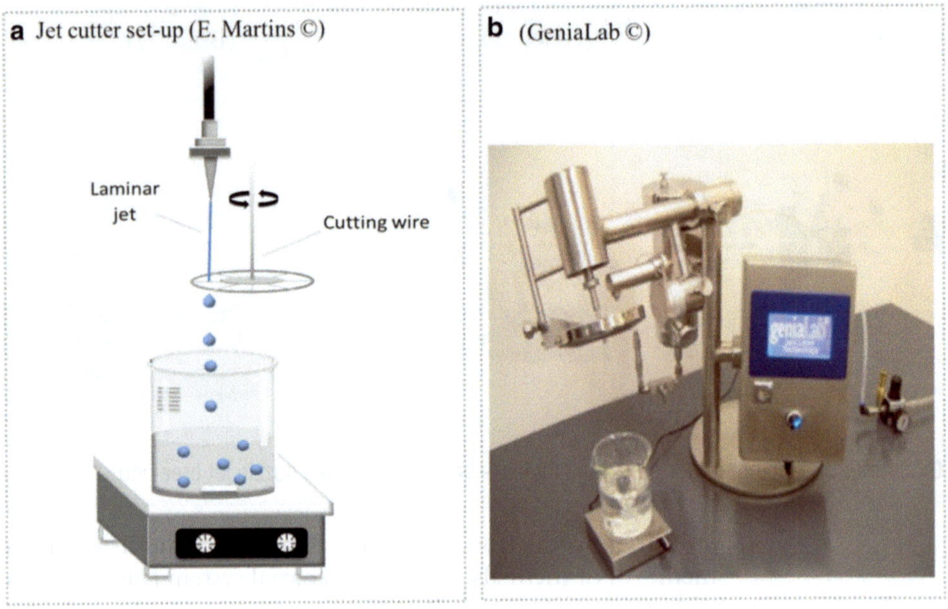

Fig. 11 Production of alginate beads using a jet-cutter

by the start-up GeniaLab. The diameter of the droplets may be predicted by:

$$d_{\mathrm{d}} = \sqrt[3]{\frac{6}{\pi}\frac{F}{N'}} \tag{11}$$

where F is the extruded liquid flow rate, N the number of wires on the wheel and Ω the rotational speed of the wheel. The liquid jet velocity must be high, similar to the maximum jet flow rate (Fig. 2) to favor a correct cutting. Some part of the liquid jet may be expelled by the wire crossing the jet. The liquid lost is mainly proportional to the wire diameter, but it may be minimized by tilting the wheel [6]. The wire diameter should preferably be lower than 80 μm (50 μm) to get yields higher than 95 % (98 %).

6.1 Equipment

GeniaLab (http://genialab.de/) sales (or sold) laboratory equipments. Recently, Nisco (http://nisco.ch/) started to propose similar equipment.

6.2 Protocol

The generic protocol may be applied. Select a nozzle diameter half the size of the expected droplets, fix the flow rate, and compute the corresponding jet velocity, fix the rotational speed to get a wire velocity at the position of the cutting similar to the jet velocity.

References

1. Steele JAM, Hallé JP, Poncelet D, Neufeld RJ (2014) Therapeutic cell encapsulation techniques and applications for diabetes. Adv Drug Deliv Rev 67–68:74–83
2. Poncelet D, Tam S (2009) Microencapsulation technologies for bioartificial endoctrine pancreas. In: Hallé JP, de Vos P, Rosenberg L (eds) The bioartificial pancreas. Transworld Research Network, Kerala, India p. 37–49
3. De Vos P, Lazarjani HA, Poncelet D, Faas MM (2014) Polymers in cell encapsulation from an enveloped cell perspective. Adv Drug Deliv Rev 67–68:15–34
4. Lindblad NR, Schneider JM (1969) Production of uniform-sized liquid droplets. J Sci Instrum 42:635–639
5. Rhodes M. (2013) Single particles in a fluid. In: Introduction to Particle Technology (2nd Ed.). Chichester: Wiley; pp 30–51
6. Poncelet D, Neufeld RJ, Goosen MFA, Bugarski B, Babak V (1999) Formation of microgel beads by electric dispersion of polymer solutions. AIChE J 45(9):2018–2023
7. Rayleigh JWS (1878) On the instability of the jet. Proc Lond Math Soc 10:4
8. Prusse U, Dalluhn J, Breford J, Vorlop KD (2000) Production of spherical beads by Jet Cutting. Chem Eng Technol 23(12):1105–1110

Chapter 4

Field Effect Microparticle Generation for Cell Microencapsulation

Brend Ray-Sea Hsu and Shin-Huei Fu

Abstract

The diameter and sphericity of alginate-poly-L-lysine-alginate microcapsules, determined by the size and the shape of calcium alginate microspheres, affect their in vivo durability and biocompatibility and the results of transplantation. The commonly used air-jet spray method generates microspheres with a wider variation in diameter, larger sphere morphology, and evenly distributed encapsulated cells. In order to overcome these drawbacks, we designed a field effect microparticle generator to create a stable electric field to prepare microparticles with a smaller diameter and more uniform morphology. Using this electric field microparticle generator the encapsulated cells will be located at the periphery of the microspheres, and thus the supply of oxygen and nutrients for the encapsulated cells will be improved compared with the centrally located encapsulated cells in the air-jet spray method.

Key words Microdroplet, Microparticle, Microsphere, Microcapsule, Microencapsulation, Alginate, Poly-L-lysine, Polycondensation, High-voltage electric field-generating circuitry, Low-current electric field

1 Introduction

1.1 Background

Microencapsulation of cells in alginate-poly-L-lysine-alginate (A-P-A) microcapsules is a commonly used procedure for immunoprotection which was originally described by Lim and Sun in 1980 [1]. Intraperitoneal implantation of A-P-A-microencapsulated xenogenetic pancreatic islets has been shown to reverse hyperglycemia in streptozotocin-induced diabetic animals for several months [2], suggesting that it is a promising approach for the treatment of insulin-dependent diabetes mellitus (IDDM) [3, 4]. Implantation of A-P-A-microencapsulated xenogenetic and allogenetic parathyroid cells [5], hepatocytes [6–8], and adrenal medulla chromaffin cells [9] has also been found to be effective in (a) correcting serum calcium levels in rats after parathyroidectomy, (b) maintaining adequate liver function in chemically-induced acute hepatic failure rats, and (c) improving involuntary movements in a

Emmanuel C. Opara (ed.), *Cell Microencapsulation: Methods and Protocols*, Methods in Molecular Biology, vol. 1479,
DOI 10.1007/978-1-4939-6364-5_4,

parkinsonism rat model. Recently, A-P-A-encapsulated cells expressing sFasL were used as a local device and found to efficiently suppress malignant Fas-sensitive tumors in a SCID mice model [10]. In addition, an A-P-A microencapsulation oral delivery system was reported to significantly enhance the gut microbiota enrichment of loaded probiotics and favorably modulate the microbiota composition of all the colonic compartments in a well-established dynamic human gastrointestinal model [11]. The A-P-A membrane provides cells with an immunoisolation environment through its selective semipermeable characteristics [12, 13]. An A-P-A membrane is formed by interfacial polycondensation between poly-L-lysine solution and calcium alginate beads [14]. The final diameter and sphericity of the microcapsules are determined mainly by the original size and shape of the calcium alginate beads. Since the size and the sphericity of the microcapsules influence their durability, biocompatibility, and effectiveness of transplantation, an effective method of preparing calcium alginate beads in which to embed donor cells or islets is important. A co-axial air-jet spray [15] is the most commonly used method to prepare microspheres. However, it produces microspheres with wide variations in sphericity and diameter, and it is difficult to obtain evenly sized capsules smaller than 500 μm in diameter. A reduction in the size of capsules is considered to be one of the most important factors to improve the microenvironment of microcapsules while allowing for the bidirectional diffusion of nutrients, oxygen, and waste [16, 17]. The size of the microcapsules is also an essential factor for the amount of insulin secreted by encapsulated islets in response to a glucose load [18], and it has been reported that smaller capsules are more biocompatible than bigger capsules [19]. Other techniques based on electrostatic force obtained by creating an electrostatic potential, either static [20] or in pulses [21], between the gelling bath and the needle have been developed and used to pull droplets off the needle and to generate more uniform and smaller microspheres [4, 16]. Therefore, we designed a field effect microparticle generator that could create a stable electric field to generate calcium alginate microspheres. By changing the generator voltage, the field distance, and the gear protruding speed, we prepared microparticles with an adjustable range of diameters between 50 and 350 μm. Herein, we present the principles and characteristics of the field effect microparticle generator and discuss its application in islet and cell transplantation.

1.2 Electric Field Microparticle Generator

The electric field microparticle generator consisted of high-voltage electric field-generating circuitry and a motor-driven worm gear piston (Fig. 1). The machine could establish a voltage of up to 10 kV. A 50-MOhm resistor was inserted into the high-voltage output circuit to prevent it from arcing between the two electrodes of the electric field. The motor-driven worm gear was clamped on

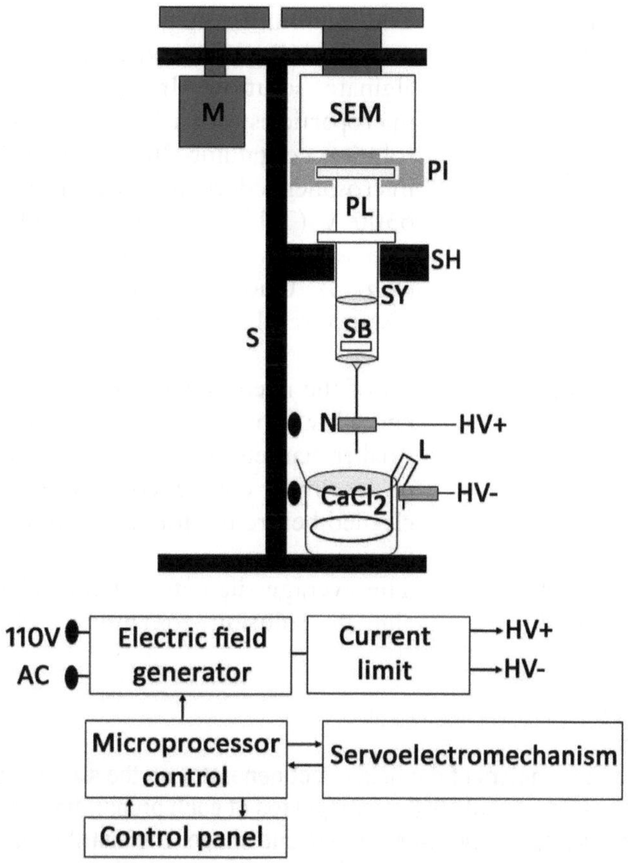

Fig. 1 Outline of installation and block diagram of the field effect microparticle generator. *M* motor, *SEM* servoelectromechanism, *PI* piston, *PL* plunger, *SH* syringe holder, *SY* syringe, *SB* stir bar, *N* needle, *L* loop, *S* stand, *HV+* positive high voltage, *HV–* negative high voltage, *AC* alternating current

top of the syringe holder, and the contents of the syringe were extruded continuously when the piston pressed on the plunger. The drops on the needle tip were separated into microdroplets before entering the calcium chloride solution and forming calcium alginate microspheres. One electrode was connected to the needle tip, and the other to a conductive loop in the beaker containing the calcium chloride solution. The distance between the syringe needle tip and the surface level of the calcium chloride solution was defined as the field distance. The field distance, voltage, and the spinning speed of the worm gear that decided the extrusion speed were all adjustable.

1.3 Factors Influencing Calcium Alginate Beads

The effects of voltage, extrusion speed, field distance, cell density, and concentration of sodium alginate on the diameter and morphology of calcium alginate beads are listed in Tables 1 and 2 (*see* **Note 1**).

1.3.1 Field Strength and Distance

The minimum voltage required to generate microdroplets varied with field distance, concentration, and cell density of the sodium alginate solution. It required at least 4000 V to generate microparticles with a 1 cm field distance and 1.5 % sodium alginate solution containing 3000 islets/mL. The average diameter of the microspheres decreased with increasing voltage between 5000 and 6500 V (Table 1). Above 6500 V, the electric field generated microspheres with a larger diameter and tailing on one pole (Fig. 2). Under similar control parameters, an increasing field distance resulted in larger microspheres (Table 1).

1.3.2 Needle Gauge

All of the needles were cut to remove the bevel evenly, as a blunt end allows the field effect to work. A 25-gauge needle produced smaller diameter microspheres in comparison to a 22-gauge needle under similar control conditions (Table 1). The needle lumen was cleaned before use for each run to ensure patency.

1.3.3 Cell Density of Sodium Alginate

The average diameter of the microspheres prepared in sodium alginate solution containing NS-1 mouse myeloma cells at a

Table 1
Effect of parameters of the field effect generator on the size and shape of calcium alginate microcapsules. About 1000 microspheres of each preparation were inspected under an inverted light microscope to evaluate the size and shape. The distribution of diameter was expressed as a percentage of the total number of microspheres examined

	Field strength effect				Gauge effect		Distance effect		Speed effect			
Needle	25				25	22	25		25		22	
Voltage	5000	5500	6000	6500	6000		6500		5500		5500	
Speed (mL/min)	0.57				0.57		0.57		0.57	0.19	0.57	0.19
Field distance (cm)	1				1		1	2	1			
Containing NS-1 (10^7/mL)	+				+		–		–			
Conc. of alginate	1.5 %				1.5 %		1.5 %		1.5 %			
Particle diameter (μm)	Percentage											
50–100									7%	27%		
100–150	10%	14%	52%	73%	52%		30%		14%	31%		
150–200	22%	54%	46%	27%	46%	10%	70%	56%	29%	16%	5%	3%
200–250	60%	32%	2%		2%	40%		25%	29%	16%	35%	30%
250–300	8%					40%		19%	21%	10%	55%	60%
300–350						10%					5%	7%

Table 2
Effect of alginate concentration on the size and shape of calcium alginate microcapsules. About 1000 microspheres of each preparation were inspected under an inverted light microscope to evaluate the size and shape. The distribution of diameter was expressed as a percentage of the total number of microspheres examined

Parameters	Effect of alginate concentration										
Needle gauge	25										
Voltage	5500	5500	5000	5000	5000	5000	5500	5500	5500	5500	5500
Speed	0.57 mL/min										
Field distance (cm)	1	1	1	2	1	1	1	1	1	1	1
NS-1	No containing cell										
Alginate (%)	1	1.2	1.2	1.2	1.3	1.35	1.4	1.45	1.5	2.25	4.5
Diameter (μm)	Percentage										
50–100	10%							4%	7%	22%	
100–150	40%	5%					8%	8%	14%	32%	
150–200	30%	40%	10%	10%	4%	5%	16%	40%	29%	14%	
200–250	20%	50%	35%	20%	8%	10%	30%	28%	29%		
250–300		5%	35%	20%	60%	64%	42%	16%	21%	32%	
300–350			15%	20%	20%	16%	4%	4%			
350–400			5%	20%	8%	5%					
400–450				10%							30%
450–500											30%
500–550											40%
Tailing	90%	80%	50%	90%	90%	5%	3%				

concentration of 10^7 cells/mL was smaller than for microspheres prepared from sodium alginate alone (Table 1).

1.3.4 Speed of Sodium Alginate Protrusion

Microspheres generated with a faster protruding speed (0.57 mL/min vs. 0.19 mL/min) had larger diameters. However, when using a 22-gauge needle, the effect of the protruding speed (0.57 mL/min vs. 0.19 mL/min) became negligible (Table 1).

1.3.5 Concentration of Alginate Solution

When using a 25-gauge needle, 5000–5500 V, 0.57 mL per minute protruding speed and 1 cm field distance, the lowest concentration of sodium alginate (without NS-1 myeloma cells) required to minimize the amount of tailing microspheres to 5% was 1.4%. The average diameter of calcium alginate microspheres decreased when the concentration of sodium alginate was increased to 2.25%.

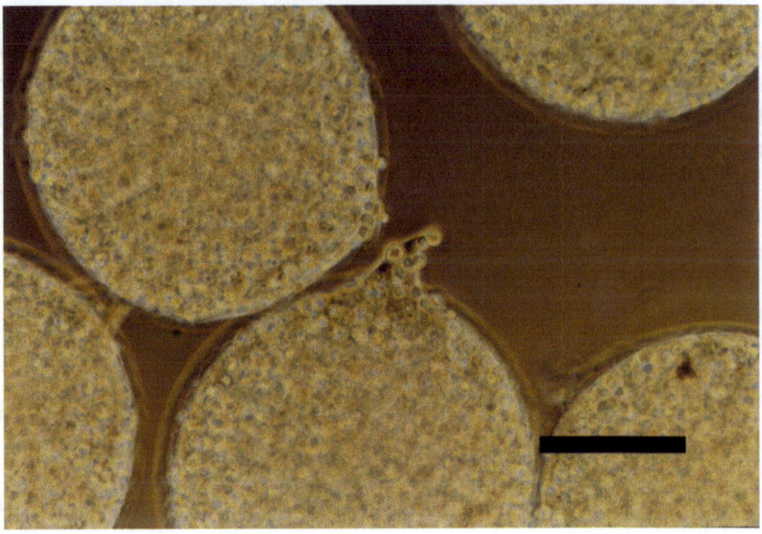

Fig. 2 Tailing effect. A combination of conditions including 1.5 % sodium alginate containing 10^7 NS-1 cells/mL, 1 cm field distance, 7000 V, 0.57 mL/min protruding speed, and 25-gauge needle was used to demonstrate the generation of tailing microspheres. The diameter of the microspheres was about 250 μm. The *black bar* indicates 100 μm

A 4.5 % alginate solution did not generate spherical beads but large, elongated, sausage-like calcium alginate droplets with a size of 400 μm × 600–800 μm. When the alginate concentration was less than or equal to 1.35 %, the percentage of tailing beads increased dramatically (Table 2).

1.3.6 Effect of Field Strength on Viability of NS-1 Cells and Rat Islets

To evaluate the effect of a high-voltage field on the viability of cells, the doubling time of NS-1 myeloma cells was estimated by counting viable cell numbers both before and after passing them through the field effect particle generator using different voltages ranging from 4000 to 6500 V, and it was found to be 14 h for all preparations. Using a 22-gauge needle, 5500 V, field distance of 1 cm, 0.57 mL/min protrusion speed and 1.5 % sodium alginate, we microencapsulated rat islets and implanted intraperitoneally 2400 encapsulated islets per mouse into diabetic BALB/c mice (Fig. 3). Fasting glucose levels decreased from above 300 mg/dL before transplantation to below 180 mg/dL within 7 days after transplantation and maintained normoglycemia for more than 4 months (data not shown). This suggested that the electric field did not exert a harmful effect on the encapsulated cells. We therefore used this field effect microparticle generator to prepare A-P-A-microencapsulated rat islets.

1.3.7 Skin Effect (Gauss' Law)

Sodium alginate (1.5 %) containing 10^7 cells was either protruded through the field effect microparticle generator or protruded

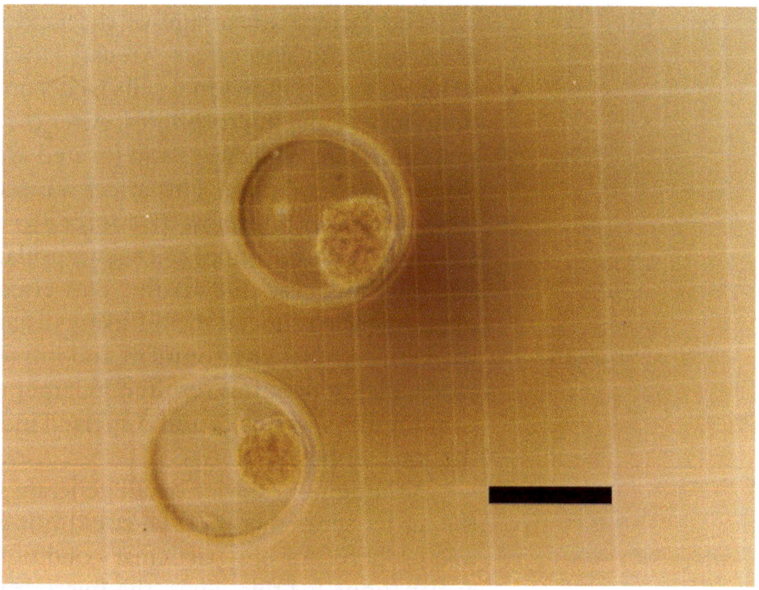

Fig. 3 A-P-A-microencapsulated Sprague-Dawley rat islets. The diameters of the microcapsules were 275–350 μm. Background is a hemacytometer grid to evaluate the capsule size. The *black bar* indicates 200 μm

manually into calcium chloride solution to generate droplets by gravity. The calcium alginate microspheres were dissected and observed under an inverted light microscope. The NS-1 cells prepared by the field effect were distributed around the periphery of the microspheres, whereas the NS-1 cells were evenly distributed in manually protruded microspheres.

1.4 The Advantages of Using an Electric Field Microparticle Generator

The first advantage of using an electric field to generate calcium alginate microspheres was its effectiveness in producing smaller and more uniform microspheres. The diameter of islet-containing microcapsules was affected by the size of calcium alginate microspheres, the length of time that the microspheres were incubated with calcium chloride solution, and the washing procedure before polylysine polycondensation. For transplantation, we used a 22-gauge needle, 5500–6500 V, field distance of 1 cm, 0.57 mL/min protrusion speed and 1.5 % sodium alginate to microencapsulate islets. With these conditions, the diameter of most islet-containing microcapsules was 300–450 μm when alginate microspheres were incubated in 100 mM $CaCl_2$ solution for a total of 6 min with 2 min of washing in normal saline. Large microspheres resulted in large microcapsules after polylysine polycondensation, however they were more fragile than small capsules. Smaller capsules had a shorter diffusing distance and thus greater concentration gradients of insulin and glucose across the capsular membrane [18]. A large-gauge needle could be used to produce smaller calcium alginate microspheres for individual microencapsulation.

The second advantage of the field effect microparticle generator was the skin effect (Gauss' Law). The skin effect of field strength forced individual cells to become distributed around the periphery of the microspheres. After the microencapsulation process, individual cells were still located around the inner surface of the A-P-A membrane. This effect was also noted by Cai et al. [6], who used an electrostatic force to generate microspheres. Eccentric localization of microencapsulated islets was always seen, and this was also due to the skin effect of field strength. This phenomenon reduced the diffusing distance of molecules between the extracapsular environment and intracapsular cells. Coaxial air-jet droplet generation resulted in larger capsules with wider variations in size and shape, and dispersed individual cells evenly inside the microspheres.

Although our device could generate smaller microspheres than the air-jet spray method, we could not generate identical sizes of microspheres for any particular condition. There are two possible explanations for this. First, the microdroplets between the needle tip and the surface of the calcium chloride solution were in line, and the field strength between each microdroplet was different. Second, the movement of the gears resulted in pulsatile but not continuous protrusion of sodium alginate. Further studies are needed to reach the goal of generating microspheres with a uniform size.

2 Materials

2.1 Microencapsulation

1. Field effect microparticle generator.
2. Normal saline.
3. Sodium alginate.
4. 100 mM Calcium chloride solution.
5. 55 mM Sodium citrate solution.
6. Poly-L-lysine hydrobromide: MW 15,000–30,000 Da.

2.2 Preparation of NS-1 Cells

1. NS-1 myeloma cell-growing media: RPMI-1640 medium supplemented with 7.5 % heat-inactivated fetal calf serum and 1 % antibiotic solution.
2. Cells were cultured at 37 °C in a humidified 5 % CO_2–95 % air atmosphere.
3. All materials must be sterile.

2.3 Preparation of Islets

1. Male Sprague-Dawley rats.
2. 100 mg/mL Ketamine

3. Islet isolation media: Modified Hanks' buffered solution (MHBS) supplemented with 1 % antibiotic solution.
4. Collagenase: Type V, 1000–1300 U/mg.
5. Ficoll: Type 400-DL, MW 40,000 Da.
6. All materials must be sterile.

2.4 Diabetic Induction

1. Male BALB/c mice.
2. Streptozotocin.
3. Citrate-buffered saline solution.

3 Methods

3.1 Preparation of the Reagents

1. Normal saline: 0.9 % NaCl, filter through a 0.22 μm filter unit. Store at 4 °C.
2. 1.5 % Sodium alginate in 0.9 % saline: Weigh out 3.0 g of sodium alginate, and slowly sprinkle into 100 mL dH_2O while stirring rapidly. Stir until dissolved, then add 100 mL of 1.8 % NaCl and mix well (*see* **Note 2**). Centrifuge at 2500 × *g* at 4 °C for 30–35 min. Filter the supernatant through a 0.22 μm filter unit. Store at 4 °C.
3. 100 mM Calcium chloride solution: Weigh out 11 g of anhydrous calcium chloride (or 14.7 g of dehydrate $CaCl_2$) and dissolve in 1 L of distilled water. Filter through a 0.22 μm filter unit. Store at 4 °C.
4. 0.15 % Sodium alginate: Dilute 30 mL of 1.5 % alginate solution with 270 mL of 0.9 % normal saline. Filter and store at 4 °C.
5. 55 mM Sodium citrate solution: Dissolve 3.23 g of citric acid trisodium salt dehydrate in 100 mL dH_2O, add 95 mL of 0.9 % saline, adjust the pH to 7.4 by adding 1 N HCl. Make up the final volume to 200 mL with 0.9 % saline. Filter and store at 4 °C.
6. 0.05 % Poly-L-lysine (PLL) solution: Dissolve 15 mg of PLL in 30 mL 0.9 % saline. Filter through a 0.22 μm filter unit. This solution must be prepared freshly for each encapsulation run.
7. Collagenase solution: Dissolve 7.8 mg collagenase V in 10 mL of MHBS and keep the solution on ice after dissolved.
8. Ficoll gradient solution: Mix Ficoll and MHBS in a 400-mL beaker to make a final concentration of 27 % (w/v). Hand-mix the mixture till wet thoroughly. Drop in a magnetic stirrer and autoclave it for 10 min (*see* **Note 3**). Then stir at room temperature till completely dissolve. Prepare different concentrations of Ficoll gradient with sterile MHBS as follows:

	27% (mL)	MHBS (mL)	Total (mL)
27%	120.00	0	120
23%	68.15	11.85	80
20%	58.25	20.75	80
11%	32.60	47.40	80

Store at 4 °C.

9. Citrate-buffered saline solution: Dissolve 2.45 g of anhydrous citric acid, 10.96 g of trisodium citrate dehydrate, and 8.77 g of NaCl in a final volume of 1000 mL dH_2O. Adjust pH to 4.5 with citric acid.
10. Streptozotocin solution: Weigh out 150 mg per kilogram body weight in 10 mL citrate buffered saline solution. Intraperitoneally inject each mouse with freshly prepared 0.01 mL streptozotocin solution per gram body weight (*see* **Note 4**).

3.2 Isolation of Rat Islets

Islets of Langerhans were isolated from the pancreases of male Sprague-Dawley rats weighing 300–350 g using a collagenase method [22].

1. Anesthetize each rat with intraperitoneal injections of 0.3 mL ketamine.
2. Place a rat in the supine position and spray the whole abdomen with 75% ethanol.
3. Use a table light and a spot light to illuminate the operation field.
4. Position a 10 mL syringe with a #25-gauge needle connected to a 20 cm PE-50 tubing filled with 10 mL of freshly prepared collagenase solution on the left-hand side of the operation field.
5. Make a cut over the low abdomen and cut along bilateral abdominal walls all the way to the diaphragm, and then flip the abdominal wall upward to expose the abdominal organs.
6. Push up the liver, and push down the intestines to the right-hand side to expose the common bile duct (CBD). Using straight Kelly forceps to clamp the entrance of the CBD to the duodenum.
7. Use two sharp-ended forceps to dissect the CBD between the hepatic ducts and the first merging pancreatic duct.
8. Use scissors to cut the inferior vena cava (IVC) and wipe away the resulting blood with tissues.
9. Make a cut over the right-hand side edge of the CBD (perpendicular to the CBD). Intubate the CBD through the small cut,

and use left-hand forceps to hold both the CBD and the inserted end of the tubing.

10. Use your right hand to slowly push the collagenase solution into the CBD/pancreatic ducts/pancreas unit.
11. After the pancreas has been fully expanded with 10 mL of collagenase solution, remove the tubing and straight Kelly forceps. Remove fat tissue from the greater curvature of the stomach. Detach the spleen, and detach the pancreas from the intestine on the right-hand side. Use two forceps to remove the pancreas and place it into a 50 mL tube.
12. Incubate the 50 mL tube in a 37 °C water bath shaker and shake for 20 min.
13. Decant the contents into a chilled 50 mL tube and rinse immediately with chilled MHBS, allow the contents to sediment by gravity, then decant the supernatant. Repeat three times and then leave one-third the volume of MHBS. Screw the cap on tightly.
14. Hand shake the tubes until most of the fragments are around 0.2 mm or smaller.
15. Filtrate immediately through a screen with a pore size of 300–350 μm in diameter.
16. Transfer the filtrate to a 50 mL tube filled up with chilled MHBS, and then centrifuge immediately at 300 × *g* for 2 min at 4 °C.
17. Remove the supernatant by suction. Resuspend the pellet gently in chilled MHBS by swirling and fill up with chilled MHBS. Centrifuge at 450 × *g* for 2 min at 4 °C.
18. Remove the supernatant. Resuspend the pellet gently in 6 mL of 27% Ficoll, and then layer over with 4 mL of 23%, 4 mL of 20%, and 4 mL of 11% Ficoll in sequence. Centrifuge at 700 × *g* for 10 min without braking. The islets will be in the top two interfaces.
19. Collect the islets from the interfaces and transfer to 50 mL tubes. Centrifuge at 240 × *g* for 5 min at 4 °C. Remove the supernatant and resuspend the pellet with chilled MHBS. Centrifuge at 130 × *g* for 5 min at 4 °C.
20. Wash with MHBS several times and transfer the islets to a sterile petri dish. Under a dissecting microscope, hand-pick islets with a 22-gauge Insyte catheter (*see* **Note 5**).

3.3 Diabetes Induction and Intraperitoneal Islet Implantation

1. Induce diabetes in male 12-week-old BALB/c mice using an intraperitoneal injection of freshly prepared streptozotocin solution in a dosage of 150 mg/kg body weight (*see* **Note 6**).

2. Intraperitoneally implant about 2000–3000 microencapsulated islets of Sprague-Dawley rats with an 18-gauge Insyte peripheral intravenous catheter into a diabetic BALB/c mouse.

3.4 Blood Glucose Monitoring

1. Determine blood glucose levels using a dry reagent method based on the glucose oxidase method (*see* **Note 7**).

3.5 Measurement of the Diameter of Calcium Alginate Microspheres

1. Extrude sodium alginate into 100 mM calcium chloride solution with the field effect microparticle generator, and leave the microspheres in the calcium chloride solution until examinations for size and shape.
2. Estimate the diameters of the microspheres under an inverted light microscope using a hemacytometer.
3. Count about 500–1000 microspheres of each preparation and express the distribution of the diameter as a percentage.

3.6 Microencapsulation of Islets

Islets were microencapsulated by the A-P-A method according to O'Shea and Sun [23].

1. Transfer a suspension of islets in 1.5 % alginate at a concentration of 2000–3000 per mL to a 10 mL syringe containing a small magnet. One milliliter of alginate produces 70,000 capsules.
2. Turn on the field effect generator and syringe pump. Extrude alginate into a beaker containing 50 mL of 100 mM $CaCl_2$ solution. Use a large magnet to rotate the small bar inside the syringe to avoid precipitation of the islets.
3. At the end of extrusion, around 3–4 min, transfer gel beads to a 50 mL conical centrifuge tube. Stand for 2 min, and then withdraw the supernatant down to 15 mL using an aspirator pump.
4. Add 15 mL 0.9 % saline. Mix briefly and stand for 2 min, and then withdraw the supernatant down to 10 mL (*see* **Note 8**).
5. React the beads with 30 mL 0.05 % PLL solution for 7 min by gently rotating for 5 min and standing for 2 min (*see* **Note 9**). Withdraw the supernatant down to 10 mL.
6. Wash twice with 30 mL 0.9 % saline, and then withdraw the supernatant down to 5–10 mL.
7. Coat with 20 mL 0.15 % alginate solution for 8 min by gently rotating for 6 min and standing for 2 min. Withdraw the supernatant down to 10 mL.
8. Wash twice with 20 mL 0.9 % saline, and then withdraw the supernatant down to 5 mL.
9. React the capsules with 20 mL 55 mM sodium citrate solution for 6 min by gently rotating for 4 min and standing for 2 min (*see* **Note 10**). Withdraw the supernatant down to 10 mL.

10. Wash twice with 20 mL 0.9% saline and resuspend in culture medium for overnight culture.

4 Notes

1. The manipulation of a field effect microparticle generator is quite straight forward and safe. However, many pitfalls may fail encapsulation of cells, such as not using blunt-end needle causing failure of field effect, lumen of needle is not cleaned causing occlusion, and contamination of working solutions causing pathogen attachment.
2. Since the viscosity of the gel is also important for the morphology of calcium alginate beads, here we use low viscosity of sodium alginate to prepare the alginate solution in this technique. It should be noted that incompletely dissolved sodium alginate results in overestimated concentration of alginate solutions.
3. Screw the cap on tightly before sterilization. While autoclave the Ficoll solution, use liquid sterilization and take it out exactly after 10-min sterilization. Do not let it go through slow exhaustion. The bottle containing Ficoll solution should be weighed before and after sterilization. After sterilization, add the lost weight back into the Ficoll solution with sterile dH_2O.
4. Wear on gloves and mask to avoid inhaling streptozotocin. Prepare ice and chilled buffer solution for streptozotocin solution. Leave weighing boat on ice while mix with buffer solution. The acidic pH value of the buffer solution is also extremely crucial for this step.
5. Pick up and discard large pieces of non-islet tissues first. Then pick up islets and transfer to a clean petri dish. Finally, pick up islets from second petri dish, transfer to a conical centrifuge tube containing normal saline and prepare for microencapsulation.
6. Only use mice that have hyperglycemia for at least two weeks and have two consecutive plasma glucose levels of more than 300 mg/dL as diabetic recipients.
7. Transplantation is deemed a failure when the plasma glucose levels exceed 200 mg/dL on two consecutive determinations.
8. Prolonged wash of calcium alginate microparticles in saline causes uneven surface of microspheres. Skipping saline wash of calcium alginate microparticles cause overexpanding and crack formation of A-P-A membrane after liquefaction of interior calcium alginate.
9. Prolonged polycondensation in PLL solution causes golf ball appearance on microcapsules.

10. Inadequate reaction with sodium citrate solution causes incomplete liquefaction of interior calcium alginate that interfering in the diffusion of oxygen, nutrients, and wastes.

References

1. Lim F, Sun AM (1980) Microencapsulated islets as bioartificial endocrine pancreas. Science 210:908–910
2. Sun AM, Vacek I, Sun YL, Ma X, Zhou D (1992) In vitro and in vivo evaluation of microencapsulated porcine islets. ASAIO J 38: 125–127
3. Soon-Shiong P, Feldman E, Nelson R, Komtebedde J, Smidsrod O, Skjak-Braek G, Espevik T, Heintz R, Lee M (1992) Successful reversal of spontaneous diabetes in dogs by intraperitoneal microencapsulated islets. Transplantation 54:769–774
4. Soon-Shiong P, Heintz RE, Merideth N, Yao QX, Yao Z, Zheng T, Murphy M, Moloney MK, Schmehl M, Harris M, Mendez R, Sandford PA (1994) Insulin independence in a type 1 diabetic patient after encapsulated islet transplantation. Lancet 343:950–951
5. Fu XW, Sun AM (1989) Microencapsulated parathyroid cells as a bioartificial parathyroid. Transplantation 47:432–435
6. Cai Z, Shi ZQ, Sherman M, Sun AM (1989) Development and evaluation of a system of microencapsulation of primary rat hepatocytes. Hepatology 10:855–860
7. Sun AM, Cai Z, Shi Z, Ma F, O'Shea GM (1987) Microencapsulated hepatocytes: an in vitro and in vivo study. Biomater Artif Cells Artif Organs 15:483–496
8. Zhang Y, Chen XM, Sun DL (2014) Effects of coencapsulation of hepatocytes with adipose-derived stem cells in the treatment of rats with acute-on-chronic liver failure. Int J Artif Organs 37:133–141
9. Aebischer P, Tresco PA, Sagen J, Winn SR (1991) Transplantation of microencapsulated bovine chromattin cells reduces lesion-induced rotational asymmetry in rats. Brain Res 560:43–49
10. Goren A, Gilert A, Meyron-Holtz E, Melamed D, Machluf M (2012) Alginate encapsulated cells secreting Fas-ligand reduce lymphoma carcinogenicity. Cancer Sci 103:116–124
11. Rodes L, Tomaro-Duchesneau C, Saha S, Paul A, Malhotra M, Marinescu D, Shao W, Kahouli I, Prakash S (2014) Enrichment of Bifidobacterium longum subsp. infantis ATCC 15697 within the human gut microbiota using alginate-poly-L-lysine-alginate microencapsulation oral delivery system: an in vitro analysis using a computer-controlled dynamic human gastrointestinal model. J Microencapsul 31:230–238
12. Darquy S, Reach G (1985) Immunoisolation of pancreatic B cells by microencapsulation. An in vitro study. Diabetologia 28:776–780
13. Lanza RP, Sullivan SJ, Chick WL (1992) Islet transplantation with immunoisolation. Diabetes 41:1503–1510
14. Arshady R (1989) Preparation of microspheres and microcapsules by interfacial polycondensation techniques. J Microencapsul 6:13–28
15. Fritschy WM, Woltens GHJ, Schilfgaarde RV (1991) Effect of alginage-polylysine-alginate microencapsulation on in vitro insulin release from rat pancreatic islets. Diabetes 40:37–43
16. Strand BL, Gåserød O, Kulseng B, Espevik T, Skjåk-Baek G (2002) Alginate-polylysine-alginate microcapsules: effect of size reduction on capsule properties. J Microencapsul 19:615–630
17. Huang X, Zhang X, Wang X, Wang C, Tang B (2012) Microenvironment of alginate-based microcapsules for cell culture and tissue engineering. J Biosci Bioeng 114:1–8
18. Chicheportiche D, Reach G (1988) In vitro kinetics of insulin release by microencapsulated rat islets: effect of the size of the microcapsules. Diabetologia 31:54–57
19. Robitaille R, Pariseau JF, Leblond F, Lamoureux M, Lepage Y, Halle JP (1999) Studies on small (<350 μm) alginate-poly-l-lysine microcapsules. III. Biocompatibility of smaller versus standard microcapsules. J Biomed Mater Res 44:116–120
20. Goosen MFA, Al-Ghafri AS, ElMardi O, Al-Belushi MIJ, Al-Hajri HA, Mahmoud ESE, Consolacion EC (1997) Electrostatic droplet generation for encapsulation of somatic tissue: assesment of high-voltage power supply. Biotechnol Prog 13:497–502
21. Halle JP, Leblond FA, Pariseau JF, Jutras P, Brabant MJ, Lepage Y (1994) Studies on small (<300 μm) microcapsules: II-parameters governing the production of alginate beads by high voltage electrostatic pulses. Cell Transplant 3:365–372
22. Lacy PE, Kostianovsky M (1967) Method for the isolation of intact islets of Langerhans from the rat pancreas. Diabetes 16:35–39
23. O'Shea GM, Sun AM (1986) Encapsulation of rat islets of Langerhans prolongs xenograft survival in diabetic mice. Diabetes 35:943–946

Chapter 5

Microfluidic Approach to Cell Microencapsulation

Varna Sharma, Michael Hunckler, Melur K. Ramasubramanian, Emmanuel C. Opara, and Kalyan C. Katuri

Abstract

Bioartificial pancreas made of insulin-secreting islets cells holds great promise in the treatment of individuals with Type-1 diabetes. Successful islet cell microencapsulation in biopolymers is a key step for providing immunoisolation of transplanted islet cells. Because of the variability in the size and shape of pancreatic islets, one of the main obstacles in their microencapsulation is the inability to consistently control shape, size, and microstructure of the encapsulating biopolymer capsule. In this chapter, we provide a detailed description of a microfluidic approach to islet cell encapsulation in alginate that might address the microencapsulation challenges.

Key words Microencapsulation, Islets, Diabetes, Microfluidic cell encapsulation

1 Introduction

Destruction of insulin producing β-cells leads to a host of complications in individuals suffering from Type-1 diabetes. Exogenous insulin is usually used as treatment methodology for Type-1 diabetes. Transplantation of insulin producing islet cells in the body to make up for insulin deficiency is considered to have potential as a viable therapeutic option. Several research groups have been working towards the development of efficient islet cell isolation techniques and encapsulation of the isolated islet cells in biopolymer materials before the cells are transplanted in the human body, a concept generally referred to as a bioartificial pancreas [1–4].

Among various candidate materials for islet cell encapsulation, alginate is the most commonly used owing to its biocompatibility and the ability to gel under mild conditions in the presence of divalent cations [5–8]. Using alginate as the encapsulating material, Opara et al. have described a method to generate permselective multilayer alginate-poly-L-ornithine-alginate microcapsules which has been successfully transplanted in an omentum pouch [1, 9].

Emmanuel C. Opara (ed.), *Cell Microencapsulation: Methods and Protocols*, Methods in Molecular Biology, vol. 1479, DOI 10.1007/978-1-4939-6364-5_5, © Springer Science+Business Media New York 2017

Among several other factors, the success of the transplantation depends on the effectiveness of cell encapsulation in preventing attack from the host immune system and oxygen availability to the transplanted islet cells, both of which are related to the overall efficiency of the encapsulation procedure and the resulting pore microstructure. Oxygen availability to encapsulated islet cells generally increases with decrease in the individual capsule size and total number of empty alginate capsules. It is challenging to devise an encapsulation technique that generates uniformed sized alginate encapsulated islet capsules at high production rates while minimizing the total number of empty capsules and satellite droplets [9].

Microfluidics-based encapsulation offers several advantages that are typically not present in other encapsulation methods such as electro-static spraying, air-syringe pump-based droplet generation. In general microfluidic device-based encapsulation leads to a smaller coefficient of variation of capsule diameters and improved sphericity of capsules.

The focus of this chapter is on generating cross-linked alginate microcapsules of consistent shape and size which can be used for encapsulating islet cells. The process of preparing islet cells for encapsulation, perm-selective coating of alginate microbeads using PLL or PLO after encapsulation, liquefaction of the alginate core using sodium citrate, and the final coating of microbeads with alginate follows the procedure that was outlined by Opara et al. [10]. We use a custom-made microfluidic device as shown in Fig. 1 to generate alginate droplets which are then cross-linked internally within the microfluidic device.

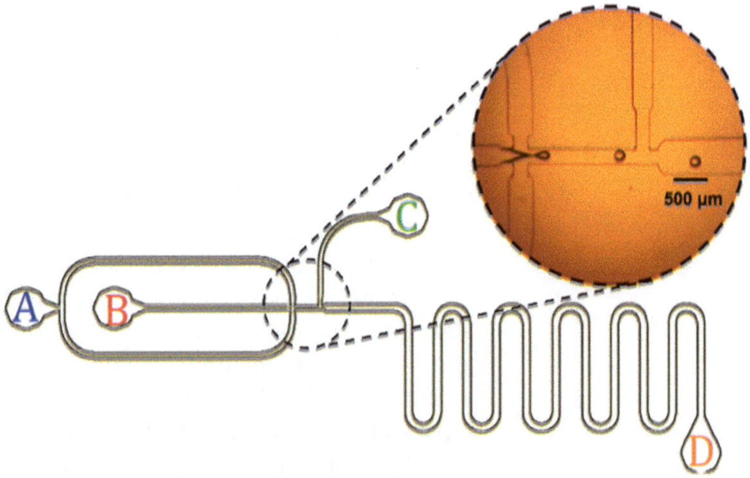

Fig. 1 Microfluidic template design with the ports labeled: (*inlet port A*) light mineral oil with 3% (V/V) Span80; (*inlet port B*) 1.5% (w/v) alginate + 100 mM Ca-EDTA (1:1); (*inlet port C*) 1% (v/v) acetic acid + mineral oil; (*outlet port D*) collection port. Production of liquid alginate microbeads suspended in light mineral oil at the flow-focusing junction is shown in the *inset*. Alginate cross-linking is initiated at the T-junction where the mineral oil with acetic acid is added to the horizontal channel

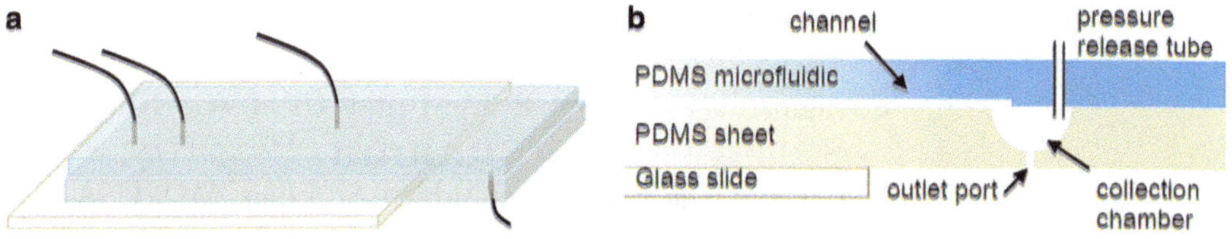

Fig. 2 Final design integration. (**a**) 3D view with tubes connected. (**b**) Cross-section of PDMS device with three layers: PDMS channel mold, PDMS sheet, and glass slide

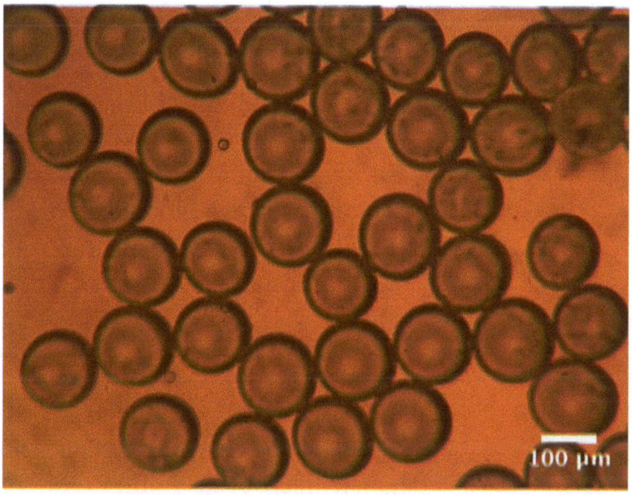

Fig. 3 Uniformly sized alginate beads formed under the following conditions: (Fig.1: (*inlet-A*) Mineral Oil + 3 % (v/v) Span 80 at 20 μl/min; (Fig.1: (*inlet-B*) 1.5 % (w/v) alginate + 100 mM Ca-EDTA (1:1) at 3 μl/min; (Fig.1: (*inlet-C*) 1 % (v/v) acetic acid + mineral oil at 12 μl/min

A 3D view, and a cross-sectional view of the encapsulation device, is shown in Fig. 2. Alginate droplets generated at the flow-focusing junction are cross-linked as they flow through the serpentine section of the device. The cross-linked capsules are collected from under the device as shown in Fig. 2. A micrograph of the collected capsules is shown in Fig. 3.

2 Materials

2.1 Chemicals

1. PDMS (Sylgard 184 Silicone Elastomer Kit, Dupont).
2. Aquapel (Aquapel 88625-47100 Glass Treatment, Aquapel Inc.).
3. Ethanol (02877 Fluka, Sigma Aldrich, St. Louis, MO, USA).
4. Sodium Citrate Dihydrate (W302600, Sigma Aldrich, St. Louis, MO, USA).

5. Light Mineral oil (M5904-500ML, Sigma Aldrich, St. Louis, MO, USA).
6. Alginate (Pronova UP LVM and UP LVG, Novamatrix, Sandvika, Norway).
7. Calcium Chloride (Sigma Aldrich, St. Louis, MO, USA).
8. Sodium-EDTA (Mallinckrodt, St. Louis, MO, USA).
9. Span80 surfactant (S6760, Sigma Aldrich, St. Louis, MO, USA).
10. Acetic acid (A6283, Sigma Aldrich, St. Louis, MO, USA).

2.2 Equipment

1. PDMS-based Microfluidic Devices (Clemson University, Clemson, SC).
2. Syringe Pumps (PHD 2000 Infusion and 11 Plus, Harvard Apparatus; NE-306, New Era Pump Systems, Inc).
3. Syringes and needles (1 and 3 ml syringes and 20 gauge blunt-tip syringe needles, Becton, Dickinson and Company).
4. Polyethylene Tubing (Intra-medic Polyethylene Tubing, PE 60, Product # 427416, Becton, Dickinson and Company).

3 Methods

Following is the procedure that is followed to generate cross-linked alginate microbeads using PDMS microfluidic device that is shown in Figs. 1 and 2.

1. Prepare a solution of light mineral oil with 3% Span80 (v/v). Addition of Span80 to mineral oil prevents bead coalescence after generation at the flow-focusing junction.
2. Prepare a solution of 1.5% (w/v) alginate mixed with 100 mM Ca-EDTA at 1:1 ratio while maintaining the concentration of alginate solution at 1.5% (w/v). Vortex the mixture prior to droplet formation. Ca-EDTA is prepared using instructions from the Supplementary Information of Utech et al. [11]. Ca-EDTA complex mixed with the alginate at a 1:1 ratio is stable at pH 7. Upon reduction of the pH with an addition of acetic acid, the calcium ions are released from the EDTA and induce internal crosslinking within the alginate.
3. To facilitate alginate droplet generation at the flow-focusing junction, the hydrophobicity of the microfluidic channel surface is increased by subjecting it to Aquapel treatment. Withdraw about 0.1 ml of Aquapel into a 1 ml syringe. Use a 20 gauge syringe needle and a small Polyethylene tube connected to the syringe needle to direct the fluid into one of the ports of the microfluidic device (*see* **Note 1**). Direct the fluid through all the holes and wait 30 s. Clean the device thoroughly with pressurized air to completely clean all the Aquapel out leaving only a thin film on the internal channel surfaces.

4. All the solutions were drawn into 1 ml BD syringes and capped with 20 gauge needles. A polyethylene tube is fitted over the needle. The solutions are placed into the syringe pumps and manually primed until the fluid flows out of the tubing. Cut the tubing at an angle (~45 °) before inserting into the device to prevent the tube ends making contact with the bottom surface of the channel. Wipe the ends of tubes or rinse with 70 % ethanol to remove any debris that could potentially get trapped within the device. Insert the tubing into the appropriate channels.
5. Start the slower flow rates first. This helps prime the channels with slower moving liquids (like alginate) and prevents the channels with faster moving liquids (mineral oil) from entering other channels and causing issues. It is especially important that the solution from Inlet C does not backflow into Inlet B, because then premature alginate crosslinking will clog the channel.
6. Steady-state operation occurs rather quickly with the device shown in Fig. 2, which includes the collection chamber. Usually within 2–3 min, the device will reach a steady state and run consistently for long periods of time until the syringes have to be refilled (*see* **Notes 2** and **3**).
7. A small cut end of a pipette tip (~5 mm long) can be snugly fit into the outlet port to direct the fluid into a dish or well or other external collection item.
8. The beads-acid oil suspension streaming out of the channel is collected in a conical tube with 1.1 % (w/w) $CaCl_2$ solution. In the collection tube acid oil will float on top of the $CaCl_2$ solution due to its lighter density while the beads will sediment at the bottom of the conical tube. Following complete partitioning of beads to the aqueous phase, the oil is discarded and the beads are then washed twice with 1 % Tween solution prior to 20-min PLO (0.1 %) coating for immunoisolation. The PLO coated beads are then finally covered with thin layer of alginate [1] prior to use in experiments.

4 Notes

1. The tubing should be thread over the 20 gauge needle for a snug fit.
2. Remaining Aquapel will discolor the PDMS channels, degrade the PDMS, and leave stains. It is found that leaving the Aquapel on for 20–30 s and then cleaning it out with pressurized air worked best. Micropipettes filled with epoxy can be used to block particular channels and ensure the aquapel or air thoroughly filled the intended channel.

3. Maintaining a clean device is important for maintaining a device that is free of debris and not degraded. After every use, it is recommended to flush the device with 55 mM Sodium Citrate and 70% ethanol solution, which helps to remove the mineral oil and remaining alginate. For storing overnight or longer, use compressed air to clean all the channels thoroughly. Ensuring that the fluid travels through each channel of the device is important in removing any foreign debris before storage of the device. Due to the presence of several channels in the device, the cleaning solution sometimes does not travel out of the desired port (redirected to a different port) and thus, does not clean the port/channel thoroughly.

References

1. Opara EC, Mirmalek-Sani SH, Khanna O, Moya ML, Brey EM (2010) Design of a bioartificial pancreas(+). J Investig Med 58(7): 831–837
2. Pareta RA, Farney AC, Opara EC (2013) Design of a bioartificial pancreas. Pathobiology 80(4):194–202
3. Kepsutlu B, Nazli C, Bal T, Kizilel S (2014) Design of bioartificial pancreas with functional micro/nano-based encapsulation of islets. Curr Pharm Biotechnol 15(7):590–608
4. Orive G, Santos E, Poncelet D, Hernandez RM, Pedraz JL, Wahlberg LU, De Vos P, Emerich D (2015) Cell encapsulation: technical and clinical advances. Trends Pharmacol Sci 36(8):537–546
5. Lee BR, Hwang JW, Choi YY, Wong SF, Hwang YH, Lee DY, Lee SH (2012) In situ formation and collagen-alginate composite encapsulation of pancreatic islet spheroids. Biomaterials 33(3):837–845
6. Lee KY, Mooney DJ (2012) Alginate: properties and biomedical applications. Prog Polym Sci 37(1):106–126
7. Richardson T, Kumta PN, Banerjee I (2014) Alginate encapsulation of human embryonic stem cells to enhance directed differentiation to pancreatic islet-like cells. Tissue Eng Part A 20(23–24):3198–3211
8. Gudmund Skjåk-Br⊣k G (1990) Alginate as immobilization matrix for cells. Trends Biotechnol 8:71–78. http://www.sciencedirect.com/science/article/pii/016777999090139O
9. Tendulkar S, Mirmalek-Sani S-H, Childers C, Saul J, Opara E, Ramasubramanian M (2012) A three-dimensional microfluidic approach to scaling up microencapsulation of cells. Biomed Microdevices 14(3):461–469
10. Opara E, McQuilling J, Farney A (2013) Microencapsulation of pancreatic islets for use in a bioartificial pancreas. In: Basu J, Ludlow JW (eds) Organ regeneration: Methods and Protocols. Humana Press, Totowa, NJ, pp 261–266. http://link.springer.com/protocol/10.1007%2F978-1-62703-363-3_21
11. Utech S, Prodanovic R, Mao AS, Ostafe R, Mooney DJ, Weitz DA (2015) Microfluidic generation of monodisperse, structurally homogeneous alginate microgels for cell encapsulation and 3D cell culture. Adv Healthc Mater 4(11):1628–1633

Part III

Biomaterials and Enabling Technologies in Cell Microencapsulation

Chapter 6

Polymeric Materials for Cell Microencapsulation

A. Aijaz, D. Perera, and Ronke M. Olabisi

Abstract

Mammalian cells have been microencapsulated within both natural and synthetic polymers for over half a century. Specifically, in the last 36 years microencapsulated cells have been used therapeutically to deliver a wide range of drugs, cytokines, growth factors, and hormones while enjoying the immunoisolation provided by the encapsulating material. In addition to preventing immune attack, microencapsulation prevents migration of entrapped cells. Cells can be microencapsulated in a variety of geometries, the most common being solid microspheres and hollow microcapsules. The micrometer scale permits delivery by injection and is within diffusion limits that allow the cells to provide the necessary factors that are missing at a target site, while also permitting the exchange of nutrients and waste products. The majority of cell microencapsulation is performed with alginate/poly-L-lysine microspheres. Since alginate itself can be immunogenic, for cell-based therapy applications various groups are investigating synthetic polymers to microencapsulate cells. We describe a protocol for the formation of microspheres and microcapsules using the synthetic polymer poly(ethylene glycol) diacrylate (PEGDA).

Key words Cell microencapsulation, Microsphere, Microcapsule, Hydrogel, PEGDA, Cell therapy

1 Introduction

Cells can be microencapsulated for a variety of applications, the most common being cell-based therapies. Cell-based therapies have numerous advantages in regenerative medicine. Cells can be grafted to differentiate and replace damaged tissues, protect other donor cells from apoptosis, promote cell activation and growth, participate in genetic and nongenetic material transfer, and provide immunomodulation [1–4]. A biological mode of drug delivery, cells can respond to their environment and secrete native factors or can be genetically modified to provide a continuous release of therapeutic factors [5–11]. An injection of a cell suspension to a target site is less invasive than a surgery to implant a device; however, a disadvantage to delivering a bolus of cells is that the injected cells may migrate from the target site. Additionally, foreign donor cells can trigger a host immune reaction that may diminish efficacy, requiring immunosuppressive therapy.

Emmanuel C. Opara (ed.), *Cell Microencapsulation: Methods and Protocols*, Methods in Molecular Biology, vol. 1479,
DOI 10.1007/978-1-4939-6364-5_6,

Microencapsulation is a technique that can be used in cell-based therapies to protect donor cells by encasing them in a material impenetrable by both the host cells and the entrapped donor cells. The process was first described in 1966 when Chang microencapsulated mammalian erythrocytes in nylon and first used therapeutically in 1980 when Lim and Sun microencapsulated pancreatic beta cells in alginate to return diabetic rats to normoglycemia [12, 13]. This isolates the donor cells from the host's immune system when they are delivered into the body for cell therapy, eliminating the need for immunosuppression. It also protects the delivered cells from injury during mechanical injection. The microencapsulated cells provide the necessary factors that are missing at a target site, while the encasing microspheres or microcapsules permit the exchange of gases, nutrients, and waste products. The microencapsulating material restricts the entrapped cells from migrating out of a target site and prevents the integration of these donor cells into host tissue. Microencapsulation specifically involves the encapsulation of cells into a material or device with at least one dimension less than 1000 μm, while macroencapsulation geometries exceed this dimension. Cells can be microencapsulated into a variety of geometries, such as cubes, spheres, prisms, and cylinders [14, 15] and can be microencapsulated within a variety of materials, such as gels, hydrogels, ceramics, and devices [15–17].

The most widely researched microencapsulated cells are pancreatic beta cells within alginate/poly-L-lysine based hydrogel microcapsules, which are currently in several clinical trials for diabetes treatment [18–21]. Hydrogels like alginate have soft tissue-like properties, such as elasticity and hydrophilicity, which make them good scaffold materials for tissue engineering purposes. For some hydrogels, the gelation process is relatively mild, which is gentle enough to permit the microencapsulation of living mammalian cells [22]. Hydrogels are available in both natural forms like collagen, gelatin, alginate, and agarose, and synthetic forms like poly(vinyl alcohol) (PVA), poly(acrylic acid) (PAA), and poly(ethylene glycol) (PEG). Research groups generally prefer either natural or synthetic polymers to form their hydrogels, depending on their applications. Each has benefits and drawbacks. Natural hydrogels can be synthesized under physiological conditions, permitting the encapsulation of living mammalian cells. Natural hydrogels also have biochemical and microarchitectural cues that can promote the viability and differentiation of entrapped cells. Synthetic hydrogels lack these cues, and can be used to introduce such cues individually to monitor cell response. Unlike natural hydrogels, the material properties of synthetic hydrogels are tunable and do not have batch-to-batch variability [23]. Conversely, synthetic polymers often have harsh polymerization conditions that are toxic to living cells. We describe a system that encapsulates living cells within an acrylated PEG, poly(ethylene glycol) diacrylate (PEGDA). Sandwiching the hydrophilic PEG, hydrophobic acrylate groups cluster in an aqueous solution and are susceptible to free-radical polymerization. Free radicals supplied by photoinitiators exposed to light initiate the polymerization of these clustered acrylate groups, resulting in a cross-linked hydrogel network. PEGDA is resistant to protein adsorption and thus cell adhesion, giving it an extremely low immunogenicity profile. It is highly hydrophilic and is generally nondegradable but can be modified to be biodegradable and/or bioactive. We have microencapsulated cells into both nondegradable and matrix metalloproteinase (MMP)-degradable PEGDA.

In this chapter, we provide protocols to microencapsulate cells into PEGDA microspheres (solid hydrogels with cells embedded into the matrix) and microcapsules (hollow hydrogels containing free-floating cells entrapped by the hydrogel membrane). The protocols are modified from those first described by Olabisi et al. and further detailed by Franco et al. and have been used to microencapsulate human, rat, ovine, and porcine mesenchymal stem cells, rat fibroblasts, MHP36 neural stem cells, murine bEnd.3 brain microvascular endothelial cells, NIH-3 T3 fibroblasts, murine MC3T3-E1 preosteoblasts, rat RIN-m beta cells, murine AtT-20ins sarcoma cells, and are currently used to microencapsulate cells for in vitro and in vivo wound healing and fracture repair applications [10, 24–27].

2 Materials

2.1 Tissue Culture Equipment

1. Incubator, humidified 37 °C, 5 % CO_2, 95 % air.
2. Tissue culture flasks: 25 cm^2, 75 cm^2, 150 cm^2, 225 cm^2.
3. Inverted microscope.
4. Automated pipettes, multichannel pipettes: 1 ml, 200 μl, 50 μl, 10 μl.
5. Centrifuge.
6. Laminar flow tissue culture hood.

2.2 Tissue Culture Reagents and Supplies

1. Cells of choice.
2. Complete cell media (including necessary antibiotics and serum).
3. Phosphate buffered saline (PBS).
4. Trypsin 0.25 % (w/v)–ethylenediaminetetraacetic acid (EDTA) 0.04 % (w/v).
5. Sterile pipette tips (1 ml, 20–200 μl, 10 μl).
6. Sterile centrifuge conical test tubes (15 ml, 50 ml).

2.3 Microencapsulation Equipment

1. Metal halide lamp (*see* **Note 1**)
   (a) Dolan Jenner MH-100 Metal Halide Illuminator
   (b) Heat absorbing glass: 50 mm diameter
   (c) Glass filter: 2 in square long pass 320 nm
   (d) Fiber optic adapter: 0.441 in ID.
   (e) Liquid light guide, 300–650 nm, 0.2 in core.
2. 2 benchtop ultravortexes (*see* **Note 2**).
3. 2 flat top vortex attachments
4. Masking tape.
5. White paper.
6. Microcentrifuge.
7. Microencapsulation station (*see* Fig. 1; *see* **Note 3**).
   (a) A support stand with rod and 8 in × 10 in base.
   (b) A medium sized dual adjustment bench clamp with 3 prongs.
   (c) A medium sized dual adjustment bench clamp with 2 prongs.
   (d) An aluminum cylinder with dimensions: OD 34 mm × ID 15 mm × H 34 mm.
   (e) A 0.2–0.5 mm thick Teflon sheet.
8. Shade 5 welding safety glasses.
9. Timer.
10. Autoclave.

2.4 Microencapsulation Reagents and Supplies

1. 10 kDa PEGDA.
2. 1-vinyl-2-pyrrolidinone (NVP; *see* **Note 4**).
3. 2,2-dimethoxy-2-phyenylacetophenone (acetophenone; *see* **Note 5**).

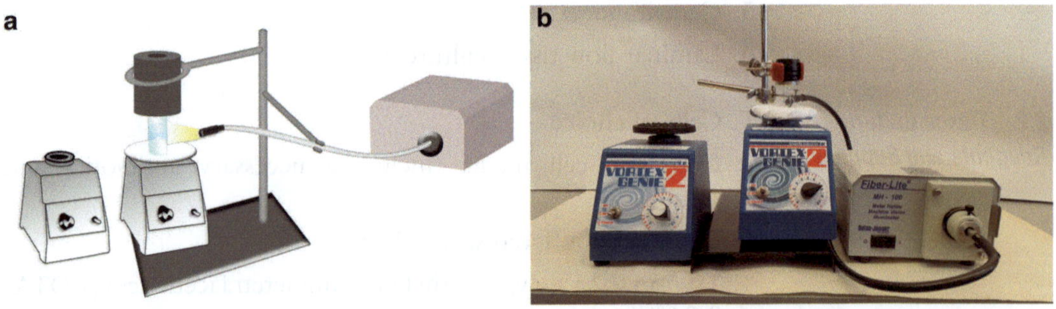

Fig. 1 (**a**) Schematic of microencapsulation station. First vortexer generates a preliminary emulsion. Second vortexer furthers the emulsion and prevents microspheres from settling prematurely. Light from the light guide is directed to the bottom of the glass test tube. In the schematic, the light is depicted further away: the light guide should directly contact the test tubes. (**b**) Image of the microencapsulation station

4. 691.9 mW eosin Y disodium salt.
5. 25 mM HEPES buffered saline (HBS, molecular biology grade).
6. 2,2′,2″-nitrilotriethanol, Tris(2-hydroxyethyl)amine (triethanolamine).
7. Pluronic F-68 solution 10%, bioreagent, sterile filtered, suitable for insect cell culture.
8. Mineral oil, sterile-filtered, suitable for mouse embryo cell culture, light oil.
9. 12 mm × 75 mm glass test tubes.
10. 15 ml amber centrifuge tubes.
11. Amber glass vials (2 ml, 5 ml, 40 ml).
12. 1.5 ml Eppendorf tubes.
13. 2 ml amber Eppendorf tubes.
14. Sterile 18 gauge needles.
15. Sterile syringes (3 ml, 10 ml).
16. Sterile 0.2 μl syringe filters.
17. Self-sealing sterilization pouches (at least 5 in × 8 in).
18. Aluminum foil.
19. Autoclave tape.
20. Millipore water.
21. Tissue culture 6-well plates.
22. Tissue culture polycarbonate membrane 6-well Transwells (8–74 μm pore size; *see* **Note 6**).
23. Tissue culture 12-well plates.
24. Tissue culture 24-well plates.
25. Tissue culture polycarbonate membrane 24-well Transwells (0.4–8 μm pore size).

2.5 Microencapsulation Working Solutions

With the exception of the mineral oil solution, after each is prepared the following solutions must be sterilized under a laminar flow tissue culture hood by filtering them through a 0.2 μm filter.

1. 1.0 mM eosin Y: in a 15 ml amber centrifuge tube, prepare a 1.0 mM eosin Y solution by combining 6.919 mg Eosin Y disodium salt in 10 ml DI water. Vortex for approximately 30 s. Store at room temperature and protect from light exposure.
2. PEGDA microsphere prepolymer: in a 5 ml amber glass vial, prepare the prepolymer solution by combining 493.025 ml HBS solution with 15 μl triethanolamine, 10 μl 1.0 mM Eosin Y solution and 3.95 μl NVP, and 0.1 g 10 kDa PEGDA, being sure to combine liquids first and the powdered PEGDA last. Vortex after each addition (*see* **Note 3**). Add 10 μl pluronic

acid to the solution and vortex briefly or mix with 1 ml pipette to avoid formation of bubbles. In a tissue culture hood, collect the prepolymer solution in a 3 ml sterile syringe using an 18 gauge sterile needle. Remove the needle and place the sterile filter on the syringe. Sterile filter the prepolymer solution by pushing it through the syringe filter and into a sterile amber 2 ml Eppendorf tube (*see* **Note 7**). Cover in foil and store at –20 °C until use.

3. PEGDA microcapsule prepolymer: in a 5 ml amber glass vial, prepare the prepolymer solution by combining 493.025 ml HBS solution with 0.1 g 10 kDa PEGDA. Vortex after each addition (*see* **Note 4**). Add 10 μl pluronic acid to the solution and vortex briefly or mix with 1 ml pipette to avoid formation of bubbles. In a tissue culture hood, collect the prepolymer solution in a 3 ml sterile syringe using an 18 gauge sterile needle. Remove the needle and place the sterile filter on the syringe. Sterile filter the prepolymer solution by pushing it through the syringe (*see* **Note 7**). Cover in foil and store at –20 °C until use.
4. Acetophenone/NVP solution: in a 2 ml amber glass vial, combine 300 mg acetophenone and 1 ml NVP. Vortex until dissolved. Protect from light and store in refrigerator. Make fresh solution every 2 weeks.
5. Microsphere mineral oil solution: work in a tissue culture hood. Combine 3 μl acetophenone/NVP solution for every 1 ml mineral oil in an amber glass vial. It is convenient to make 30 ml solution in a 40 ml amber glass vial. Vortex until acetophenone/NVP solution completely dissolves in mineral oil. Cover vial in foil and store at –20 °C until use.
6. Microcapsule mineral oil solution: work in a tissue culture hood. Combine 15 μl acetophenone/NVP solution for every 1 ml mineral oil in an amber glass vial. It is convenient to make 30 ml solution in a 40 ml amber glass vial. Vortex until acetophenone/NVP solution completely dissolves in mineral oil. Cover vial in foil and store at –20 °C until use.

2.6 Solutions to Assess Microencapsulated Cell Viability and Encapsulation Efficiency

These solutions are prepared and used immediately on samples that will be subsequently discarded and therefore do not need to be sterilized.

1. Calcein AM/ethidium homodimer 1 LIVE/DEAD® Viability/Cytotoxicity stain: work in a tissue culture hood. In a 15 ml amber centrifuge tube, add 10 ml of warmed media. Retrieve ethidium homodimer-1 dye and calcein AM dye from the LIVE/DEAD® Viability/Cytotoxicity Kit for mammalian cells (Life-Technologies). Minimize exposure to light by keeping dyes wrapped in aluminum foil. In hood, pipette 20 μl of

ethidium homodimer-1 dye and 5 μl of calcein AM dye into a 15 ml amber centrifuge tube.

2. CellTiter 96® AQueous One Solution Reagent (Promega Corporation, Madison, WI): [3-(4,5-dimethylthiazol-2-yl)-5-(3-carboxymethoxyphenyl)-2-(4-sulfophenyl)-2H-tetrazoliu m, inner salt; MTS]

3 Methods

3.1 Formation of Microspheres or Microcapsules

1. Put 5 glass 12 mm × 75 mm test tubes into sterilization pouches or wrap in foil and place autoclave tape on test tubes. Five test tubes will make approximately 1 ml of microspheres.
2. Autoclave test tubes. Sterilize as many 5-test tube pouches as needed.
3. Turn on metal halide lamp 30 min prior to use to allow it to warm and achieve its full intensity. Ensure aperture is fully open and cover end of light guide with foil to protect others' vision.
4. Work in tissue culture hood and gently pipette 1 ml mineral solution into each autoclaved glass 12 × 75 mm test tubes (To make microspheres, use the microsphere prepolymer solution with the microsphere mineral oil solution. To make microcapsules, use the microcapsule prepolymer solution with the microcapsule mineral oil solution.).
5. Harvest cells as usual (remove tissue culture flasks from incubator, aspirate media, rinse with PBS, incubate with trypsin, deactivate trypsin with complete media and pipette detached cells into 50 ml test tubes. Count cells as usual).
6. Spin cells down in a centrifuge.
7. Aspirate off media.
8. Suspend desired number of cells for microencapsulation in 1 ml media (or PBS, if to be injected in animal; *see* **Note 8**). If cell pellet is greater than 1 ml, split into multiple aliquots.
9. Pipette cell suspension into 1.5 ml Eppendorf tubes. Keep on ice to prevent hypoxic cell death.
10. Spin down cells in Eppendorf tubes at 800 rpm in a microcentrifuge at 4 °C.
11. Aspirate off media (or PBS, if to be injected in animal), until it just barely touches the top edge of the cell pellet (Fig. 2).
12. Add identical amount of prepolymer solution to a separate Eppendorf tube, holding them side by side to compare amounts (Fig. 3).
13. Withdraw all of prepolymer from the matched Eppendorf tube and add to the cell pellet tube.

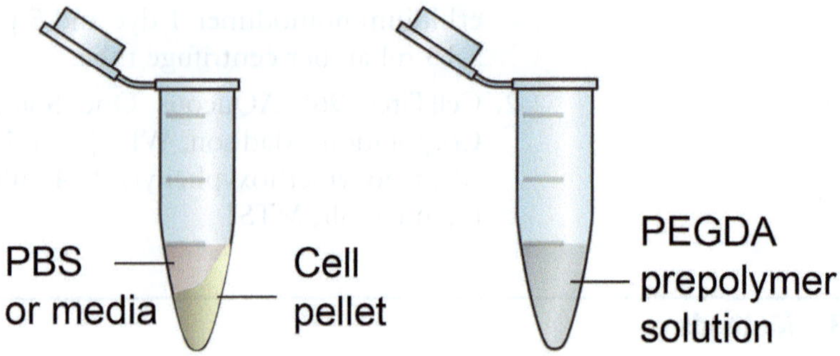

Fig. 2 The method to measure the amount of cells and prepolymer is based on cell volume rather than cell number, since cells vary in size. When spun in a microcentrifuge, the cell pellet will form a "tail" that creeps up the Eppendorf tube. Media or PBS is filled to the top of this tail, then precursor solution is filled to the same height in an identical Eppendorf. The two solutions are mixed well with care taken not to form bubbles immediately prior to forming microspheres

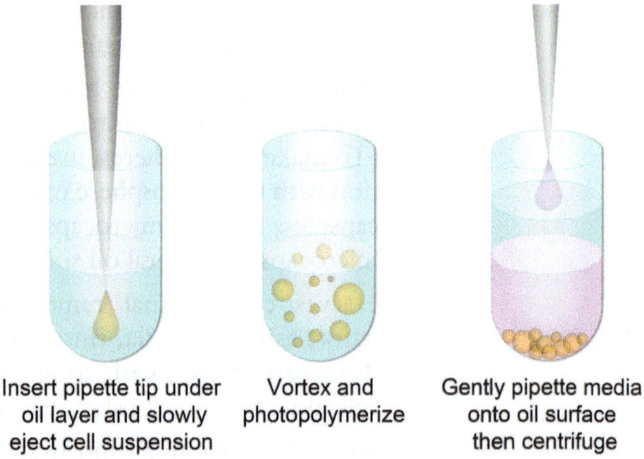

Fig. 3 The PEGDA precursor solution containing cells is pipetted just below the mineral oil solution in a glass test tube. The solution is vortexed, not to mix but to generate an emulsion, briefly before and during exposure to a crosslinking light source. Microdroplets in the emulsion containing cells and prepolymer are cross-linked into microparticles. A dual-initiator method with a hydrophobic photoinitiator in the oil phase and a hydrophilic photoinitiator in the aqueous phase forms microspheres while a single-initiator method with the hydrophobic photoinitiator in the oil phase and no photoinitiators in the aqueous phase forms microcapsules. Slowly adding media or buffer to the surface of the oil layer allows the media to fall through the oil, pulling with it some of the microparticles. Remaining microparticles are isolated from the oil by centrifugation and aspiration of the oil layer

14. Mix well without forming bubbles, careful to fully resuspend cells.
15. Draw up no more than 200 μl cell–prepolymer suspension and carefully expel the contents into the mineral oil solution by inserting the pipette tip just below the oil layer. Greater than

200 μl and microdroplets coalesce. Smaller than 200 μl is fine, but less efficient.

16. Turn both vortexers on and to high. Do not use the "Touch" feature. Work in dim or ambient light. Emulsify the cell–prepolymer suspension by pulsing at the first vortexer. The goal is to create an emulsion, not to mix the two solutions. Stop when mineral oil just begins to look cloudy (approximately 4 s; *see* **Note 9**).
17. Move to the white light vortex-clamp stand setup.
18. Place on welding glasses.
19. Uncover the white light.
20. Drop test tube through hole in metal holder and vortex for 3 s under white light. The text tube should bounce around freely, while touching the light. Turn vortexer off.
21. With the vortexer off, continue to expose the entire emulsion to white light for 20 s by rotating slowly with fingers and with a vortex pulse at 10 s.
22. Remove test tube from apparatus.
23. Slowly add 1 ml complete culture media to emulsion to draw microspheres or microcapsules from the oil and into the media (*see* **Note 10**).
24. Gently drop glass test tube into regular 15 ml conical centrifuge tube.
25. Finish making the desired amount of microspheres in 200 μl increments, following **steps 15–23**.
26. When microspheres are complete, centrifuge the 15 ml conical centrifuge tubes at 300×*g*. Microspheres or microcapsules (Fig. 4) will settle at the bottom.

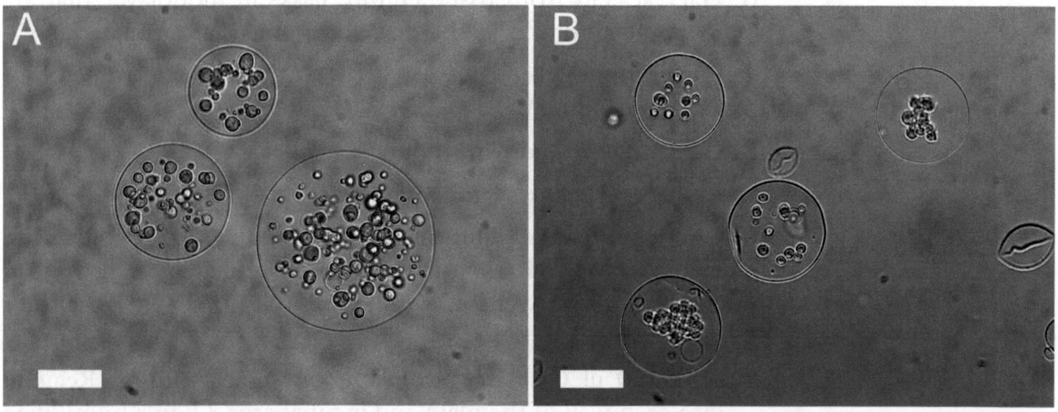

Fig. 4 Representative images of microspheres (**a**) and microcapsules (**b**) containing AtT-20ins cells. Scale bars are 100 μm. Microspheres embed cells in place within the polymeric mesh. Cells within microcapsules are free to cluster or float freely

27. Move to tissue culture hood.
28. Remove the oil layer by aspiration. Remove some of the media to ensure all of the oil is removed. Stay clear of microspheres as these are settled, but will not pellet like cells.
29. Resuspend settled microspheres or microcapsules in complete culture media and divide equally for culture in the top Transwell of 6-well plates with 4–5 ml complete culture media. The number of plates used will depend on the amount of cells microencapsulated. Ensure there is enough media per well for the number of cells in microspheres.
30. Maintain microencapsulated cells in a humidified incubator at 37 °C/5 % CO_2.
31. To feed entrapped cells aspirate media from the bottom well—microspheres will stay settled in the Transwell and media will drain. Replace media in the top well.

3.2 Determining Encapsulation Efficiency

3.2.1 MTS Assay

1. Determine encapsulation efficiency on the same day of encapsulation, but at least 4 h after microencapsulation (*see* **Note 11**).
2. In hood, dispense 1 ml of complete media into the desired number of wells of 24-well plate. If evaluating 8 samples, fill all 24 wells to run samples in triplicate. Label wells as necessary.
3. Retrieve microspheres or microcapsules from incubator.
4. Keep microsphere wells still, careful not to disturb microspheres. Layer of microspheres should be apparent on the Transwell. Drag a pipette tip through the layer of microspheres, drawing up 20 μl of the microspheres.
5. Dispense 20 μl of microspheres per well into the complete culture media in the 24-well plate. Do not reuse pipette tip between samples.
6. Harvest cells from tissue culture flasks in monolayer that are the same kind of cells that have been microencapsulated.
7. Count monolayer cells as usual (with a hemocytometer or Coulter counter).
8. Centrifuge cells down to obtain a pellet.
9. Resuspend cells in 1 ml complete media.
10. Conduct a serial dilution by thoroughly mixing cells with media in well A1, then draw up 1 ml of media from well A1 and repeat with wells A2–A5. Leave well A6 devoid of cells for use as a blank.
11. Add 200 μl CellTiter 96® AQueous One Solution Reagent into each well of the plate and incubate for 3 h in a humidified incubator at 37 °C/5 % CO_2 (*see* **Note 6**).

3.2.2 Analysis of Results

1. Measure the amount of soluble formazan produced by cellular reductionofthetetrazoliumcompound[3-(4,5-dimethylthiazol-2-yl)-5-(3-carboxymethoxyphenyl)-2-(4-sulfophenyl)-2H-tetrazolium, inner salt; MTS] by reading the absorbance of the medium at 490 nm in a plate reader. The amount of the formazan product is proportional to the number of live cells.
2. Plot a standard curve of absorbance from the serial dilution of known cell numbers.
3. Using the standard curve calculate the number of live cells within microspheres or microcapsules.
4. Calculate the encapsulation efficiency using the values calculated values for live cells and the starting live cell number used for encapsulation.

3.3 Determining Cell Viability

3.3.1 Live-Dead Assay

1. In hood, dispense 1 ml of calcein AM/ethidium homodimer 1 solution into the desired number of wells of 12-well plate. If evaluating 12 samples, fill all 12 wells. Label wells as necessary.
2. Retrieve microspheres or microcapsules from incubator.
3. Label wells of 12-well plate.
4. Keep microsphere wells still, careful not to disturb microspheres. Layer of microspheres should be apparent on the Transwell. Drag a pipette tip through the layer of microspheres, drawing up 10–50 μl of the microspheres. Dispense these into the calcien AM/ethidium homodimer 1 solution already in the 12-well plate. Do not reuse pipette tip if it has been immersed in dye solution during microsphere transfer.
5. Incubate 12-well plate for 10 min.

3.3.2 Imaging

1. Keep 12-well plate covered in foil to minimize exposure to light as it is transferred for imaging. Image immediately, as dye solution only lasts 30 min–1 h.
2. Use a fluorescent microscope to obtain both phase contrast images and fluorescent images in both red and green channels. Image at 5× magnification. Live cells fluoresce green (excitation/emission: 495/515), dead cells fluoresce red (excitation/emission: 528/617).
3. Export images.

3.3.3 Analysis

Analysis can be qualitative, semiquantitative, or quantitative. For quantitative analysis, count cells in each channel to get percentages of live and dead cells. Semiquantitative analysis can be performed if all images were obtained with the same settings.

1. Using ImageJ software, open the green channel image.
2. Set the image type to 16 Bit. (Image > Type > 16 Bit).

3. Adjust the threshold by moving the upper slider bar to the right. Click Apply. (Image > Adjust > Threshold…) Choose a threshold that minimizes artifact but does not eliminate any positively stained cells.
4. Between each analysis, select a region of interest that has approximately the same number and size of microspheres from one image to the next.
5. Click Analyze > Analyze Particles…
6. Ensure the "Summarize" option is checked on. Click OK.
7. Record the count value for live cells.
8. Repeat the procedure for the red channel image to count dead cells, using the same region of interest.

4 Notes

1. Other sources of intense white light can be used, but these protocols were developed and validated using the Dolan Jenner metal halide illuminator, Edmund Optics glass filters, and a Newport liquid crystal light guide. Polymerization times and cell viabilities will differ with different illuminators. Take the illuminator, unscrew the front assembly, remove the existing filter, and replace with the heat absorbing and long pass filters as follows:
   (a) There should be a slot holding a 2 in circular filter. Replace the original 2 in circular filter with the circular heat absorbing glass. The replacement filter will absorb the infrared waves and dissipate the heat so that the other filters and liquid crystal cable do not overheat.
   (b) Next place the 2 in square filter in the available space behind the metal plate containing the slot for the circular filter. There should be four bolts holding this metal plate. Loosen these to fit the square filter behind the metal plate (it will be slightly askew) and then screw the assembly back together. The purpose of this square filter is to cut off short wavelength UV (below 320) which would damage the liquid crystal cable and would also harm the cells. With the final assembly, the light from the lamp should enter the heat absorbing circular filter first, then the square 320 long pass filter, then the liquid crystal cable.
   (c) Place and secure the fiber optic adapter into the opening on the front of the lamp.
   (d) Place and secure the liquid crystal light guide into the fiber optic adapter opening.
2. Wrap one of the flat top attachments with white paper to reflect as much light from the lamp as possible. The white paper

reflects more of the light from the light guide than the black rubber flat-top attachment, in addition to limiting the amount the glass test-tube heats up from friction against the rubber. Hydrogel polymerization and cell viability are improved with this piece of paper.

3. To assemble the microencapsulation station:
   (a) Cut a Teflon sheet to a height and length such that it can line the inside of the aluminum cylinder.
   (b) Tape the Teflon sheet into the inside of the aluminum cylinder such that it lines the inside of the cylinder in one layer. This will protect glass test tubes from rattling against the aluminum during vortexing.
   (c) Place the vortex with the paper covered flat top attachment on the ring stand base.
   (d) Clamp the aluminum cylinder with the 3 prong clamp and secure to the support stand such that the base of the cylinder sits 30–50 mm above the center of the paper covered attachment. The height should be adjusted to allow the 12 mm × 75 mm glass test tubes to bounce freely.
   (e) Clamp the liquid light guide at the light end and arrange such that the end just barely touches the glass test tubes that are placed into the aluminum cylinder. Microspheres will not polymerize if the light guide is too far from the test tube (even less than 1 cm can make a difference). Ensure the light guide touches the test tube and that it is angled to illuminate as much of the mineral oil as possible.
4. Do not use NVP more than 2–3 weeks after opening as it loses its potency and can crash out of solution.
5. Cover the acetopheonone bottle with foil to protect it from light and after each use blanket with argon to preserve its potency.
6. Microparticles are cultured in Transwells since the polymers do not adhere to tissue culture plastic. Media can be changed by aspirating the bottom well. If microencapsulation efficiency is poor, there will be free cells that will fall through larger pore Transwells and adhere to the bottom of the well plates, which can then be replaced. With smaller pore Transwells, cells can form a monolayer in the Transwell, which can affect changing the media, imaging, and results. Selecting larger pore Transwells is a tradeoff between losing smaller microparticles and discarding unwanted free cells.
7. Since small volumes are used, much of the solution will be retained in the syringe. To retrieve the maximum solution from the filter, remove the filter from the syringe and keep the syringe sterile. Draw air into the syringe. Reattach syringe to filter and force air through the syringe filter. Repeat until no

more fluid can be extracted. This may introduce bubbles into the prepolymer solution. These bubbles will dissolve with time or can be eliminated by centrifuging.

8. The liquid containing the cells will become part of the microparticles. If performing in vitro work only, it does not matter if serum proteins from complete media are entrapped within microparticle surfaces. If performing in vivo work, the serum proteins can trigger an immune response, therefore PBS should be used.
9. The longer the solutions are vortexed, the smaller the microparticles will be. This part is highly operator dependent, and times should be optimized per individual if a specific size range is desired. If not, the stated times produce 50–300 μm sized microparticles.
10. Even if doing in vivo work, this step should involve media. It is best to allow encapsulated cells to recover in media for at least 4 h since the photopolymerization process affects the permeability of the cell membrane. When ready for in vivo work, centrifuge cells and resuspend in PBS.
11. The MTS assay measures the metabolic activity of cells. Each cell type has a different level of metabolic activity, therefore, the cell number used for serial dilution and the incubation times depend on the type of cells encapsulated. Incubation times can be optimized by carefully monitoring the color change of the MTS reagent and measuring absorbance at different time intervals (e.g., every 30 min up to 3 h). The cell number should be chosen such that the absorbance read remains in the linear range of the assay. Incubation time also depends on diffusion of the reagent into the cell-laden microspheres, which depends on the size of the hydrogel mesh, which in turn is governed by the molecular weight of the PEGDA used.

References

1. Zhang JJ (2013) Mechanisms of cell therapy for clinical investigations: an urgent need for large-animal models. Circulation 128:92–94
2. Kim JK et al (2010) Autologous bone marrow infusion activates the progenitor cell compartment in patients with advanced liver cirrhosis. Cell Transplant 19:1237–1246
3. Camussi G, Dereqibus MC, Tetta C (2010) Paracrine/endocrine mechanism of stem cells on kidney repair: role of microvesicle-mediated transfer of genetic information. Curr Opin Nephrol Hypertens 19(1):7–12
4. Abumaree M, Al Jumah M, Pace RA, Kalionis B (2012) Immunosuppressive properties of mesenchymal stem cells. Stem Cell Rev 8(2): 375–392
5. Moutsatsos IK et al (2001) Exogenously regulated stem cell-mediated gene therapy for bone regeneration. Mol Ther 3:449–461
6. Goren A, Dahan N, Goren E et al (2010) Encapsulated human mesenchymal stem cells: a unique hypoimmunogenic platform for long-term cellular therapy. FASEB J 24(1): 22–31
7. Ausländer S, Wieland M, Fussenegger M (2012) Smart medication through combination of synthetic biology and cell microencapsulation. Metab Eng 14(3):252–260

8. Tomaro-Duchesneau C, Saha S, Malhotra M et al (2013) Microencapsulation for the therapeutic delivery of drugs, live mammalian and bacterial cells, and other biopharmaceutics: current status and future directions. J Pharm 2013:103527
9. Teramura Y, Minh LN, Kawamoto T, Hiroo Iwata H (2010) Microencapsulation of islets with living cells using PolyDNA-PEG-lipid conjugate. Bioconjugate Chem 21(4):792–796
10. Olabisi RM, Lazard ZW, Franco CL et al (2010) Hydrogel microsphere encapsulation of a cell-based gene therapy system increases cell survival of injected cells, transgene expression, and bone volume in a model of heterotopic ossification. Tissue Eng Part A 16(12): 3727–3736
11. Afkhami F, Yves Durocher Y, Satya Prakash S (2010) Investigation of antiangiogenic tumor therapy potential of microencapsulated HEK293 VEGF165b producing cells. J Biomed Biotechnol 2010:645610
12. Chang TMS, Macintosh FC, Mason SG (1966) Semipermeable aqueous microcapsules: I. Preparation and properties. Can J Physiol Pharmacol 44:115–128
13. Lim F, Sun AM (1980) Microencapsulated islets as bioartificial endocrine pancreas. Science 210:908–910
14. Olabisi RM (2015) Cell microencapsulation with synthetic polymers. Tissue Eng Pt A 103: 846–859
15. Hong JS, Shin SJ, Lee S, Wong E et al (2007) Spherical and cylindrical microencapsulation of living cells using microfluidic devices. Korea Aust Rheol J 19(3):157–164
16. Hunt NC, Grover LM (2010) Cell encapsulation using biopolymer gels for regenerative medicine. Biotechnol Lett 32(6):733–742
17. Liu L, Shujun Gao S, Yu Y et al (2006) Bioceramic hollow fiber membranes for immunoisolation and gene delivery: I: membrane development. J Membrane Sci 280:375–382
18. Baker J. A phase I/IIa open-label investigation of the safety and effectiveness of DIABECELL(R) [immunoprotected (alginate-encapsulated) porcine islets for Xenotransplantation] in patients with type I diabetes mellitus (study NCT00940173). Available at: http://www.clinicaltrials.gov. Accessed 18 Sept 2015
19. Dufrane D. A monocentre phase 1 trial to assess a monolayer cellular device in the treatment of type 1 diabetes (study NCT00790257). Available at: http://www.clinicaltrials.gov. Accessed 18 Sept 2015
20. Keymeulen B. Functional survival of beta cell allografts after transplantation in the peritoneal cavity of non-uremic type 1 diabetic patients (study NCT01379729). Available at: http://www.clinicaltrials.gov. Accessed 18 Sept 2015
21. Schwartz S, Mulgrew P. A single-center phase I/II study of PEG-encapsulated islet allografts implanted in patients with type I diabetes (study NCT00260234). Available at: http://www.clinicaltrials.gov. Accessed 18 Sept 2015
22. Bryant SJ, Anseth KS (2002) Hydrogel properties influence ECM production by chondrocytes photoencapsulated in poly(ethylene glycol) hydrogels. J Biomed Mater Res 59:63–72
23. Tibbitt MW, Anseth KS (2009) Hydrogels as extracellular matrix mimics for 3D cell culture. Biotechnol Bioeng 103:655–663
24. Aijaz A, Faulknor R, Berthiaume F, Olabisi RM (2015) Hydrogel microencapsulated insulin-secreting cells increase keratinocyte migration, epidermal thickness, collagen fiber density, and wound closure in a diabetic mouse model of wound healing. Tissue Eng Part A. doi:10.1089/ten.TEA.2015.0069
25. Franco CL, Price J, West JL (2011) Development and optimization of a dual-photoinitiator, emulsion-based technique for rapid generation of cell-laden hydrogel microspheres. Acta Biomater 7(9):3267–3276
26. Mumaw J, Jordan ET, Sonnet C et al (2012) Rapid heterotrophic ossification with cryopreserved poly(ethylene glycol-) microencapsulated BMP2-expressing MSCs. Int J Biomater 2012:861794
27. Sonnet C, Simpson CL, Olabisi RM et al (2013) Rapid healing of femoral defects in rats with low dose sustained BMP2 expression from PEGDA hydrogel microspheres. J Orthop Res 31(10):1597–1604

Chapter 7

Polymeric Materials for Perm-Selective Coating of Alginate Microbeads

William F. Kendall Jr. and Emmanuel C. Opara

Abstract

Application of microencapsulation to the immunoisolation of pancreatic islets holds promise for expanding the use of islet transplantation as a treatment option for Type 1 diabetes. It is generally believed that successful development of a reliable methodology will ideally allow for transplantation of pancreatic islets that are protected from the immune system, thereby obviating the need for the use of immunosuppressive drugs and their attendant side effects. In addition, this technology has the potential to expand the donor pool as islets from nonhuman donors could be used as xenografts in human patients. The complex polysaccharide, alginate, has been the most widely used polymer for microencapsulation of islets. However, it is known that alginate lacks appreciable permselectivity to confer immunoisolation of encapsulated islets, thus necessitating the routine permselective coating of alginate microbeads with polymers of amino acids, mainly, poly-L-lysine (PLL) and poly-L-ornithine (PLO). The protocol described in this chapter outlines the steps we have used in our studies on perm-selective coating of alginate microbeads for islet transplantation.

Key words Diabetes, Microencapsulation, Islets, Immunoisolation, Polymers

1 Introduction

Diabetes continues to plague our society. The worldwide prevalence of diabetes, for all age groups was estimated to be 2.8% (171 million) in 2000, and projected to increase to 4.4% in 2030 (366 million) according to a previous study by Wild et al. [1]. In 2012, it was estimated that 29.1 million Americans (9.3% of the population) had diabetes, with approximately 1.25 million American adults and children having Type 1 diabetes [2]. The incidence of new cases of diabetes in the USA ranged from 1.7 to 1.9 million new cases per year in 2010 and 2012. In 2010 diabetes was the seventh leading cause of death in the USA with 234,051 death certificates listing diabetes as underlying or contributing cause of death [2]. The estimated total cost of diagnosed diabetes in the USA, in 2010, was 245 billion, with 176 billion for direct medical

Emmanuel C. Opara (ed.), *Cell Microencapsulation: Methods and Protocols*, Methods in Molecular Biology, vol. 1479, DOI 10.1007/978-1-4939-6364-5_7, © Springer Science+Business Media New York 2017

costs (including treatment for hypoglycemia, hypertension, dyslipidemia, cardiovascular disease, stroke, eye problems, blindness, kidney disease, and limb amputations) [2]. It is estimated that there was about $69 billion in reduced productivity, and that the average medical expenditure for people with diagnosed diabetes was 2.3 times higher than for those without diabetes [2].

Pancreas transplantation was first performed in 1966 by William Kelly and Richard Lillehei [3]. Since then, it has emerged as a viable treatment option for patients with type 1 diabetes and select patients with type 2 diabetes [4]. It can help to delay or decrease some of the potential sequelae that arise from the metabolic complications attributable to hyperglycemia in diabetes, and between 1966 and 2000 over 15,000 pancreas transplants were performed worldwide [4]. Pancreas transplantation requires the use of potent immunosuppressants in order to prevent organ rejection, and these anti-rejection drugs have their own attendant complications [3]. In 1967, Lacy and Kostianovsky described a method that allowed them to isolate intact islets of Langerhans from the rat pancreas [5]. Progressive advances in islet isolation techniques ultimately lead to the successful clinical transplantation of human pancreatic islets using the Edmonton immunosuppression protocol [6].

In the quest to develop an immunosuppression-free transplantation procedure, numerous techniques and devices have been evaluated through the years, with the goal to achieve similar success as pancreas transplantation with immunoisolation of islets by microencapsulation prior to transplantation [6, 7]. Despite the occasional individual successes, an optimal microencapsulation technique that requires no use of immunosuppressive drugs in clinical islet transplantation is yet to be developed. Immunoisolation of islets encapsulated with alginate remains a viable option due to alginate's ease of gelling and biocompatibility [8]. Our protocol described in this chapter provides an opportunity to explore the use of durable permselectively coated alginate microbeads in clinical islet transplantation.

2 Materials

2.1 Equipment

1. 2-Channel air jacket microencapsulator (Duke Med Ctr. Instrument Shop).
2. Rotor (Clay Adams Nutator, Becton Dickinson, Sparks, MD).
3. Inverted light microscope (Olympus CK40), linked to the UTHSCA image tool computer program (University of Texas, Austin).

2.2 Chemicals/Biochemicals

1. (1.56 mM) Dithizone.
2. 0.5% (w/v) Trypan blue.
3. Ultrapurified sodium alginate [EX-8085] (3% w/v) with a high content of mannuronic acid, obtained from Kelco/Monsanto (San Diego, CA).
4. FITC-Ricinus Communis I (RCA-I).
5. FITC-Sambucas Nigra (SNA), from EY Lab Inc.
6. FITC _Maackia Amurensis I, from Vector Laboratories (Burlingame, CA).
7. Poly-lysine
8. Poly-L-ornithine, from Sigma Chemical (St Louis, MO).
9. 6 mM Sodium sulfate.
10. Nylon mesh with a pore size of 100 μm.

2.3 Solutions

1. 0.24 Na-alginate solution.
2. Dissolve 8 mL of sodium alginate in 100 ml of normal saline.
3. Place 10 cc aliquots into 15 mL centrifuge tubes. Store at −75 °C.
4. 6 mM Na_2So_4.
5. 0.756 g Na_2SO_4 diluted in 1 l of distilled water.
6. Cross-linking solution (make 500 cc).
7. 10 mM HEPES (260 mg/100 ml).
8. 100 $CaCl_2$ (1.11% = 1.11 g/100 ml).
9. 500 mL of distilled water.
10. Adjust to pH 7.4. Make up weekly.
11. 0.10% Poly-lysine solution.
12. Dissolve 0.100 g of poly-lysine in 100 ml of normal saline.
13. Make up monthly.
14. 55 mM Sodium citrate solution in normal saline (make 500 cc).
15. Dissolve 1.6 g of sodium citrate per 100 mL of normal saline.
    (a) Adjust to pH 7.4.
    (b) Make up monthly.
16. Hepes buffer solution.
    (a) 25 mM HEPES (651 mg = 0.651 g/100 mL).
    (b) 118 mM NaCl (690 mg = 0.690 g/100 mL).
    (c) 5.6 mM KCl (41.7 mg = 0.042 g/100 mL).
    (d) 2.5 mM $MgCl_2$ (23.8 mg = 0.024 g/100 mL).
    (e) 100 mL of distilled water.
    (f) Adjust to pH 7.4.
    (g) Make up monthly (prn).

17. Dithizone solution (*see* **Note 1**).
    (a) 400 mg DTZ (Dithizone).
    (b) 28 mM DMSO.
    (c) 160 mL Hanks solution.

2.4 Animals

1. Sprague-Dawley rats.

3 Methods

3.1 Islet Microencapsulation with the 2-Channel Air-Jacket Device

1. Centrifuge the 15 cm centrifuge tube, containing the hand-picked islets, obtained as previously described in our islet isolation manuscript [9] suspended in RPMI solution, for 2 min at 4 °C.
2. Carefully, pipette off the supernatant as much as possible, without disturbing the pellet.
3. Resuspend the pellet in the residual RPMI solution (*see* **Note 2**).
4. Add 1 ml of pure alginate to the centrifuge tube (or modify volume as experiment or pellet size dictates).
5. Suspend the islets by hand-vortexing.
6. Prepare the microencapsulator.
    (a) Ensure that the needle, metal tip, syringe, and funnel (be sure filter is in funnel) are clean.
    (b) Turn on the air compressor until it hisses.
    (c) Attach needle to syringe.
    (d) Place white hard plastic ring around syringe, with screw beds facing upward.
    (e) Place black ring around syringe.
    (f) Connect white ring to silver syringe holder (be sure that it is appropriately connected to the silver holding arm).
    (g) Adjust syringe so that tip of needle is approximately 1–2 mm from the end of the exit lumen from the silver syringe holder.
    (h) Tighten the syringe apparatus as high as possible on the silver pole with the holding arm (to avoid splashing onto the lumen).
    (i) Place the funnel onto its holder.
    (j) Turn on and test gauges (adjust with black knobs).
    (k) Set gauges to appropriate settings.
    (l) After testing and adjusting gauges, leave gas on; turn off air jacket and alginate switches.
    (m) Put 5 ml of cross-linking solution into the syringe (be sure gas is on).

(n) Adjust gas so that slow bubbling is occurring in crosslinking solution (No large bubbling).
(o) Re-vortex tube containing islets suspended in alginate.
(p) Using pipettman, place islet/alginate mixture into syringe.
(q) Attach white connector on tubing to the syringe.
(r) Turn on air jacket switch.
(s) Turn on alginate switch.

7. Monitor droplet formation.
   (a) Add additional cross-linking solution as needed.
   (b) Turn on air compressor as needed.
8. After capsule formation is completed, empty contents of funnel into sterile capsule-washing-container (use sterile funnel); turn gas up slightly to assist with blowing remaining capsules out from bottom of funnel. Use additional cross-linking solution as needed to rinse capsules into the container.
   (a) Turn off switches.
   (b) Open exhaust switch on capsule-washing-container, and hook air to input switch on capsule-washing-container.
   (c) Incubate for 15 min, in capsule-washing container, placed on ice on rotator.
   (d) Close air input switch on capsule-washing container, then close exhaust switch on capsule-washing container.
   (e) Let sit for 1 min on ice, and then remove supernatant with disposable bulb pipet.
   (f) Add saline to tube to 30 ml mark (If transplant planned use culture medium), gently invert capsule-washing-container; let tube sit in ice for 1 min; remove supernatant carefully with disposable bulb pipet (do not disturb capsules).
   (g) Add saline to tube to 30 ml mark, gently invert capsule-washing container. Let tube sit in ice for 1 min. Remove supernatant carefully with disposable bulb pipet (do not disturb capsules).
9. Coat the alginate bead with 0.05 % polylysine in saline for 6 min.
   (a) Fill container (containing capsules) with polylysine to 20 ml mark.
   (b) Gently invert.
   (c) Open exhaust switch on capsule-washing container, and hook air to input switch on capsule-washing container.
   (d) Incubate for 4 min, in capsule-washing container, placed on ice on rotator.
   (e) Close air input switch on capsule-washing container, and then close exhaust switch on capsule-washing container.

- (f) Let sit for 2 min on ice, and then remove supernatant with disposable bulb pipet.
- (g) Carefully remove supernatant with a disposable bulb pipet.
- (h) Fill tube to 30 ml mark with saline.
- (i) Gently invert.
- (j) Sit upright in ice bucket for 2 min.
- (k) Carefully remove supernatant with a disposable bulb pipet.

10. Apply outer alginate layer by suspension of capsules in 0.06% (wt/vol) sodium alginate in saline for 4 min.
    - (a) Fill tube to 20 ml with 0.06% sodium alginate.
    - (b) Open exhaust switch on capsule-washing container, hook air to input switch on capsule-washing-container.
    - (c) Incubate for 2 min, in capsule-washing container, placed on ice on rotator.
    - (d) Close air input switch on capsule-washing container, and then close exhaust switch on capsule-washing-container.
    - (e) Let sit for 2 min on ice, and then remove supernatant with disposable bulb pipet.
    - (f) Carefully remove supernatant with a disposable bulb pipet.
    - (g) Fill tube to 30 ml mark with saline.
    - (h) Gently invert.
    - (i) Sit upright in ice bucket for 2 min.
    - (j) Carefully remove supernatant with a disposable bulb pipet.
11. Liquefaction of the inner calcium alginate core by suspension in 50 mM sodium citrate for 7 min.
    - (a) Fill tube to 20 ml mark with sodium citrate.
    - (b) Open exhaust switch on capsule-washing container, and hook air to input switch on capsule-washing container.
    - (c) Incubate for 5 min, in capsule-washing container, placed on ice on rotator.
    - (d) Close air input switch on capsule-washing container, and then close exhaust switch on capsule-washing container.
    - (e) Let sit for 2 min on ice, and then remove supernatant with disposable bulb pipet.
    - (f) Carefully remove supernatant with a disposable bulb pipet.
    - (g) Fill tube to 30 ml mark with saline.
    - (h) Gently invert.
    - (i) Sit upright in ice bucket for 2 min.
    - (j) Carefully remove supernatant with a disposable bulb pipet.

(k) Fill tube to 30 ml mark with saline.

(l) Gently invert.

(m) Sit upright in ice bucket for 2 min.

(n) Carefully remove supernatant with a disposable bulb pipet (two washings total with saline).

12. Treat capsule with treatment solution as dictated by experiment being conducted (four subgroups: (1) PLL 6-min coating; (2) PLL 20-min coating; (3) PLO 6-min coating; (4) PLO 20-min coating).

(a) Add treatment solution to tube to 20 ml mark (if using sodium sulfate incubation for planned transplantation, add 5 mM glucose to the sodium sulfate solution).

(b) Gently invert.

(c) Open exhaust switch on capsule-washing container, and hook air to input switch on capsule-washing container.

(d) Incubate for 30 min (or modify as dictated by experiment), in capsule-washing container, placed on ice on rotator.

(e) Close air input switch on capsule-washing container, and then close exhaust switch on capsule-washing-container.

(f) Let sit for 2 min on ice, and then remove supernatant with disposable bulb pipet.

(g) Add saline to tube to 30 ml mark, and gently invert container. Sit upright in ice bucket for 2 min. Remove supernatant carefully with disposable bulb pipet (do not disturb capsules). Repeat this entire step once (two washings with saline).

3.2 Generation of Empty Alginate Microbeads with the Microencapsulator

1. Prepare the microencapsulator.

(a) Ensure that the needle, metal tip, syringe, and funnel (be sure that filter is in funnel) are clean.

(b) Turn on the air compressor until it hisses.

(c) Attach needle to syringe.

(d) Place white hard plastic ring around syringe, with screw beds facing upward.

(e) Place black ring around syringe.

(f) Connect white ring to silver syringe holder (be sure that it is appropriately connected to the silver holding arm).

(g) Adjust syringe so that the tip of the needle is approximately 1–2 mm from the end of the exit lumen from the silver syringe holder.

(h) Tighten the syringe apparatus as high as possible on the silver pole with the holding arm (to avoid splashing onto the lumen).

(i) Place the funnel onto its holder.

(j) Turn on and test gauges (adjust with black knobs).

(k) Set gauges to appropriate settings.

(l) After testing and adjusting gauges, leave gas on; turn off air jacket and alginate switches.

(m) Put 5 ml of cross-linking solution into the syringe (be sure gas is on).

(n) Adjust gas so that slow bubbling is occurring in cross-linking solution (no large bubbling).

(o) Attach white connector on tubing to the syringe.

(p) Turn on air jacket switch.

(q) Turn on alginate switch.

2. Monitor droplet formation.

(a) Add additional cross-linking solution as needed.

(b) Turn on air compressor as needed.

3. After capsule formation is completed, empty contents of funnel into sterile capsule-washing-container (use sterile funnel); turn gas up slightly to assist with blowing remaining capsules out from bottom of funnel. Use additional cross-linking solution as needed to rinse capsules into the container.

(a) Turn off switches.

(b) Open exhaust switch on capsule-washing container, and hook air to input switch on capsule-washing-container.

(c) Incubate for 15 min, in capsule-washing container, placed on ice on rotator.

(d) Close air input switch on capsule-washing container, and then close exhaust switch on capsule-washing container.

(e) Let sit for 1 min on ice, and then remove supernatant with disposable bulb pipet.

(f) Add saline to tube to 30 ml mark, and gently invert capsule-washing container. Let tube sit in ice for 1 min. Remove supernatant carefully with disposable bulb pipet (don't disturb capsules).

(g) Add Saline to tube to 30 ml mark, gently invert capsule-washing-container; Let tube sit in ice for 1 min; Remove supernatant carefully with disposable bulb pipet (do not disturb capsules).

4. Coat the alginate bead with 0.05 % polylysine in saline for 6 min.

(a) Fill container (containing capsules) with polylysine to 20 ml mark.

(b) Gently invert.

   (c) Open exhaust switch on capsule-washing container, and hook air to input switch on capsule-washing container.
   (d) Incubate for 4 min, in capsule-washing container, placed on ice on rotator.
   (e) Close air input switch on capsule-washing container, and then close exhaust switch on capsule-washing container.
   (f) Let sit for 2 min on ice, and then remove supernatant with disposable bulb pipet.
   (g) Carefully remove supernatant with a disposable bulb pipet.
   (h) Fill tube to 30 ml mark with saline.
   (i) Gently invert.
   (j) Sit upright in ice bucket for 2 min.
   (k) Carefully remove supernatant with a disposable bulb pipet.
5. Apply outer alginate layer by suspension of capsules in 0.06% (wt/vol) sodium alginate in saline for 4 min.
   (a) Fill tube to 20 ml with 0.06% sodium alginate.
   (b) Open exhaust switch on capsule-washing container, and hook air to input switch on capsule-washing-container.
   (c) Incubate for 2 min, in capsule-washing container, placed on ice on rotator.
   (d) Close air input switch on capsule-washing container, and then close exhaust switch on capsule-washing container.
   (e) Let sit for 2 min on ice, and then remove supernatant with disposable bulb pipet.
   (f) Carefully remove supernatant with a disposable bulb pipet.
   (g) Fill tube to 30 ml mark with saline.
   (h) Gently invert.
   (i) Sit upright in ice bucket for 2 min.
   (j) Carefully remove supernatant with a disposable bulb pipet.
6. Liquefaction of the inner calcium alginate core by suspension in 50 mM sodium citrate for 7 min.
   (a) Fill tube to 20 ml mark with sodium citrate.
   (b) Open exhaust switch on capsule-washing container, hook air to input switch on capsule-washing container.
   (c) Incubate for 5 min, in capsule-washing container, placed on ice on rotator.
   (d) Close air input switch on capsule-washing container, and then close exhaust switch on capsule-washing container.
   (e) Let sit for 2 min on ice, and then remove supernatant with disposable bulb pipet.

(f) Carefully remove supernatant with a disposable bulb pipet.

(g) Fill tube to 30 ml mark with saline.

(h) Gently invert.

(i) Sit upright in ice bucket for 2 min.

(j) Carefully remove supernatant with a disposable bulb pipet.

(k) Fill tube to 30 ml mark with saline.

(l) Gently invert.

(m)Sit upright in ice bucket for 2 min.

(n) Carefully remove supernatant with a disposable bulb pipet (two washings total with saline).

7. Treat capsule with treatment solution as dictated by experiment being conducted (four subgroups: (1) PLL 6-min coating; (2) PLL 20-min coating; (3) PLO 6-min coating; (4) PLO 20-min coating).

(a) Add treatment solution to tube to 20 ml mark.

(b) Gently invert.

(c) Open exhaust switch on capsule-washing container, and hook air to input switch on capsule-washing container.

(d) Incubate for 30 min (or modify as dictated by experiment), in capsule-washing container, placed on ice on rotator.

(e) Close air input switch on capsule-washing container, and then close exhaust switch on capsule-washing container.

(f) Let sit for 2 min on ice, and then remove supernatant with disposable bulb pipet.

(g) Add saline to tube to 30 ml mark, and gently invert container. Sit upright in ice bucket for 2 min. Remove supernatant carefully with disposable bulb pipet (do not disturb capsules). Repeat this entire step once (two washings with saline).

3.3 Evaluation of Microbead Permeability with Fluorescent Lectins [9]

1. Pipette 150 μL of encapsulated pig islets (batch of encapsulated islets as outlined in Subheading 3.2), from each subgroup into new and appropriately labelled containers for incubation with one of four lectins of various molecular weight: triticum vulgare (WGA, MW 36,000), Maackia Amurensis (MAL-I, MW 75,000), Ricinus Communis (RCA-I, MW 120,000) and Sambuca Nigra (SNA, MW 150,000).

(a) Add 15 μL (1 mg/μL) of the appropriate lectin to each labeled container: triticum vulgare (WGA, MW 36,000), Maackia Amurensis (MAL-I, MW 75,000), Ricinus Communis (RCA-I, MW 120,000) and Sambuca Nigra (SNA, MW 150,000).

(b) Incubate capsules in each subgroup, with mixing, for 48 h at 4 °C.

(c) Then examine the capsules on an Olympus BH2 fluorescent microscope for fluorescence activity.

3.4 Analysis of Changes in Microbead Size [10]

1. Make a batch of empty microcapsules as outlined in: Subheading 3.2, and then suspend the microcapsule in saline.
2. Periodically perform daily diameter measurements of microcapsules over the course of 14 days using an inverted light microscope (Olympus CK40), linked to the UTHSCA image tool computer program (University of Texas, Austin).

3.5 Implantation of PLL- and PLO-Coated Microbeads in Normal Rats for the Assessment of Fibrotic Response [11]

1. Make batches of empty microcapsules as outlined in Subheading 3.2 (one batch per rat).
2. In **step 19**, treat half of the batches with PLL for 6 min; and place in an appropriately labeled container. Treat the other batches with PLO for 6 min and place in an appropriately labeled container.
3. Suspend each batch in 10 ml of normal saline.
4. Anesthetize each rat sequentially and place 10 ml aliquots of the appropriate batch of microcapsules into the peritoneal cavity, through a small (approximately 1 cm abdominal skin opening, which is then closed with watertight skin closure using vicryl suture.
5. Place rats into appropriately labeled cages (*see* **Note 3**).
   (a) Sequentially sacrifice animals at 1, 2, 3, and 4 weeks when each animal is sacrificed, retrieve the capsules. Assess each capsule for degree of fibrotic response using an inverted light microscope (Olympus CK40).

3.6 Results from Our Studies

1. After incubation, unencapsulated, naked porcine islets contained 100% fluorescence while empty PLL or PLO-coated capsules contained no fluorescence. Encapsulated porcine islets fluoresced with the smaller molecular weight lectins (mal-1, mw 75 kDa; WGA, mw 36 kDa). No fluorescence was noted in the capsules incubated with the larger lectins (SNA, mw 150 kDa; and RCA-1, mw 120 kDa).
2. The microcapsules coated with PLL for 6 min were similar to those coated with PLL after 20 min of incubation.
3. We also observed that in contrast to the bead swelling phenomenon that is commonly observed in PLL-coated microcapsules, the PLO-coated microcapsules did not significantly change in size when incubated over 14 days in normal saline (Table 1, Figs. 1 and 2).

Table 1
PLL vs. PLO microcapsule swelling summary data

Treatment	Pre-incubation	Post-36 h incubation	4 days incubation	6 days incubation	8 days incubation	10 days incubation	12 days incubation	14 days incubation
PLL	718.06 ± 17.59	809.03 ± 27.03	817.55 ± 20.17	819.49 ± 15.15	824.03 ± 15.10	819.78 ± 16.49	819.01 ± 18.59	821.72 ± 17.37
PLO	689.35 ± 17.15	701.56 ± 15.76	711.81 ± 13.39	706.40 ± 16.26	706.35 ± 15.89	707.37 ± 14.92	704.83 ± 7.94	708.98 ± 13.44

Data represents: mean ± SD, $n = 20$ capsules

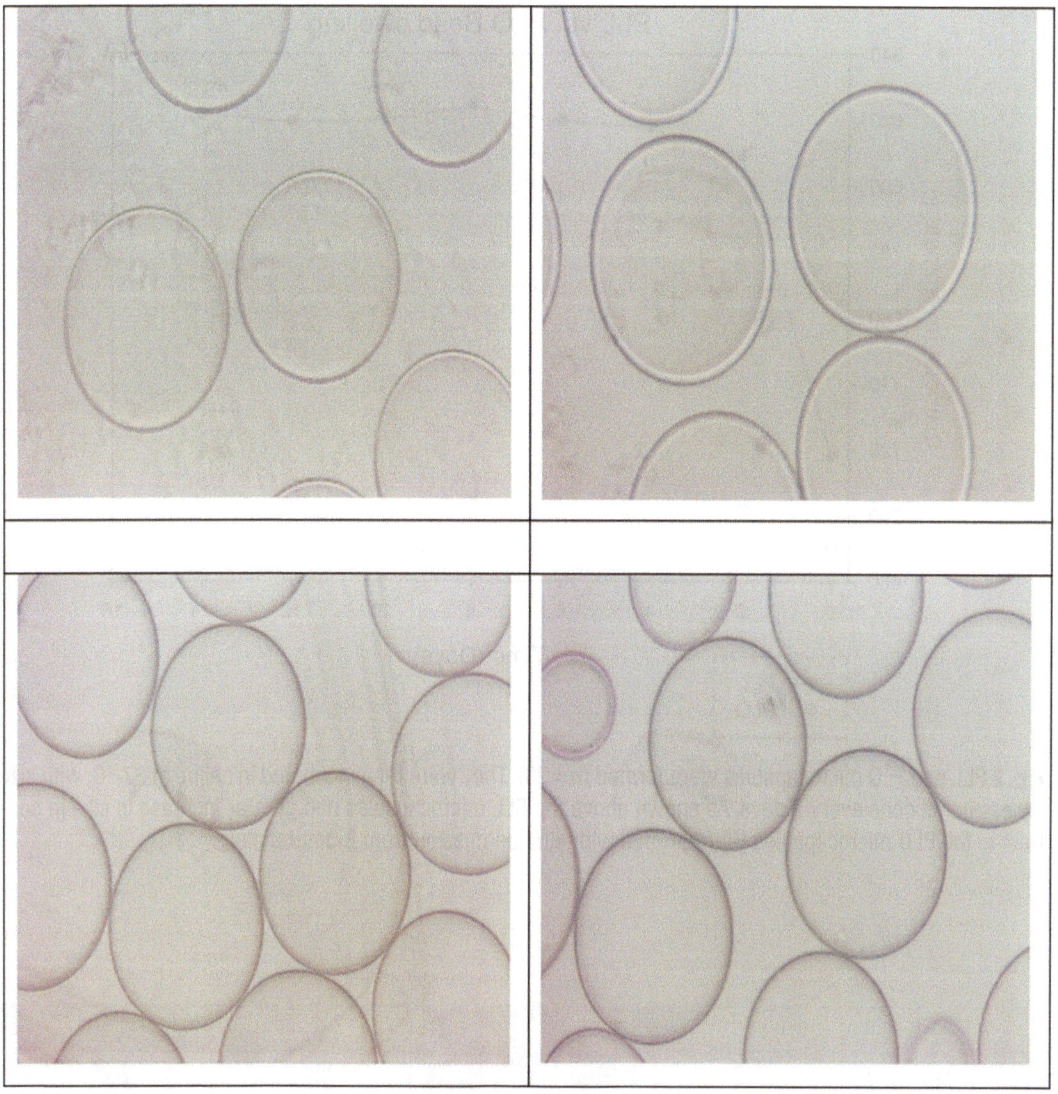

Fig. 1 *Top*: PLL day 0 (after formation), then PLL day 14 (after incubation in saline). *Bottom*: PLO day 0 (after formation), then PLO day 14 (after incubation in saline)

4. Capsules that were implanted in Sprague-Dawley rats were retrieved at 1, 2, 3, and 4 weeks post-implantation. The PLO-coated capsules appeared to have less fibrosis, suggesting better biocompatibility (Fig. 3 and Table 2).

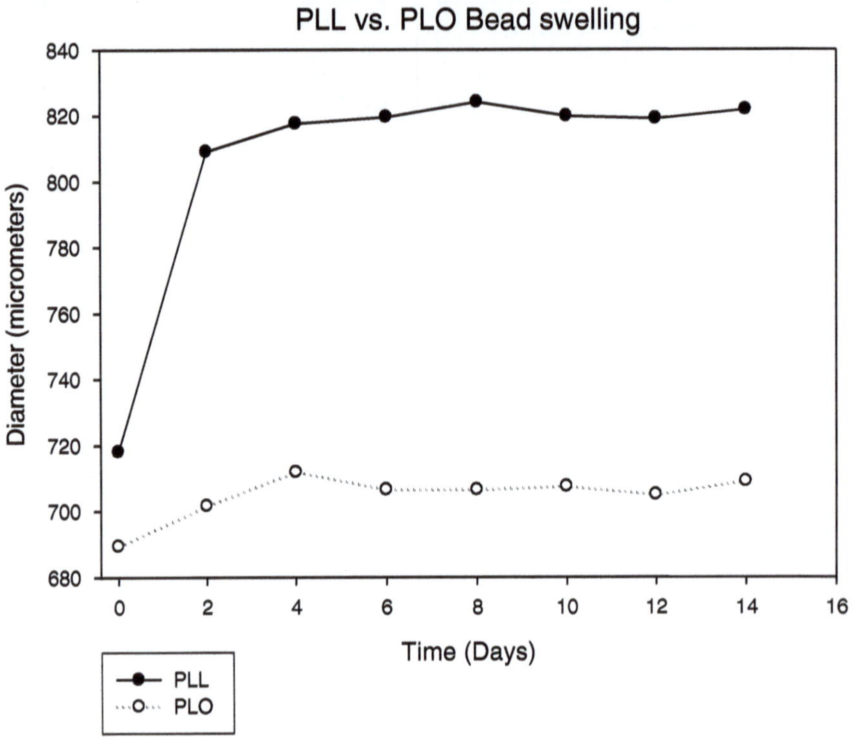

Fig. 2 PLL and PLO microcapsules were formed ($n = 20$). They were then incubated in saline at 37 °C, with size assessments done every 2 days. As shown above the PLL microcapsules had greater increase in size in contrast to the PLO microcapsules [9] (reprinted with kind permission from Biomaterials)

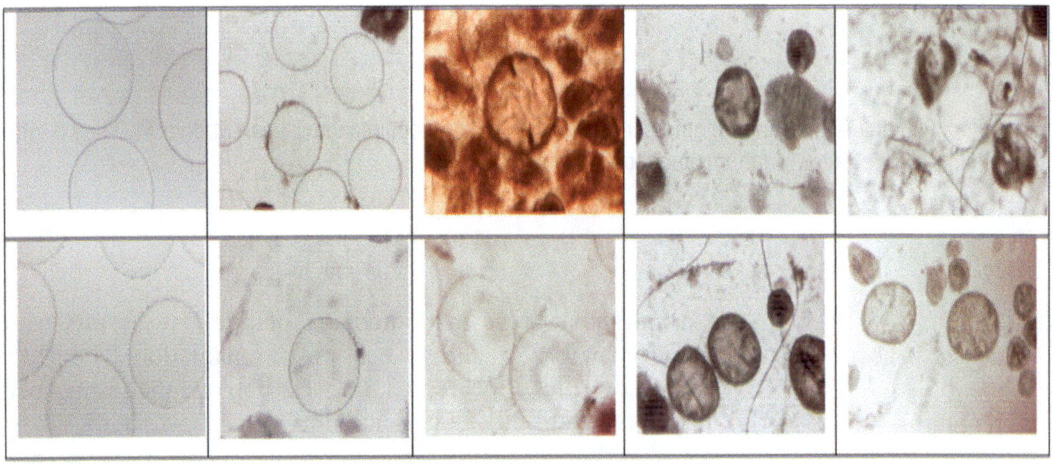

Fig. 3 *Top*: PLL preimplantation; PLL 7 days post-implantation into Sprague-Dawley rat's peritoneal cavity; PLL 12 days post-implantation; PLL 19 days post-implantation; PLL 28 days post-implantation. *Bottom*: PLL preimplantation; PLL 7 days post-implantation into Sprague-Dawley rat's peritoneal cavity; PLL 12 days post-implantation; PLL 19 days post-implantation; PLL 28 days post-implantation

Table 2
PLO and PLL microcapsules were formed. Samples of capsules were injected through a small ventral midline incision made into the peritoneum of male rats. The rats were followed for a 4-week period. At 1-week intervals, representative rats from each group were sacrificed. The capsules were then retrieved via gentle peritoneal lavage with slight suction. The capsule were examined for degree of fibrosis, following peritoneal lavage (*n* = 8). Fibrosis was confirmed by microscopic examination

	PLL capsules	PLO capsules
Mean capsule diameter in μm (SD)	743.27 (23.13)	743.27 (23.13)
Pore size in kDa	120	120
% Fibrosis at 1 month	75 %	25 %

4 Notes

1. Store in sterile 200 mL centrifuge tube. Pour into 30 cc syringe when ready to use. Place blue filter on the tip of the syringe. Place needle on the tip of the blue filter.
2. Finger-flick to resuspend.
3. Be sure to check on animals daily. We placed no more than two animals per cage.

References

1. Wild S, Roglic G, Green A, Sicree R, King H (2004) Global Prevalence of Diabetes. Estimates for the year 2000 and projections for 2030. Diabetes Care 27(5):1047–1053
2. www.diabetes.org Data from the National Diabetes Statistics Report, 2014 (released June 10, 2014).
3. Duck JH, Sutherland DER (2010) Gut Liver 4(4):450–465
4. Hakim NS (2002) Pancreatic transplantation for patients with Type I diabetes. HPB (Oxford) 4(2):59–61
5. Lacy PE, Kostianovsky M (1967) Method for the isolation of intact islets of Langerhans from the rat pancreas. Diabetes 16(1):35–39
6. Shapiro AM, Lakey JR, Ryan EA et al (2000) Islet transplantation in seven patients with type 1 diabetes mellitus using a glucocorticoid-free immunosuppressive regimen. N Engl J Med 343(4):230–238
7. Kendall WF, Collins BH, Opara EC (2001) Islet cell transplantation for the treatment of diabetes mellitus. Expert Opin Biol Ther 1(1):109–119
8. Opara EC, McQuilling JP, Farney AC (2013) Microencapsulation of pancreatic islets for use in a bioartificial pancreas. Methods Mol Biol 1001:261–266
9. Ching DC, Harland RC, Collins BH, Kendall W, Hobbs H, Opara EC (2001) A reliable method for isolation of viable porcine islet cells. Arch Surg 136:276–279
10. Darrabie MD, Kendall WF, Opara EC (2005) Characteristics of poly-L-ornithine-coated alginate microcapsules. Biomaterials 26(34): 6846–6852
11. Hobbs H, Kendall W, Darrabie M, Collins B, Bridges S, Opara EC (2000) Substitution of polyornithine for polylysine in alginate microcapsules. Diabetes 49(Suppl 1):A111

Chapter 8

Determination of the Mechanical Strength of Microcapsules

Marcus D. Darabbie and Emmanuel C. Opara

Abstract

The promise of pancreatic islet transplantation is hindered by organ shortage, and the need for immunosuppression of transplant recipient in order to prevent rejection. Alginate microencapsulation can overcome these hurdles; however further optimization of this technique is required. Among the critical factors to be optimized is the durability of alginate microcapsules, which can be determined by their mechanical strength tests. Here we describe several simple and reliable methods to assist in assessing the mechanical strength of alginate beads.

Key words Alginate, Microencapsulation, Islet transplantation, Encapsulated islets, Osmotic stress, Bead agitation

1 Introduction

1.1 Background

Despite improvements in immunosuppressive regimens pancreatic islet transplantation is still hindered by graft failure and organ shortage [1]. Encapsulated pancreatic islets obviate the need for immunosuppression through the use of semipermeable hydrogel capsules. The original alginate poly-L-lysine-alginate microcapsule developed by Lim and Sun over 30 years ago consisted of an alginate hydrogel with enclosed pancreatic islet tissue [2]. The pore-size exclusion was restricted with a poly-L-lysine layer after which, liquefaction of the inner core is achieved with sodium citrate prior to a final alginate coating. Very few changes have been made to the principle components of capsule formation. Islet cell viability is solely dependent on the durability and stability of the encapsulated islet tissue. To address this key requirement the ideal microcapsule must withstand internal sheer stress and osmotic pressure generated from capsule formation and implantation. Hence, there is still a need to further optimize alginate gel microcapsules, particularly with regard to improved mechanical strength. Among a number of important variables, the main factors that influence mechanical strength

Emmanuel C. Opara (ed.), *Cell Microencapsulation: Methods and Protocols*, Methods in Molecular Biology, vol. 1479, DOI 10.1007/978-1-4939-6364-5_8, © Springer Science+Business Media New York 2017

include alginate composition, morphometry, gelling cation, and polyamino coating. The balance between these factors and capsule performance is delicate. Therefore the ideal microcapsule should be mechanically durable to withstand sheer internal forces without compromising perme-selectivity or biocompatibility.

1.2 Microcapsule Composition

Sodium alginate is the most popular hydrogel polymer used in islet cell encapsulation and is comprised of monomeric units of 1,4-linked-D-mannuronic acid and L-guluronic acid [3]. Alterations in the chemical structure and molecular composition of alginate have significant effects on microcapsule size and mechanical strength. Specifically, using Ca^{++} as cross-linking cation, alginate preparations that have high guluronic acid content have been shown to be stiffer than low guluronic acid content alginate, while higher mannuronic acid content is more elastic [4, 5]. Additionally, microencapsulation with highly purified mannuronic acid-rich alginate has been shown to result in smaller initial microspheres with reduced morphometric changes; however mannuronic acid-rich alginate micro-beads also swell more than guluronic acid capsules [6], as illustrated in Fig. 1. Alginate molecular structure directly influences its affinity for cation binding. Guluronic acid residues bind divalent cations with greater affinity than mannuronic acid. Upon cross-linking with divalent cations, alginate forms

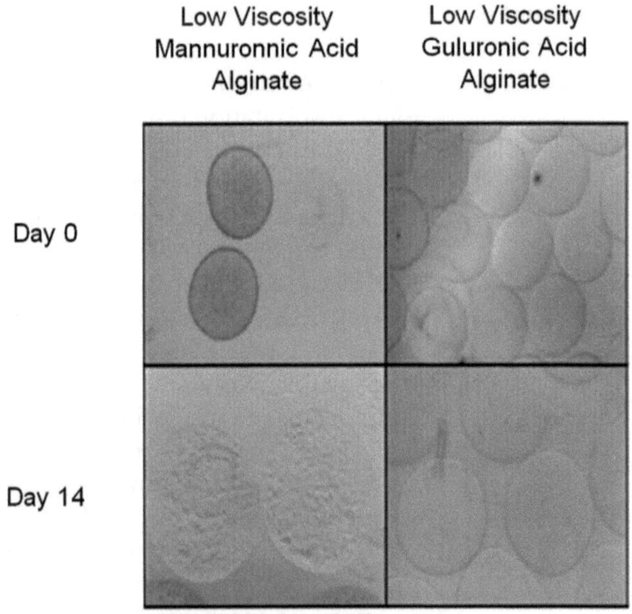

Fig. 1 Microcapsule swelling and alginate composition. Alginate microcapsules composed of low-viscosity mannuronic acid (*top left*) have increased swelling after 14 days of saline incubation (*bottom left*) in comparison to low-viscosity guluronic alginate capsules (*top* and *bottom right*)

an egg-box-shaped lattice, with guluronic-guluronic residues. Mannuronic-mannuronic residues form in a less efficient manner and less dense network. It is therefore predicted that higher percentage guluronic content increase mechanical strength [7]. A previous study evaluating the degradation patterns of alginate-polyornithine-alginate microcapsules also demonstrated the influence of the mannuronic:guluronic ratio on stability [8].

1.3 Cross-Linking Cation

Cross-linkage of alginate microcapsules is achieved through binding of the carboxyl units of alginate with divalent cations [9]. Though calcium is the most widely used component, other divalent cations that have been investigated include barium and strontium. It has been reported that strontium and barium bind specifically to mannuronic and guluronic acid blocks [10]. Barium binds to G–G and M–M blocks, calcium binds to G–G and M–G blocks, and strontium binds to G–G blocks [10]. Initial studies with barium demonstrated alginate microspheres that were able to withstand more mechanical stress than calcium capsules which was attributed to its greater affinity for alginate and capacity to form two bonds to the carboxyl groups of alginate [11]. Both strontium and barium bind alginate with greater affinity than calcium [9]. These capsules were initially felt to be less biocompatible as the potential leakage of barium was thought to lead to enhanced fibrotic reaction. However, a subsequent study with barium capsules showed similar proliferation rates and fibrosis to calcium cross-linked capsules after peritoneal implantation in rats [12].

Barium's high affinity for alginate leads to reduced incorporation of polycations into the final alginate capsule leading to reduced perme-selectivity [3, 13]. Also, liquefaction of the inner core is not possible with barium cross-linkage; therefore, the solid gel core of barium capsules would be expected to reduce oxygen content [14]. Finally, there are remaining concerns regarding the potential toxicity in using Ba^{++} -cross-linked alginate microbeads [15].

1.4 Polyamino Coating

Application of polyamino coating is not only vital to immunoisolation and perme-selectivity of encapsulated tissue but also for stabilization of the alginate membrane. Polyamino coatings allow added thickness which is an important component of membrane strength. Barium capsules with poor polyamino coating swell more rapidly and rupture quickly during osmotic stress testing [13]. While early applications utilized poly-L-lysine [2], additional polymers have been investigated and include poly-L-ornithine, poly-D-lysine, poly (methylene-co-guanide) hydrochloride, and poly-L-arginine [16, 17]. Studies have demonstrated increased strength of poly-L-ornithine capsules which was attributed to its molecular structure which is shorter by one methyl group as well as reduction in bead swelling [18, 19]. In general, polyamino acids bind alginate via electrostatic interactions between the NH3 functional groups and the –COOH groups of the alginate. Additional interactions add to the increased stability of poly-L-ornithine and poly-L-arginine compared to poly-L-lysine [20]. Other investigators have described using an alginate-aminopropyl-silicate/alginate which forms a ≡Si–O–Si≡ bond as well as electrostatic bonds with alginate. The additional molecular interactions allow for the generation of a firm ceramic like matrix in these capsules. These capsules have been studied in vitro and demonstrate increased durability compared to alginate-poly-lysine capsules. In vivo studies demonstrate similar functionality to traditional alginate poly-L-lysine capsules [21].

1.5 Microcapsule Morphometry

Capsule size and morphometry play an essential role in microcapsule durability. Optimally encapsulated cells are spherical and free from tissue protrusion. A spherical alginate microcapsule with a small initial diameter is theoretically more capable of withstanding the forces that cause swelling and polymorphism. These conditions would lower shear forces, increase biocompatibility, extend in vivo durability, and ultimately increase islet viability and insulin release kinetics [22–24]. Previous studies have demonstrated that viscosity of the alginate as well as the ratio of mannuronic/guluronic acid are important factors in fabricating spherical capsules [20]. Osmotic swelling from liquefaction of the inner core has also been shown to cause capsule polymorphism [25]. Additional morphometric changes can be influenced by device settings, distance to cross-linking solution and size of the principle components of droplet formation [26].

1.6 Assessment of Microcapsule Mechanical Strength

A number of methods to assess membrane physical and mechanical properties have been developed including the probing of capsules with fine tip tweezers [27]. Several studies have examined the mechanical properties of alginate microcapsules using micropipette techniques. It was determined that sensitivity to micropipette aspiration was an indicator of overall gel weakening and calcium loss [21]. A micropipette vacuum system has also been utilized to assess microcapsule elastic properties [28]. These methods allow finite measurements of the mechanical properties of capsules but may not directly assess primary membrane strength in settings similar to in vivo conditions.

Direct measurement of capsule burst strength has been described using uniaxial compression, which was used to evaluate the correlation between membrane deformation and capsule volume [29]. Similar devices were used to measure force of compression and the time to compression or elasticity of alginate microcapsules [20]. Qualitative methods that incorporated dextran FITC into the alginate were later developed as well as a method of determining the osmotic strength of capsules [30, 31]. A modification of the osmotic pressure test consists of the explosion assay, which involves placement of capsules in water and recording the number of intact capsules over time [9]. The prime component of osmotic stress is attributed to pressure generated between the inner alginate core and surrounding tissues, and its influence on the rigid capsule membrane. This process is exacerbated by aggressive liquefaction of the inner core which results in residual unbound divalent cations which cause capsule swelling [25]. Other reports have assessed capsule integrity using advanced microscopy techniques after implantation [8, 32]. In-vivo methods that combine assessment of islet performance after implantation into biological compartments likely provide the most direct assessment of functional stability and durability. However, the in-vitro assessments described in this chapter provide high-throughput, simple, and reliable assessment of mechanical strength and facilitate systematic optimization of encapsulated islets.

2 Materials

2.1 Chemicals and Solutions

1. 1.2–1.8% Alginate (Pronova UP LVM and UP LVG, Novamatrix, Sandvika, Norway).
2. 0.1% Poly-L-lysine (PLL) (P4957, Sigma-Aldrich, St. Louis, MO, USA).
3. 0.1% Poly-L-ornithine (PLO) (P5061, Sigma-Aldrich).
4. 100 mM $CaCl_2$ solution (C614-10, Fischer Scientific, Waltham, MA, USA).
5. 55 mM Sodium citrate solution (S467-3, Fischer Scientific).
6. 0.9% Sodium chloride solution (normal saline) (71376-5KG, Sigma-Aldrich).
7. 10 mM HEPES solution (H3375-2KG, Sigma-Aldrich).
8. 0.5% Trypan Blue (T6146, Sigma-Aldrich).

2.2 Equipment

1. Air-syringe pump droplet generator.
2. Microfluidic devices (Clemson University, SC).
3. Labline Orbital Shaker.
4. 9. 3 mm Soda Lime Beads (26396-508, VWR).
5. Stereomicroscope (Swift Instruments, Schertz, TX).

3 Methods

3.1 Generation of Alginate Microcapsules

1. 1.2–1.8% alginate microspheres are generated and allowed to gel into microbeads in a bath of 100 mM $CaCl_2$ dissolved in 10 mM HEPES solution at 4 °C, pH 7.4.
2. Following two washes with normal saline, the microbeads are perm-selectively coated with either 0.1% PLL or 0.1% PLO for variable duration of time depending on the desired pore-size exclusion limit.

3. Liquefaction of the alginate core of the microcapsules is achieved by a brief (2 min) incubation in 55 mM sodium citrate solution at 4 °C.
4. Wash three times with normal saline; this is accomplished by allowing the capsules to settle in the bottom of a 100 mL beaker over the course of approximately 2 min.
5. External coating with a lower concentration (routinely about 10% of the concentration used in generating the initial microspheres) which is accomplished by incubating the capsules with alginate for 5 min or 45 min if angiogenic protein is incorporated into the outer layer at 4 °C.
6. To cross-link the external alginate coat, normal saline may be used with or without 22 mM calcium chloride.

3.2 Induction of Mechanical Stress Using Osmotic Pressure

1. Equal aliquots of alginate microcapsules are placed in 10 mL of H_2O and incubated at 37 °C for 2 h (*see* **Note 1**).
2. Capsules are then washed twice with saline and stained with 0.5% (w/v) Trypan blue.
3. A final washing with saline is then performed prior to assessment of broken and intact capsules as illustrated in Fig. 2 using an inverted light microscope.

3.3 Induction of Mechanical Stress Using Bead Agitation

1. Equal aliquots of capsules are placed into T-25 tissue culture flasks containing approximately 6.5 g of 3 mm inert soda lime beads and 30 mL of normal saline (*see* **Note 2**).
2. The capsules are then subjected to mechanical stress for 36 h by shaking at approximately 300 RPM using a Lab Line Orbital Shaker.
3. The percentages of broken and unbroken capsules are determined manually by visual analysis and handpicking under a stereomicroscope (*see* **Note 3**).

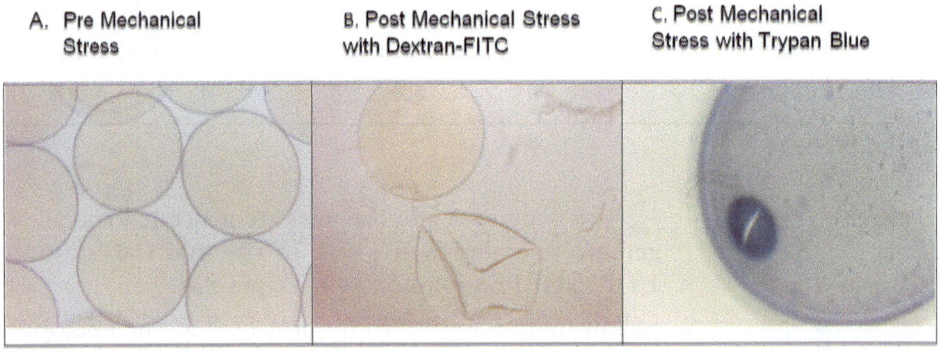

Fig. 2 Assessment of broken and intact capsules. (**a**) Intact alginate microcapsules with incorporated dextran-FITC prior to mechanical stress. (**b**) Broken and intact capsules after bead agitation. (**c**) Broken capsules after induced osmotic stress and stained with Trypan blue

4 Notes

1. The sensitivity of this assay can be increased by using an osmometer and placing capsules in successive hypotonic solutions generated from dilution of serum free media.
2. We have found that this size and quantity of beads provide the optimal bead to surface ratio for force generation in our studies. Smaller beads were less likely to cause enough generated force. Capsules were less likely to collide with larger beads. In our studies, adjustment of the shaker speed was the easiest way to scale the amount of force generated.
3. An alternative to manual counting involves inclusion of dextran-FITC into alginate preparations at a final concentration of 0.2 % (w/v). After induction of mechanical stress, quantification of intact capsules is performed by dilution of the residual capsules and supernatant in glycine buffer.

References

1. Hatziavramidis DT, Karatzas TM, Chrousos GP (2013) Pancreatic islet cell transplantation: an update. Ann Biomed Eng 41:469–476
2. Lim F, Sun AM (1980) Microencapsulated islets as bioartificial endocrine pancreas. Science 210:908–910
3. Smidsrod O, Skjak-Braek G (1990) Alginate as immobilization matrix for cells. Trends Biotechnol 8:71–78
4. Amsden B, Turner N (1999) Diffusion characteristics of calcium alginate gels. Biotechnol Bioeng 65:605–610
5. De Vos P, De Haan B, Van Schilfgaarde R (1997) Effect of the alginate composition on the biocompatibility of alginate-polylysine microcapsules. Biomaterials 18:273–278
6. Kendall WF Jr, Darrabie MD, El-Shewy HM, Opara EC (2004) Effect of alginate composition and purity on alginate microspheres. J Microencapsul 21:821–828
7. Haug AMS, Larsen B, Smidsrod O (1967) Correlation between chemical structure and physical properties of alginates. Acta Chem Scand 21:768–778
8. Thanos CG, Bintz BE, Bell WJ, Qian H, Schneider PA, MacArthur DH, Emerich DF (2006) Intraperitoneal stability of alginate-polyornithine microcapsules in rats: an FTIR and SEM analysis. Biomaterials 27:3570–3579
9. Thu B, Bruheim P, Espevik T, Smidsrod O, Soon-Shiong P, Skjak-Braek G (1996) Alginate polycation microcapsules. I Interaction between alginate and polycation. Biomaterials 17:1031–1040
10. Morch YA, Donati I, Strand BL, Skjak-Braek G (2006) Effect of Ca2+, Ba2+, and Sr2+ on alginate microbeads. Biomacromolecules 7:1471–1480
11. Peirone M, Ross CJ, Hortelano G, Brash JL, Chang PL (1998) Encapsulation of various recombinant mammalian cell types in different alginate microcapsules. J Biomed Mater Res 42:587–596
12. Li HB, Jiang H, Wang CY, Duan CM, Ye Y, Su XP, Kong QX, Wu JF, Guo XM (2006) Comparison of two types of alginate microcapsules on stability and biocompatibility in-vitro and in-vivo. Biomed Mater 1:42–47
13. Tam SK, Bilodeau S, Dusseault J, Langlois G, Halle JP, Yahia LH (2011) Biocompatibility and physicochemical characteristics of alginate-polycation microcapsules. Acta Biomater 7:1683–1692
14. Schrezenmeir J, Kirchgessner J, Gerö L, Kunz LA, Beyer J, Mueller-Klieser W (1994) Effect of microencapsulation on oxygen distribution in islets organs. Transplantation 57:1308–1314
15. Zimmermann U, Mimietz S, Zimmermann H, Hillgartner M, Schneider H, Ludwig J, Hasse C, Haase A, Rothmund M, Fuhr G (2000) Hydrogel-based non-autologous cell and tissue therapy. Biotechniques 29:564–572, 574, 576 passim

16. van Raamsdonk JM, Cornelius RM, Brash JL, Chang PL (2002) Deterioration of polyamino acid-coated alginate microcapsules in-vivo. J Biomater Sci Polym Ed 13:863–884
17. Wang T, Lacik I, Brissova M, Anilkumar AV, Prokop A, Hunkeler D, Green R, Shahrokhi K, Powers AC (1997) An encapsulation system for the immunoisolation of pancreatic islets. Nat Biotechnol 15:358–362
18. De Castro M, Orive G, Hernandez RM, Gascon AR, Pedraz JL (2005) Comparative study of microcapsules elaborated with three polycations (PLL, PDL, PLO) for cell immobilization. J Microencapsul 22:303–315
19. Darrabie MD, Kendall WF Jr, Opara EC (2005) Characteristics of Poly-L-Ornithine-coated alginate microcapsules. Biomaterials 26:6846–6852
20. Bhujbal SV, Paredes-Juarez GA, Niclou SP, de Vos P (2014) Factors influencing the mechanical stability of alginate beads applicable for immunoisolation of mammalian cells. J Mech Behav Biomed Mater 37:196–208
21. Sakai S, Ono T, Ijima H, Kawakami K (2000) Control of molecular weight cut-off for immunoisolation by multilayering glycol chitosan-alginate polyion complex on alginate-based microcapsules. J Microencapsul 17:691–699
22. Chicheportiche D, Reach G (1988) In vitro kinetics of insulin release by microencapsulated rat islets: effect of the size of the microcapsules. Diabetologia 31:54–57
23. Poncelet D, Neufeld RJ (1989) Shear breakage of nylon membrane microcapsules in a turbine reactor. Biotechnol Bioeng 33:95–103
24. Robitaille R, Pariseau JF, Leblond FA, Lamoureux M, Lepage Y, Hallé JP (1999) Studies on small (<350 microm) alginate-poly-L-lysine microcapsules. III Biocompatibility Of smaller versus standard microcapsules. J Biomed Mater Res 44:116–120
25. Hobbs HA, Kendall WF Jr, Darrabie M, Opara EC (2001) Prevention of morphological changes in alginate microcapsules for islet xenotransplantation. J Invest Med 49:572–575
26. Chan ES, Lee BB, Ravindra P, Poncelet D (2009) Prediction models for shape and size of ca-alginate macrobeads produced through extrusion-dripping method. J Colloid Interface Sci 338:63–72
27. Goosen MF, King GA, McKnight CA, Marcotte N (1989) Animal cell culture engineering using alginate polycation microcapsules of controlled membrane molecular weight cut-off. J MembrSci 41:323–343
28. Jay AW, Edwards MA (1968) Mechanical properties of semipermeable microcapsules. Can J Physiol Pharmacol 46:731–737
29. Rehor A, Canaple L, Zhang Z, Hunkeler D (2001) The compressive deformation of multicomponent microcapsules: influence of size, membrane thickness, and compression speed. J Biomater Sci Polym Ed 12:157–170
30. Leblond FA, Tessier J, Halle JP (1996) Quantitative method for the evaluation of biomicrocapsule resistance to mechanical stress. Biomaterials 17:2097–2102
31. Van Raamsdonk JM, Chang PL (2001) Osmotic pressure test: a simple, quantitative method to assess the mechanical stability of alginate microcapsules. J Biomed Mater Res 54:264–271
32. van Hoogmoed CG, Busscher HJ, de Vos P (2003) Fourier transform infrared spectroscopy studies of alginate-PLL capsules with varying compositions. J Biomed Mater Res A 67:172–178

Chapter 9

The Diffusive Properties of Hydrogel Microcapsules for Cell Encapsulation

D.M. Lavin, B.E. Bintz, and C.G. Thanos

Abstract

Hydrogel microcapsules have been used for decades to encapsulate cells and treat diseases ranging from neurodegenerative disorders to more systemic applications like Type I Diabetes. This cell encapsulation modality has been developed through more cumulative experiments than perhaps any other, owing to the relative ease of accessing the required materials, the commercial availability of droplet-generating instrumentation, and the mild microenvironment and unique permeability properties of hydrogels that are difficult to attain with alternative encapsulation systems employing thermoplastic materials. Because of their size and shape, microcapsules have an inherent advantage over macroencapsulation devices due to the more favorable surface area to volume ratio, which allows for greater efficiency in the amount of cellular cargo that is entrapped and enhanced nutrient exchange and efflux of secreted products. Unfortunately, with this significant positive benefit comes the caveat of difficult or impractical retrievability, highlighting the paradox that is particularly relevant as differentiated stem cell sources become more readily available. This chapter focuses on several techniques that can be used to evaluate the permeability and pore structure of hydrogel microcapsules, including a simplistic model for describing the diffusive behavior of alginate-polycation-alginate (APA) microcapsules with a liquid core, and an ancillary method to evaluate the ultrastructure of the APA membrane including morphometric analysis.

Key words Microcapsule, Diffusion, Alginate, Encapsulation, Cell transplantation

1 Introduction

1.1 Hydrogel Encapsulation

Hydrogel microcapsules have long been studied [1] for their ability to deliver therapeutic cells to locations throughout the body, including the subcutaneous space, the peritoneal cavity, the brain parenchyma, and others. As a universal tool, hydrogel microencapsulation is perhaps the most widely used cell encapsulation modality in research, with some iterations of this technology having developed into GMP-grade therapies that have advanced into clinical trials [2]. Because of the inherent advantages in the efficiency of diffusion radially from a small sphere and the ability to distribute the dose of microcapsules throughout a large space such as the peritoneal cavity that vastly exceeds the delivered volume, cell-encapsulated hydrogel

Emmanuel C. Opara (ed.), *Cell Microencapsulation: Methods and Protocols*, Methods in Molecular Biology, vol. 1479, DOI 10.1007/978-1-4939-6364-5_9,

microcapsules remain one of the most promising delivery paradigms in cell therapy.

The permeability of hydrogel microcapsules can be controlled by the properties of the primary materials from which they are made [3], through chemical modifications of the polymer backbone or as pendant side chains [4], or through ionic layering of additional materials to form multilayered membrane walls [5]. Other chemical modifications to the hydrogel itself have been implemented to improve biocompatibility [6]. Hydrogel microcapsules have a dynamically changing pore structure that responds to fluctuations in osmotic or oncotic pressure differentials by swelling or shrinking, thereby expanding or compressing the wall structure and impacting permselectivity [7]. Because of this, microcapsules are often overlooked for certain applications in favor of non-degradable [8] thermoplastic membranes made via solvent inversion, track etching, or biaxial stretching due to their stability with defined pore structures. Moreover, these non-degradable polymers can be retrieved in the event of an adverse event or failure. As with all implanted biomaterials, interaction with the host inevitably causes some degree of change between initial pre- and post-implanted functional properties. These changes may be due to protein adsorption, cell coating, and matrix deposition. As such, in order to adequately characterize the functionality of either type of encapsulation system, it is important to quantify these properties under a variety of conditions to adequately predict in vivo performance.

The current chapter focuses on the characterization of the diffusive properties of hydrogel microcapsules in vitro, with a focus on method development for evaluation of the diffusion coefficient and extrapolation of the nominal molecular weight cutoff (MWCO) in a simplistic encapsulation and release paradigm. These basic methods can be used to study the diffusive properties of a formulation of interest in a release buffer, as shown here, or in a more physiologically relevant medium incorporating proteins or serum, mimicking the surface passivation that would occur in vivo (*See* **Note 1**). Additionally, we describe ultrastructural analysis using transmission electron microscopy. The principles employed in these applications can be applied to other types of hydrogel systems such as conformal coatings or slabs.

1.2 Microcapsule Formulations and Diffusion

Cell suspensions in alginate, cellulose, chitosan, and modified versions thereof can form droplets through any number of techniques that can be crosslinked to form stiff hydrogel beads that may serve as a protective matrix for the study of cells in vitro or in vivo. Depending on the properties of the hydrogel material employed, the solid, three-dimensional bead can be used as a final delivery vehicle, or it can be further coated with another set of polymers or polypeptides to achieve a higher degree of permselectivity [9]. The finalized microcapsule can contain a solid or liquid core, with the encapsulated cells separated from the external environment by a thin wall on the order of microns [10], or by a continuous hydrogel with a radius of millimeters [11]. The size, shape, and structure of the microcapsules all play a part in the biocompatibility, longevity, porosity, and effective diffusion coefficient associated with the encapsulation system.

Immunoprotection refers to the ability of a cell encapsulation system to prevent or restrict molecules of the host immune system from penetrating through the membrane and eliciting an immune response against the encapsulated cells, or restricting the ability of shed antigens to exit the capsule. In theory, the pore size cutoff that enables a material to be immunoprotective must be small enough to restrict passage of complement components, immunoglobulins, and other small immune protein fragments. To prevent antibody transport, the pores should be smaller than about 5.9 nm, the Stokes radius of IgG [12], but large enough to allow outward diffusion of therapeutic products such as insulin (about 2 nm [13]), growth factors (about 1.5–4.5 nm), and other proteins.

These theoretical values are certainly useful as a guide, but in practice, the dynamic in vivo environment and unpredictable composition of shed antigens make designing the pore characteristics of a membrane a task that requires empirical evaluation. Macroencapsulation devices containing thermoplastic membranes made by phase inversion, track etching, or MEMs techniques can all achieve relatively narrow pore size distributions within a finite, theoretically immunoprotective range owing to the relatively stable nature of the polymers employed and precision associated with current manufacturing techniques [13]. Depending on the technique utilized, the resulting thermoplastic membrane can achieve relatively consistent porosity with either a homogenous pore distribution throughout the thickness of the membrane wall (reticular), or with a distinct limiting skin (trabecular), typically occupying a very thin region on the inner or outer wall edge.

The nominal pore size of a hydrogel matrix can be dictated by a limiting skin in certain configurations, but more frequently it is comprised of a continuous high-density barrier wherein the limitation on diffusion is imparted by its cumulative tortuosity and thickness. Alginate, a polysaccharide derived from seaweed, has been used extensively in this role in a solid format [11], in a multi-layered capsule integrated with a diffusion-controlling polycation [14, 15], and in hybrid applications with other polysaccharides such as chitosan [16]. In a solid bead, diffusion of solutes is dependent on the density of the matrix and distance to the surface. The properties of alginate vary significantly depending on the composition of the alginate, both in aqueous solutions [17] as well as in cross-linked matrices [18]. For example, calcium-cross-linked alginate with a high proportion of guluronate residues forms a stiffer matrix than cross-linked high-mannuronate alginate. The molecular weight, block length, and ratio of the monomers contribute to the overall mechanical and diffusive properties of the system.

Alginate-polycation-alginate (APA) microcapsules contain a multi-layered wall comprised of an inner core of alginate bonded to a polycationic layer typically comprised of poly-L-ornithine, poly-L-lysine, or other polypeptides [19]. In this configuration, the solid calcium-crosslinked alginate core is coated with one or more layers of polycation, which is further coated with an outer layer of biocompatible alginate. It has been demonstrated that while the polycation contributes significantly to diffusive control and potentially immunoprotection, its exposure in vivo is associated with inflammation, leading to immune involvement and, potentially, one of the primary failure modes of this encapsulation paradigm: fibrosis. Therefore it is important that the outermost layer of the APA membrane be stable and resistant to degradation that could occur in the transplant site [10]. Once formed, the APA membrane is a permselective multilayer composite wherein the flux of solutes can be controlled by manipulation of the concentration of each component, the timing of the coating reactions, and the molecular weights and other physicochemical properties of each membrane constituent.

1.3 Permselective Hydrogel Membrane Diffusion Testing Models

Solute diffusion studies are used throughout the pharmaceutical industry to quantify the flux properties of flat-sheet, tubular, and other membrane geometries fabricated from both thermoplasts and hydrogels. To illustrate, a flat sheet membrane can be placed in a Franz Diffusion cell [20] between an open receptacle (the donor chamber), and a closed receptacle with a sampling port (the receptor chamber). A solute maintained at a high concentration (but well below its solubility limit) is placed in the top chamber against a pre-wetted test membrane that effectively creates a semi-permeable seal between the two chambers, while the lower collection chamber is filled with a release buffer in which the test solute has confirmed solubility. Depending on the molecular weight of the test solute and the diffusion coefficient of the membrane, samples can be taken for minutes, hours, or days. Dextrans linked to fluorophores are produced in a variety of relatively monodisperse molecular weight ranges, and they make convenient test solutes for this type of investigation. However, as dextrans are typically linear or branched compared to their protein counterparts with complex three-dimensional structures, the diffusion behavior of test solutes of dextrans or proteins of equivalent molecular weight can be considerably different [21].

Simplistically, APA microcapsules can be studied in a similar manner wherein the core of liquefied alginate can represent the donor chamber, containing a test solute that is incorporated during the encapsulation process, and the APA wall represents the permselective membrane. Known volumes of APA microcapsules can be aliquoted into tubes filled with a diffusion buffer (receptor chamber), and the supernatant sampled over time to measure the presence of the test solute. In practice, however, sampling of supernatant is associated with several sources of experimental error related to incomplete washing of the capsule surface and inhomogeneous solute distribution. Rather, we propose that analyzing individual, unique sample aliquots for remaining solute in the APA capsule core (donor chamber) is associated with an improvement in the signal to noise ratio, and reduces confounding artifact related to capsule washing.

As a system, including the fluid components within the capsule wall as well as the surrounding supernatant, the total volume must be able to accommodate the complete dissolution of the test solute to a level above the solubility limit of the donor compound. While the resulting diffusion coefficient can be described as concentration dependent [22], for the purpose of the simplistic method described here, it is important to maintain equivalent conditions when comparing changes in formulations.

1.4 Microencapsulation of FITC-Dextrans

Due to their high solubility in aqueous solutions, wide availability in discrete size ranges, and ease of detection, FITC-dextran solutions are ideal test solutes for APA diffusion testing. They can be incorporated into the liquid alginate solution at concentrations that are compatible with gelation conditions, and that are suitably high to measure levels released even at very early time points. As the APA capsule is formed, the solute remains with the primary alginate droplet, localizing mostly to the core of the capsule. One caveat to this approach is that due to the high solubility of FITC-dextrans in all of the crosslinking and coating solutions employed in the APA encapsulation process, even with high encapsulation efficiency, there are still significant losses of the test solute as the APA membrane is crosslinked, coated, and chelated. It is important that these losses from theoretical loading levels be quantified so that the concentration within the APA microcapsule is known as the experiment is initiated. In the experiment described below, we created terminal samples for each time point that could be incubated for different durations and terminally sampled by extraction. In this manner, we have reduced the variability frequently encountered during supernatant sampling due to unknown final loading, incomplete sampling, and losses due to swelling.

2 Materials

2.1 Buffers and Solutions

0.9% (W/V) NaCl (pH 7.4) (Sigma-Aldrich, St. Louis, MO).

Sodium alginate (High-M: Proprietary; High-G: Pronova) (*see* **Note 2**).

1× Calcium and magnesium-free Dulbecco's phosphate-buffered saline (DPBS): (Thermo Fisher, Pittsburgh, PA).

Polycation (Sigma-Aldrich, St. Louis, MO).

109 mM Calcium chloride (Sigma-Aldrich, St. Louis, MO).

55 mM Sodium citrate (Sigma-Aldrich, St. Louis, MO).

3-(*N*-morpholino) propanesulfonic acid (MOPS) buffer (Sigma-Aldrich, St. Louis, MO).
2.5 % (W/V) FITC-dextran dissolved in sterile-filtered 1.7% (W/V) alginate (Sigma-Aldrich, St. Louis, MO).
0.1 M Sodium cacodylate buffer (Electron Microscopy Sciences, Hatfield, PA).
1 % Osmium tetroxide (Electron Microscopy Sciences, Hatfield, PA).

2.2 Kits

BCA Assay Kit (Pierce Biotechnology).

2.3 Special Equipment

Air knife (Kineteks, Warwick, RI).
Fluorescent plate reader (Beckman Coulter, Danvers, MA).
Transmission electron microscope (Hitachi).
ImageJ software.
Viscometer (Brookfield).

3 Methods

3.1 Viscometry

Viscosity was measured using a Brookfield Cone/Plate Viscometer at 25 °C.

(a) 1 mL of 1.0 % (W/V) alginate sample was added to the sample cup and distributed in a thin layer across the bottom of the cup and the spindle was inserted to assess rotational resistance at speeds ranging from 1 to 20 rpm.

(b) Torque ranged from 25 to 95 % at the shear speeds within the optimum working range of the viscometer.

(c) Dynamic viscosity was calculated by the change in resistance compared to the speed of the probe.

3.2 Characterization of Contaminants

A Micro BCA Protein Assay (Pierce, USA) was used to measure total protein levels in alginate samples.

(a) 1 mL 1.7 % (W/V) alginate samples were diluted to 2-, 5-, 10-, and 20-fold and were incubated with the working reagent for 2 h at 37 °C.

(b) The developed reagent was detected on a 96-well plate with a UV–Vis spectrophotometer at 562 nm and quantified against a linear standard curve with bovine serum albumin. The Limulus Amebocyte Lysate (LAL) CL-1000 Chromogenic LAL Endpoint Assay (Cambrex, USA) was used to quantify the total endotoxin content of the alginates under study.

(c) 1.7 % (W/V) samples were incubated at a tenfold dilution in dH2O for 18 h at 50 °C for endotoxin extraction and reacted over a defined time course against standard concentrations [23].

(d) Endpoint product was analyzed on a Beckman-Coulter DTX-880 UV–VIS Spectrophotometer. The assay had a detection range of 1–50 EU/mL.

3.3 Micro-encapsulation

(a) Sterilized alginate containing 2.5 mg/mL of FITC-dextran (20, 70, or 150 kDa) was passed through a custom air knife (Kineteks, USA) and into a bath of 109 mM CaCl2 stirring without vortex.

(b) After cross-linking for 5 min, capsules were removed and immediately reacted with polycation for 10 min followed by two washes in MOPS buffer.

(c) A subsequent reaction with 0.05 % (W/V) PLO and washing followed. An outer alginate coat was applied by stirring the capsules in low concentration alginate for 6 min and the coated capsules were washed twice in MOPS buffer.

3.4 Release Studies

(a) Release studies were initiated by aliquoting 1.0 mL of MOPS buffer containing 100 μL APA capsules into microcentrifuge tubes. The total MOPS volume excluding capsules was 900 μL.

(b) As a releasate, MOPS buffer was selected to resist ionic exchange with microcapsules during the extended incubation period. Samples were placed on an end-over-end rotating rack, incubated at 37 °C, and sampled at time intervals between 15 min and 16 h.

(c) At each sampling point, the extracapsular release buffer was removed, the APA capsules were washed transiently with 1.0 mL of MOPS buffer, and the entire APA capsule volume was stored at −20 °C.

(d) After the last sampling interval, frozen APA capsule samples were thawed at room temperature, and 1.5 mL of sodium citrate buffer was added to each vial.

(e) Using a probe sonicator with a microtip operating at medium intensity, samples were sonicated for 30 s taking care to avoid overheating or sample losses. Confirmation of complete capsule rupture utilizing this method was assessed under bright-field microscopy (Fig. 1).

(f) Samples were centrifuged at 2000 × g for 5 min, and the supernatant (intracapsular retentate) was removed and analyzed for fluorescent intensity on a Beckman-Coulter DTX-880 UV–VIS Spectrophotometer with excitation at 488 nm and emission at 520 nm. Samples were also taken prior to initiation of the release study at $t=0$ for comparison of release relative to actual encapsulated FITC.

(g) The diffusion coefficient was calculated for each solute and capsule type according to Fickian diffusion (*see* **Note 3**).

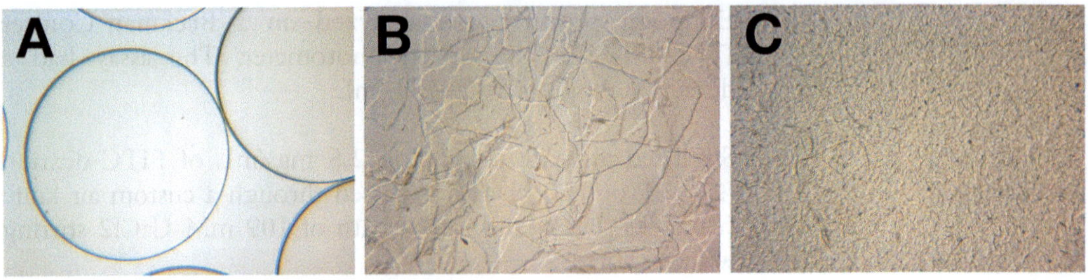

Fig. 1 Representative bright-field images of (**a**) APA microcapsules (**b**) during, and (**c**) after post-sonication (magnification, 50×). Sonication was used to disrupt capsule and liberate remaining FITC-dextran for diffusion studies

Table 1
Comparison of alginate material properties and the resulting APA membrane morphometry

		Purified alginate	Raw alginate
	Viscosity @ 1.7 % W/V (cP)	131.3 ± 1.3	108.5 ± 1.5
	Endotoxin content (EU/mL)	<0.2 EU/mL	1200 EU/mL
Average wall thickness (μm)	Inner wall	2.4	1.4
	Inner-mid layer	4.7	3.4
	Mid-layer	1	1.1
	Mid-outer layer	9.1	3.9
	Outer layer	0.3	0.4
	Full wall thickness	17.4	10.2
Pore size (nm)	Outer	75.2	91.2
	Distribution (%CV)	41	26
	Inner	31.4	51.4
	Distribution (%CV)	19	24

3.5 Transmission Electron Microscopy to Evaluate APA Wall Ultrastructure

APA microcapsules or Ca-Alginate beads can be processed using conventional techniques for visualization with transmission electron microscopy (TEM) with minimal swelling artifact or deformation of the wall structure [3]. With high-magnification TEM in combination with digital image analysis, it is possible to measure the porosity of the matrix or membrane by measuring the void area within the continuous structure.

(a) APA microcapsules were formed as previously described from a base material of purified high-M Keltone LVCR alginate. The purified alginate differed from the unpurified material in endotoxin concentration, as well as dynamic viscosity (Table 1).

(b) Immediately after fabrication of APA microcapsules, samples were rinsed twice for 10 min each with cacodylate buffer and post-fixed with 1% osmium tetroxide in buffer for 30 min at room temperature.

(c) Specimens were rinsed twice with 0.1 M sodium cacodylate buffer. Samples were dehydrated through a series of graded ethyl alcohols for 30 min each followed by three changes of 100% ethanol for 10 min each.

(d) The samples were first infiltrated overnight with a 1:1 solution of 100% ethanol, and Spurr embedding medium (Electron Microscopy Science, Hatfield, PA), then infiltrated with fresh Spurr overnight, embedded in molds, and polymerized at 60 °C.

(e) Blocks were sectioned at a thickness of 90 nm on a Reichert Ultramicrotome with a diamond knife, placed on copper grids, and viewed on a Phillips 410 transmission electron microscope equipped with an Advantage HR CCD camera.

(f) Images were acquired with Advanced Microscopy Techniques imaging software, and image analysis was performed with ImageJ software (*see* **Note 4**).

4 Notes

1. These in vitro methods do not introduce the variable of encapsulated cells, the media that they are maintained in, or the in vivo environment to which they are introduced. While we do not detail the impact of protein adsorption on the diffusivity of APA membranes, it plays a significant role in the long-term function and stability of APA microcapsules, and the propensity for proteins to stick on the outer alginate surface can cause an inflammatory response associated with implantation [26]. Perifusion can be used to compare encapsulated cells to free islets in a convective flow chamber that allows for controlled stimulus of the cells with glucose, and repeated sampling for detection of insulin. It has been shown the some configurations of the APA capsule have an impact of this real-time diffusion, such as those with a solid core, while others are nearly identical to unencapsulated islets [27]. It is without question that these functional assays are enormously valuable, and should serve as a final method of characterization following appropriate analysis of diffusion and material characterization. In this system and other broad hydrogel formulations, the methods described here may help to further elucidate the effects of material on ultrastructure and how they impact the structure-function relationship as a component of diffusion.

2. High-M sodium alginate was purchased from FMC and purified in-house to rid the material of contaminating proteins and endotoxin.

3. The FITC-dextrans in the present example have a maximal solubility of 25 mg/mL in water. Therefore, all release conditions were selected such that the highest detected FITC-dextran concentration did not exceed 2.5 mg/mL. These examples modeled mass transport kinetics of membrane and alginate materials. Our studies involve multilayered alginate capsules with a liquid alginate core and alginate-PLO-alginate membrane surrounding an internal calcium-crosslinked alginate sheath. Kwok and colleagues [24] modeled the diffusion of protein from a bulk solution of albumin into alginate-PLL-alginate microcapsules by solving for the radial dissolution rate of solute in the membrane in combination with Fick's law (the assumption that the diffusion coefficient through the PLL membrane is constant) (Fig. 2). Therefore, our system may be similarly modeled using the governing equation for diffusion characteristics in the membrane:

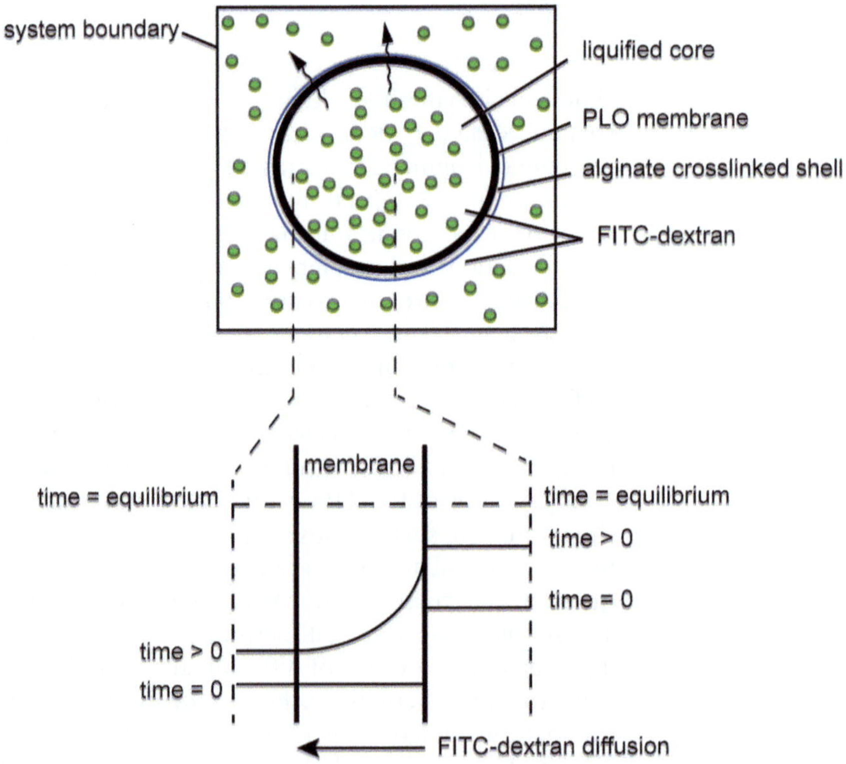

Fig. 2 FITC-dextran diffusion model for limited volume bulk solution and liquefied APA microcapsule core (adapted from ref. 24)

$$\frac{\partial C_{\mathrm{m}}}{\partial t}=\frac{1}{r^2}\frac{\partial}{\partial r}\left(D_{\mathrm{m}}\, r^2\, \frac{\partial C_{\mathrm{m}}}{\partial r}\right) \tag{1}$$

where $\frac{\partial C_{\mathrm{m}}}{\partial t}$ represents the rate of change in the concentration of solute in the membrane, D_{m} the diffusivity of solute in the membrane, r the radial position (direction of solute flux originating from the center of the sphere), $\frac{\partial C_{\mathrm{m}}}{\partial r}$ the solute concentration gradient in the membrane, and t time. In the case of a well-mixed liquid core intracapsular solution, the rate of solute exiting the membrane from the APA core solution into the bulk solution becomes

$$\frac{\mathrm{d}M_{\mathrm{i}}}{\mathrm{d}t}=-A_{\mathrm{m}}^{'}\, D_{\mathrm{m}}\left(\frac{\partial C_{\mathrm{m}}}{\partial r}\right)_1 \tag{2}$$

where the total surface area of the microcapsules $A_{\mathrm{m}}^{'}=NA_{\mathrm{m}}$; N being the number of microcapsules in 100 μL volume, and A_{m} the surface area of an individual microcapsule. In Eq. 2, M_{i} denotes the amount of solute in the chelated alginate core, and $\left(\frac{\partial C_{\mathrm{m}}}{\partial r}\right)_1$ the solute concentration gradient at the membrane-core solution interface. The APA membrane flux, J, can then be obtained from Eq. 2 by dividing by the total surface area of microcapsules yielding

$$J=-D_{\mathrm{m}}\left(\frac{\partial C_{\mathrm{m}}}{\partial r}\right)_1 \tag{3}$$

The experimental data obtained from release studies was used to determine the molar flux and diffusion coefficient of 20, 70, and 150 kDa dextrans from microcapsules containing high M (keltone, FMC Biopolymer) and high G (SLG20, Novamatrix) alginates. A linear regression analysis was then used to calculate the molecular weight cutoff of APA capsules.

The diffusion profiles of 20, 70, and 150 kDa FITC-dextrans from high M and high G alginate microcapsules (1.1 mm ± 0.1 mm in diameter) showed a difference in diffusivity. After 16-h incubation, high M-block alginates released 88, 54, and 81 % of encapsulated 20, 70, and 150 kDa FITC-dextran, respectively. In contrast, high G-block alginates released 80, 79, and 39 % of encapsulated 20, 70, and 150 kDa FITC-dextran, respectively (Figs. 3, 4, and 5). The differences in diffusive flux from APA microcapsules could be explained by M-block and G-block content of alginates and FITC-dextran molecular weight, as only the G-blocks of alginates are involved

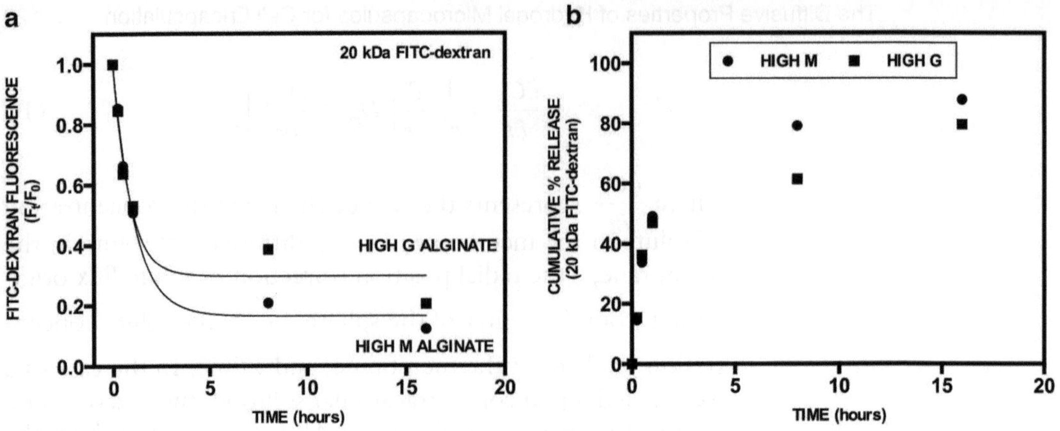

Fig. 3 Comparison of experimental (**a**) diffusion and (**b**) % cumulative release profiles of 20 kDa FITC-dextran from APA microcapsules formulated using high G-block (*squares*) and high M-block (*circles*) alginates

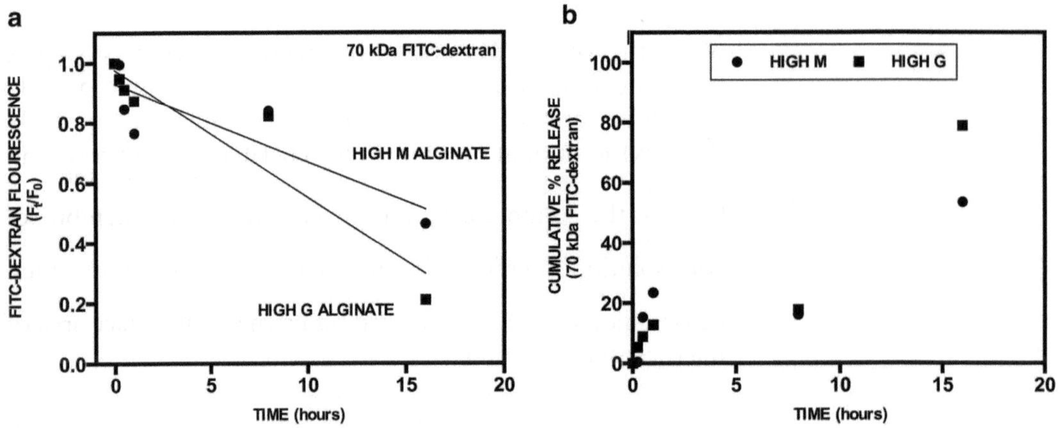

Fig. 4 Comparison of experimental (**a**) diffusion and (**b**) % cumulative release profiles of 70 kDa FITC-dextran from APA microcapsules formulated using high G-block (*squares*) and high M-block (*circles*) alginates

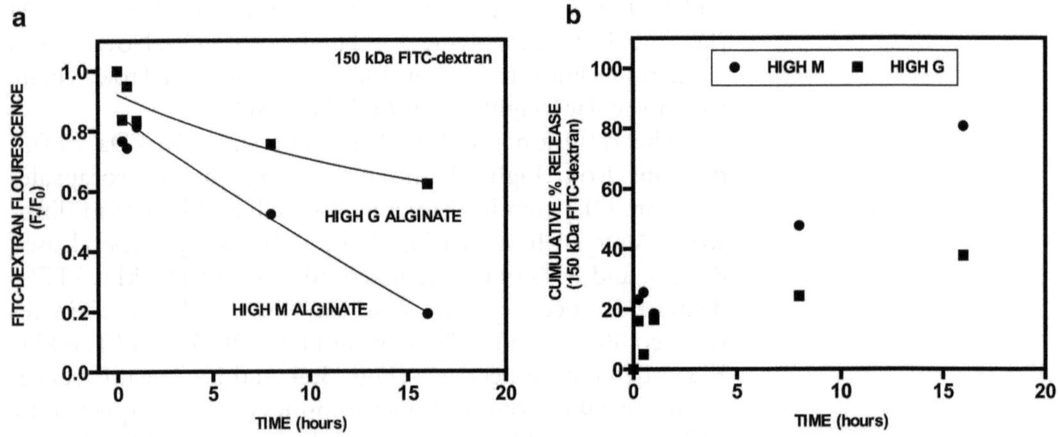

Fig. 5 Comparison of experimental (**a**) diffusion and (**b**) % cumulative release profiles and of 150 kDa FITC-dextran from APA microcapsules formulated using high G-block (*squares*) and high M-block (*circles*) alginates

in cross-linking with calcium to form stiff hydrogels [25]. This could explain why the release kinetics of FITC-dextrans from high G alginate capsules was not as rapid as those comprised of high M alginates at high FITC-dextran molecular weight.

Calculated flux values for high M-block alginate microcapsules were found to be 1.4×10^{-10} cm^2/s, 1.4×10^{-10} cm^2/s, and 3.4×10^{-12} cm^2/s for 20, 70, and 150 kDa FITC-dextran, respectively. High G-block alginate microcapsules produced diffusion coefficients of 1.3×10^{-11} cm^2/s, 8.7×10^{-12} cm^2/s, and 4.4×10^{-13} cm^2/s for 20, 70, and 150 kDa FITC-dextran, respectively. Calculated diffusion coefficients are representative of differences observed in molar flux of the APA membrane, with higher flux rates of 20 kDa in comparison to 70 kDa and 150 kDa FITC-dextran. Diffusional flux values from the experiments shown here with dextran markers indicate that these membranes may have greater permselectivity than calcium-alginate slabs [3], where bovine serum albumin (BSA, 66 kDa) was associated with a diffusion coefficient of 1.8×10^{-6} cm^2/s, several orders of magnitude greater than the 70 kDa diffusion values reported here (1.14×10^{-10} cm^2/s). A linear regression analysis of diffusion coefficients over the three molecular weights evaluated here revealed molecular weight cutoff (MWCO) values of 165 and 156 kDa for high-M and high-G APA microcapsules (Fig. 6).

4. Capsule walls were measured using Motic image analysis software from TEM images acquired at 7100× magnification. In all of

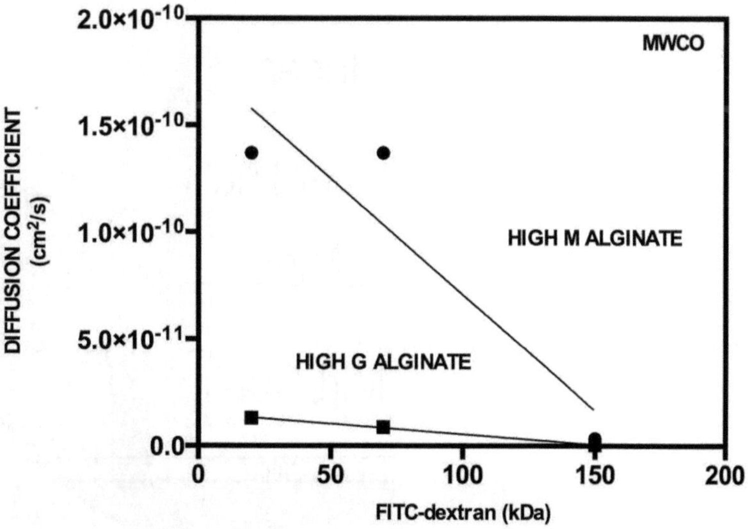

Fig. 6 Linear regression analysis of 20, 70, and 150 kDa FITC-dextran diffusion coefficients to determine molecular weight cutoff (MWCO) of high M-block (MWCO, 165 kDa) and high G-block alginates (MWCO, 156 kDa)

the capsule formulations, there appeared to be five separate layers that constituted the complete wall, as shown in Fig. 7. In general, capsule walls varied in thickness from about 10 to 17 μm, with the purified APA microcapsules having a thicker wall as well as a higher starting viscosity (Table 1).

The layers of the capsule wall vary hypothesized PLO-rich areas, the inner-mid layer and the mid-outer layer. These formulations, made from the same base material, could vary so significantly because of the influence of residual contaminants (such as protein) on the crosslinking or coating kinetics.

The wall structure of the alginate formulations both have a similar layered appearance, but differ dramatically in contrast due to electron density, and distribution of PLO at the layer interfaces. For example, at 4400× magnification, the central layer of the purified APA capsules appears without contrast, while capsules made from raw alginate has very high contrast in this area. The pore shape is also circular in the purified group and more striated in the raw alginate cohort, perhaps resulting from differences in wall thickness and swelling within the process.

Porosity was measured using ImageJ analysis software at 31,000× magnification in the zones depicted in Fig. 7. Figure 8 shows a sample image that was used with this technique. The pore size in the outer zone of the capsule wall was significantly

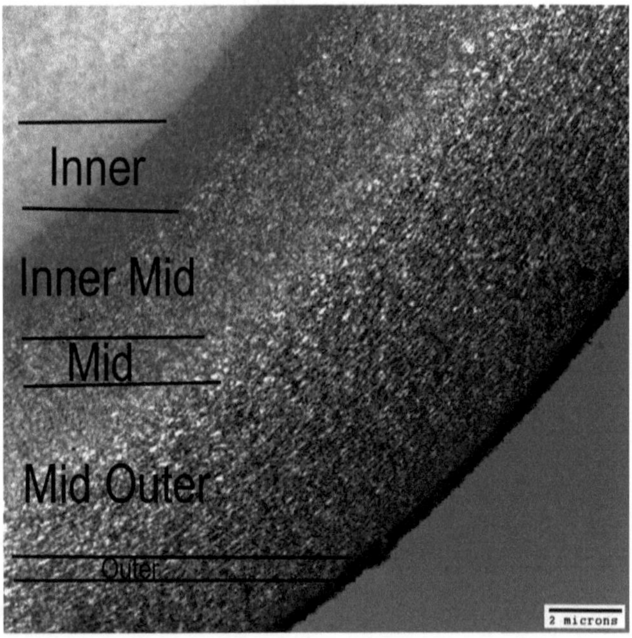

Fig. 7 Example of a TEM micrograph captured at 7100× magnification. The interior core of the APA microcapsule is shown in the *upper left*, while the exterior boundary is on the *bottom right*

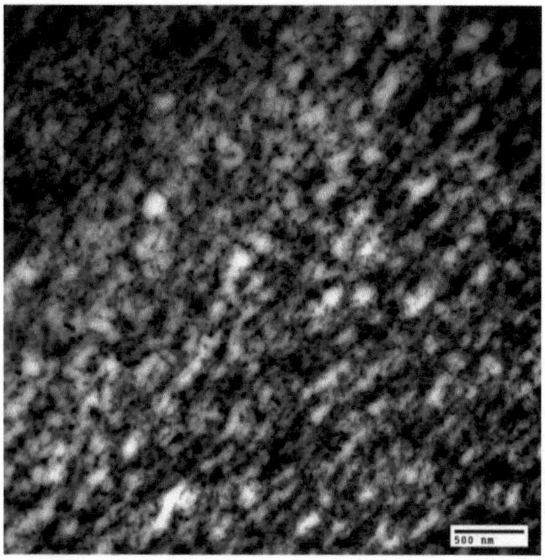

Fig. 8 Sample image used for pore measurement acquired at 33,000× magnification

higher overall when compared to the inside wall (Table 1), with a consistent gradient of porosity observed in groups made from either purified or raw alginates. The outer pores in capsules formed from raw keltone LVCR are over 90 nm in diameter, while the purified cohort showed a tighter outer pore structure at 75.2 nm. This carried over to the inner pores as well, with the purified group showing a 39 % reduction in pore size.

References

1. Chang T (1964) Semipermeable microcapsules. Science 146:524–525
2. Elliott R, Tan P, Wilson R, Wolf L (2008) PRESS RELEASE: living cell technologies reports clinical benefit for all patients in diabetes trial. http://www.lct.com.au
3. Martinsen A, Storro I, Skjak-Braek G (1992) Alginate as immobilization material: III. Diffusional properties. Biotechnol Bioeng 39(2):186–194
4. Bouhadir KH, Alsberg E, Mooney DJ (2001) Hydrogels for combination delivery of antineoplastic agents. Biomaterials 22(19):2625–2633
5. Lim F, Sun A (1980) Microencapsulated islets as bioartificial endocrine pancreas. Science 210(4472):908–910
6. King A, Strand B, Rokstad A, Kulseng B, Andersson A, Skjak-Braek G, Sandler S (2003) Improvement of the biocompatibility of alginate/poly-L-lysine/alginate microcapsules by the use of epimerized alginate as a coating. J Biomed Mater Res A 64(3):533–539
7. Liu X, Xue W, Liu Q, Yu W, Fu Y, Xiong X, Ma X, Yuan Q (2004) Swelling behaviour of alginate–chitosan microcapsules prepared by external gelation or internal gelation technology. Carbohydr Polym 56(4):459–464
8. Altman JJ, Houlbert D, Chollier A, Leduc A, McMillan P, Galletti PM (1984) Encapsulated human islet transplants in diabetic rats. Trans Am Soc Artif Intern Organs 30:382–386
9. Thu B, Bruheim P, Espevik T, Smidsrod O, Soon-Shiong P, Skjak-Braek G (1996) Alginate polycation microcapsules. II. Some functional properties. Biomaterials 17(11):1069–1079
10. Thanos C, Bintz B, Bell W, Qian H, Schneider P, MacArthur D, Emerich D (2006) Intraperitoneal stability of alginate-polyornithine microcapsules in rats: an FTIR and SEM analysis. Biomaterials 27(19):3570–3579

11. Veiseh O, Doloff J, Ma M, Vegas A, Tam H, Bader A, Li J, Langan E, Wyckoff J, Loo W, Jhunjhunwala S, Chiu A, Siebert S, Tang K, Hollister-Lock J, Aresta-Dasilva S, Bochenek M, Mendoza-Elias J, Wang Y, Qi M, Lavin D, Chen M, Dholakia N, Thakrar R, Lacik I, Weir G, Oberholzer J, Greiner D, Langer R, Anderson D (2015) Size- and shape-dependent foreign body immune response to materials implanted in rodents and hon-human primates. Nat Mater 14(6):643–651
12. Kang J, Erdodi G, Kennedy J (2007) Third-generation amphiphilic conetworks. III. Permeabilities and mechanical properties. J Polym Sci A Polym Chem 45(18): 4276–4283
13. Schweicher J, Nyitray C, Desai TA (2014) Membranes to achieve immunoprotection of transplanted islets. Front Biosci (Landmark Ed) 19:49–76
14. Lim F, Moss RD (1981) Microencapsulation of living cells and tissues. J Pharm Sci 70(4): 351–354
15. Leung YF, O'Shea GM, Goosen MF, Sun AM (1983) Microencapsulation of crystalline insulin or islets of Langerhans: an insulin diffusion study. Artif Organs 7(2):208–212
16. Sakai S, Ono T, Ijima H, Kawakami K (2000) Control of molecular weight cut-off for immunoisolation by multilayering glycol chitosan-alginate polyion complex on alginate-based microcapsules. J Microencapsul 17(6): 691–699
17. Matsumoto T, Kawai M, Masuda T (1992) Influend of concentration and mannuronate/guluronate ratio on steady flow properties of alginate aqueous systems. Biorheology 29(4): 411–417
18. van Schilfgaarde R, de Vos P (1999) Factors influencing the properties and performance of microcapsules for immunoprotection of pancreatic islets. J Mol Med 77(1):199–205
19. Wang T, Lacik L, Brissova M, Anilkumar AV, Prokop A, Hunkeler D, Green R, Shahrokhi K, Powers AC (1997) An encapsulation system for the immunoisolation of pancreatic islets. Nat Biotechnol 15(4):358–362
20. Addicks WJ, Flynn GL, Weiner N (1987) Validation of a flow-through diffusion cell for use in transdermal research. Pharm Res 4(4): 337–341
21. Pluen A, Netti P, Jain R, Berk D (1999) Diffusion of macromolecules in agarose gels: comparison of linear and globular configurations. Biophys J 77:542–552
22. O'Leary TJ (1987) Concentration dependence of protein diffusion. Biophys J 52(1):137–139
23. Nakagawa Y, Murai T, Hasegawa C, Hirata M, Tsuchiya T, Yagami T, Haishima Y (2003) Endotoxin contamination in would dressings made of natural biomaterials. J Biomed Mater Res 66B:347–355
24. Kwok WY, Kiparissides C, Yuet P, Harris TJ, Goosen MFA (1991) Mathematical-modeling of protein diffusion in microcapsules - a comparison with experimental results. Can J Chem Eng 69(1):361–370
25. Lee KY, Mooney DJ (2012) Alginate: properties and biomedical applications. Prog Polym Sci 37(1):106–126. doi:10.1016/j.progpolymsci.2011.06.003
26. Tam SK, de Haan BJ, Faas MM, Halle J-P, Yahia LH, deVos P (2007) Adsorption of human immunoglobulin to implantable alginate-poly-L-lysine microcapsules: effect of microcapsule composition. J Biomed Mater Res A 89(3):609–615
27. Garfinkel MR, Harland RC, Opara EC (1998) Optimization of the microencapsulated islet for transplantation. J Surg Res 76:7–10

Chapter 10

Methods for Incorporating Oxygen-Generating Biomaterials into Cell Culture and Microcapsule Systems

John Patrick McQuilling and Emmanuel C. Opara

Abstract

A major obstacle to long-term performance of tissue construct implants in regenerative medicine is the inherent hypoxia to which cells in the engineered construct are exposed prior to vascularization of the implant. Various approaches are currently being designed to address this problem. An emerging area of interest on this issue is the use of peroxide-based materials to generate oxygen during the critical period of extended hypoxia that occurs from the time cells are in culture waiting to be used in tissue engineering devices through the immediate post-implant period. In this chapter we provide protocols that we have developed for using these chemical oxygen generators in cell culture and tissue constructs as illustrated by pancreatic islet cell microencapsulation.

Key words Oxygen-generating biomaterials, Calcium peroxide, Sodium percarbonate, Alginate, Microencapsulation

1 Introduction

One hurdle for many cell- and tissue-based therapies is overcoming hypoxia during the immediate posttransplant period prior to the onset of angiogenesis [1, 2]. This is particularly challenging with cells or tissue with exceptionally high metabolic demands such as skeletal muscle or pancreatic islets [3, 4]. There are several approaches to reducing hypoxic injury following implantation including the pre-vascularization of the transplantation site, [5] reducing the time to engraftment with the localized delivery of angiogenic growth factors, [6, 7] the use or incorporation of oxygen carriers within biomaterial constructs, [8] and locally supplying oxygen with oxygen-generating materials.

Compared to pre-vascularization strategies, oxygen-generating materials have the advantage of requiring only a single surgery. In contrast to strategies which utilize a localized delivery of angiogenic growth factors to decrease the time to engraftment, oxygen-generating materials immediately provide implanted cells or tissues with oxygen.

Emmanuel C. Opara (ed.), *Cell Microencapsulation: Methods and Protocols*, Methods in Molecular Biology, vol. 1479, DOI 10.1007/978-1-4939-6364-5_10, © Springer Science+Business Media New York 2017

Unlike oxygen carrying strategies such as hemoglobin-based materials, oxygen-generating materials do not require a vasculature supply or oxygen reservoir. Oxygen-generating materials include calcium peroxide (CPO) and sodium percarbonate (SPO) which have both been applied to several tissue engineering applications.

Sodium percarbonate (SPO) is an adduct of sodium carbonate and hydrogen peroxide which in the presence of water will rapidly decompose to generate oxygen as shown in Eqs. 1 and 2 below:

$$2[Na_2CO_3 \cdot 1.5H_2O_2] \rightarrow 2Na_2CO_3 + 3H_2O_2 \tag{1}$$

$$2H_2O_2 \rightarrow 2H_2O + O_2 \tag{2}$$

SPO particles, similar to those illustrated in Fig. 1a, have previously been utilized within a poly(D,L-lactide-co-glycolide) (PGLA) scaffold to reduce necrosis and cellular apoptosis in an ischemic flap model and has also been administered via intramuscular injection to an ischemic muscle injury which resulted in improved functional outcomes [9, 10].

CPO decomposes in the presence of water to form H_2O_2 which then decomposes to oxygen and water, as shown in Eqs. 3 and 4.

$$CaO_2 + 2H_2O \rightarrow Ca(OH)_2 + H_2O_2 \tag{3}$$

$$2H_2O_2 \rightarrow 2H_2O + O_2 \tag{4}$$

CPO particles, illustrated in Fig. 1b, decompose at a slower rate than SPO. These particles have been utilized within a PLGA scaffold and were shown to improve fibroblast proliferation over a 10-day period in hypoxic conditions [11]. When incorporated into a silicone construct, CPO was able to generate oxygen for up to a 6 week period. Using this CPO-silicone system for pancreatic islets culture, a significant reduction in hypoxia-induced cell death and dysfunction was observed for up to 48 h. It has also been shown to improve MIN6 cell survival for up to 3 weeks [12].

While there is growing evidence that oxygen-generating materials can be utilized to prevent hypoxic damage to cells and tissues, this is a novel technology which is still under development. One challenge to implementing this technology is controlling the rate of oxygen generation so that there is enough oxygen to prevent hypoxia, but not too much that the angiogenic process is disrupted [13]. Additionally these materials often have a burst release, which results in the buildup of hydrogen peroxide which is cytotoxic, and this buildup in turn increases the chance of hydroxyl radical formation with further damage to tissues [14]. There are several factors which play a role in controlling the duration and rate of oxygen release including the local pH [14], particle size, the purity of either the SPO or CPO, the applications of hydrophobic coatings

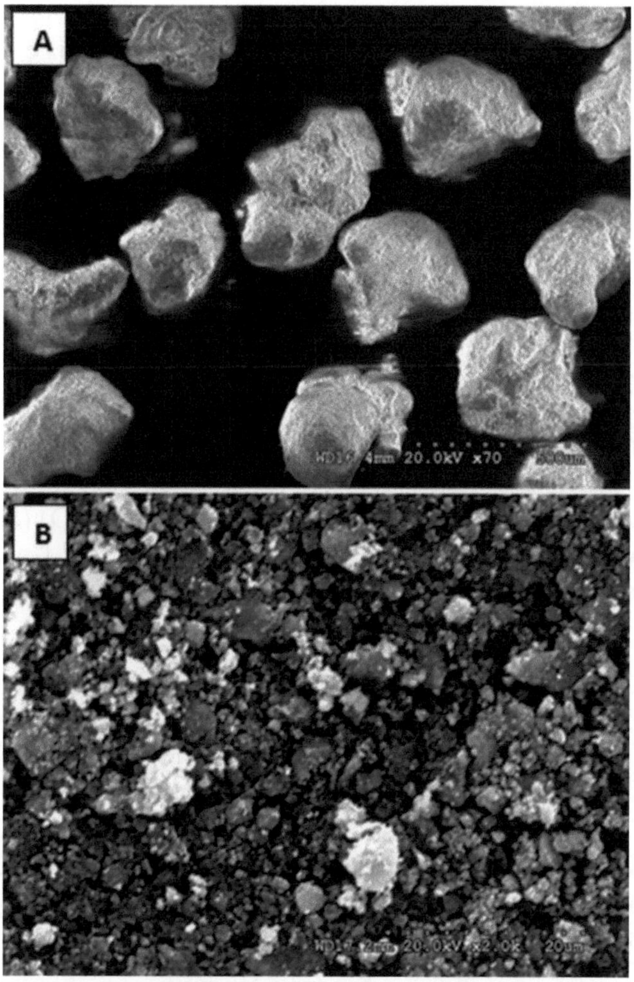

Fig. 1 SEM images of oxygen-generating materials. (**a**) Representative image of SPO particles with an average diameter of 330 ± 8 μm. (**b**) Representative image of CPO particles with an average diameter of 2.78 ± 1.83 μm

such as steric acid or paraffin, and the materials used to encapsulate the oxygen-generating material. Additionally, the use of antioxidants, in particular, catalase is recommended to mitigate cytotoxicity caused by oxidative stress from these materials.

Here we present two different methods for utilizing these two oxygen-generating materials. The first method outlines the process for producing a culture system designed to generate oxygen from sodium percarbonate encapsulated within a layer of polydimethylsiloxane (PDMS). While SPO typically reacts rapidly in the presence of water and catalase under physiological conditions and can completely degrade within minutes, the addition of the hydrophobic PDMS greatly slows this reaction enabling oxygen generation to occur over a period of days as water slowly comes into contact with the SPO between the layers of PDMS. The second method outlines the process for fabricating alginate microspheres with CPO to generate oxygen to encapsulated islets. In order to prevent cell death as a result of the initial burst release of CPO, CPO was pre-incubated with PBS overnight and then washed with PBS to remove smaller particles. Both systems have been tested with pancreatic islets under relatively hypoxic conditions and have been shown to reduce damage from oxidative stress; however it is recommended that the concentration and particle size of these oxygen-generating materials as well as the antioxidant loads be experimentally determined for each cell or tissue type tested with these systems.

2 Materials

2.1 Materials for Oxygen-Generating Culture System

1. Sodium Percarbonate (SPO) (Sigma-Aldrich)
2. Sylgard 184 Silicone elastomer kit (World Precision Instruments, Sarasota FL)
3. 150 mm × 20 mm plastic Petri dish
4. 90 mg Sodium Percarbonate particles ~1 mm in diameter (Sigma-Aldrich)
5. Bovine Catalase (Sigma-Aldrich)
6. Vacuum Oven

2.2 Materials for Oxygen-Generating Microcapsules

1. Calcium Peroxide (Sigma-Aldrich)
2. Phosphate Buffered Saline (without Ca^{2+} or Mg^{2+})
3. 1.5 % Alginate (Pronova UP LVM and UP LVG, Novamatrix, Sandvika Norway)
4. 0.1 % Poly-L-Ornithine (PLO) (P5061, Sigma-Aldrich)
5. 100 mM $CaCl_2$ solution (C614-10, Fischer Scientific, Waltham, MA, USA)
6. 55 mM Sodium citrate solution (S467-3, Fischer Scientific)
7. 0.9 % Sodium chloride solution (normal saline) (71376-5KG, Sigma-Aldrich)
8. 10 mM HEPES solution (H3375-2KG, Sigma-Aldrich)
9. Eight channel microfluidic encapsulation device (Department of Mechanical Engineering, Clemson University, SC (*see* **Note 1**).

3 Methods

3.1 Oxygen-Generating Culture System

1. Following the manufacturer's instructions, prepare a 1:10 ratio (by weight) of Catalyst to PDMS.
2. Add enough of the PDMS solution to form a 1 mm thick layer on the bottom of the petri dish.

3. Remove air bubbles in PDMS by placing in a vacuum for 1 h.
4. Allow to cure at room temperature for 48 h (*see* **Note 2**).
5. Evenly distribute SPO particles across the cured layer of PDMS.
6. Prepare a second 1:10 mixture of catalyst and PDMS and evenly apply over SPO and bottom layer of silicone to form a top layer of silicone approximately 1 mm thick.
7. Remove air bubbles via vacuum for 1 h at room temp.
8. Allow 48 h for final layer of silicone to cure.
9. Sterilize via ethylene oxide or gamma irradiation prior to use (*see* **Note 3**).
10. We recommend supplementing standard culture media with 100 U/mL of Catalase prior to use.

3.2 Oxygen-Generating Microcapsules

1. Add 10 mg of CPO to 50 mL of sterile PBS and incubate over night at 37 °C.
2. Allow CPO to settle by gravity to the bottom of the conical tube and aspirate PBS.
3. Wash with 50 mL PBS three times and allow particles to settle by gravity prior to aspirating.
4. Suspend remaining CPO particles in 1 mL alginate (*see* **Note 4**).
5. Suspend islets in sodium alginate/CPO suspension Islets are typically suspended in a solution of sodium alginate (usually from 1.2 to 1.8 % w/v made up in normal saline) with 5000 islets suspended in 1 mL of alginate.
6. Microspheres of alginate containing one or two islets (depending upon the alginate-islet ratio in the suspension), as illustrated in Fig. 2, are generated and allowed to gel into microbeads in a solution of 100 mM $CaCl_2$ and 10 mM HEPES at 4 °C, pH 7.4 for 5–10 min.
7. Collect microspheres into a 50 mL conical tube and allow the capsules to settle with gravity.
8. Aspirate calcium chloride solution and wash three times in normal saline.
9. Incubate microspheres in 0.1 % PLO for 30 min on ice.
10. Liquefaction of the alginate core of the microcapsules is achieved by a brief (2 min) incubation in 55 mM sodium citrate solution.
11. Wash with normal saline three times, as previously described.
12. Coat outer layer of capsule with 1.25 % LVM for 5 min.
13. Separate excess alginate from microcapsules and crosslink outer layer.
14. Place microcapsules in culture using standard culture media supplemented with 100 U/mL Catalase.

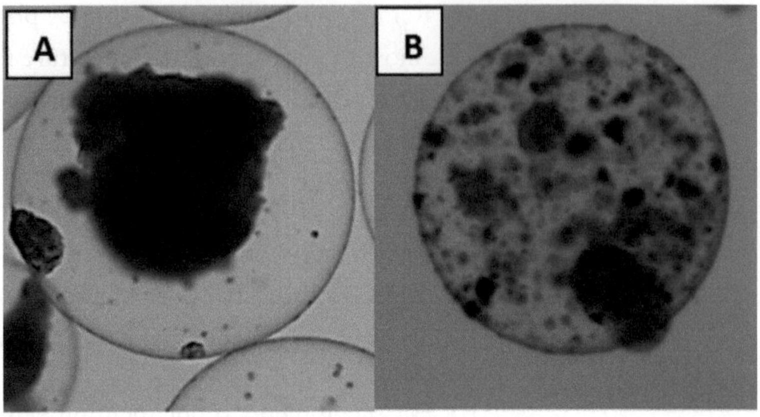

Fig. 2 (**a**) Phase contrast image of microencapsulated islets in alginate microcapsules with poly-L-orthinine semipermeable coatings (control). (**b**) Phase contrast images of islets co-encapsulated with CPO

4 Notes

1. The microfluidic device used here is constructed with in-house materials as previously described [15]. This eight nozzle device uses standard syringe needles (gauges 20–27) and generates capsules at air jacket pressures at 2.5–10 psi and with alginate input flow rates of 0.5–2.5 mL/min.
2. SPO is temperature sensitive; curing the PDMS at high temperatures may result in the early degradation of SPO particles.
3. This system was originally designed to culture islets and since this is a suspension culture we have not included instructions for treating the cell culture surface for attachment cultures prior to sterilization.
4. The addition of CPO will begin to crosslink the alginate immediately after mixing. While pre-washing the particles with PBS will slow this process, encapsulation should occur quickly after preparing the alginate/CPO suspension.

Acknowledgements

The authors would like to acknowledge financial support from the National Institutes of Health (RO1 DK080897) and the Vila Rosenfeld Estate, Greenville NC for the work in Dr. Opara's laboratory at the Wake Forest Institute for Regenerative Medicine.

Also, research reported in this publication was supported by the National Institute of Biomedical Imaging and Bioengineering (NIBIB) of the National Institutes of Health under award # T32EB014836. The content is solely the responsibility of the authors and does not necessarily represent the official views of the National Institutes of Health.

References

1. Malda J, Klein TJ, Upton Z (2007) The roles of hypoxia in the in vitro engineering of tissues. Tissue Eng 13:2153–2162
2. Bland E, Dreau D (2013) Overcoming hypoxia to improve tissue-engineering approaches to regenerative medicine. J Tissue Eng Regen Med 4:505–514
3. Blaisdell FW (2002) The pathophysiology of skeletal muscle ischemia and the reperfusion syndrome: a review. Cardiovasc Surg 10:620–630
4. Davalli AM, Scaglia L, Zangen DH et al (1996) Vulnerability of islets in the immediate post-transplantation period. Dynamic changes in structure and function. Diabetes 45:1161–1167
5. Pileggi A, Molano RD, Ricordi C et al (2006) Reversal of diabetes by pancreatic islet transplantation into a subcutaneous, neovascularized device. Transplantation 81:1318–1324
6. McQuilling JP, Arenas-Herrera J, Childers C et al (2011) New alginate microcapsule system for angiogenic protein delivery and immunoisolation of islets for transplantation in the rat omentum pouch. Transplant Proc 43(9):3262–3264
7. Zisch AH, Lutolf MP, Ehrbar M et al (2003) Cell-demanded release of VEGF from synthetic, biointeractive cell in growth matrices for vascularized tissue growth. FASEB J 17:2260–2262
8. Espes D, Lau J, Quach M et al (2015) Cotransplantation of polymerized hemoglobin reduces β-cell hypoxia and improves β-cell function in intramuscular islet grafts. Transplantation 99:2077–2082
9. Harrison BS, Eberli D, Lee SJ et al (2007) Oxygen producing biomaterials for tissue regeneration. Biomaterials 28:4628–4634
10. Ward CL, Corona BT, Yoo JJ et al (2013) Oxygen generating biomaterials preserve skeletal muscle homeostasis under hypoxic and ischemic conditions. PLoS One 8, e72485
11. Oh SH, Ward CL, Atala A et al (2009) Oxygen generating scaffolds for enhancing engineered tissue survival. Biomaterials 30:757–762
12. Pedraza E, Coronel MM, Fraker CA et al (2012) Preventing hypoxia-induced cell death in beta cells and islets via hydrolytically activated, oxygen-generating biomaterials. Proc Natl Acad Sci 106:4245–4250
13. Steg H, Buizer AT, Woudstra W et al (2015) Control of oxygen release from peroxides using polymers. J Mater Sci Mater Med 26:207
14. Northup A, Cassidy D (2008) Calcium peroxide (CaO_2) for use in modified Fenton chemistry. J Hazard Mater 152:1164–1170
15. Tendulkar S, McQuilling JP, Childers C et al (2011) A scalable microfluidic device for the mass production of microencapsulated islets. Transplant Proc 43:3184–3187

Chapter 11

Noninvasive Tracking of Alginate-Microencapsulated Cells

Genaro A. Paredes-Juarez, Brad P. Barnett, and Jeff W.M. Bulte

Abstract

This chapter presents a description of standardized techniques used routinely in our laboratory to encapsulate different cell types using the alginate-PLL-alginate immunoisolation system. Given the importance of noninvasive tracking of encapsulated cell transplants, we present a detailed guidance to achieve maximum efficiency and functionality of the capsule preparations for optimal tracking posttransplantation. The provided protocols cover tracking of encapsulated cells using magnetic resonance (MR), X-ray, computed tomography (CT), and ultrasound (US) imaging. Practical suggestions to optimize each method with specific references to recommended suppliers are included.

Key words Alginate, Encapsulation, Noninvasive tracking, MRI, X-ray, Ultrasound

1 Introduction

Encapsulation of cells for transplantation purposes is a technique intended to protect cells from the innate and adaptive immune system of the host, while allowing diffusion of essential nutrients and oxygen required for graft survival [1, 2]. Encapsulation of cells is an established technique, first described by Lim and Sun in 1980 [3]. Since then, it has been applied with different rates of success in different laboratories [4–7]. In the case of allografts and xenografts, encapsulation can minimize or eliminate the need for immunosuppression thereby sparing the transplant recipient from the deleterious effects of immunosuppressive agents [8]. The use of xenografts is especially attractive due to the shortage of allograft tissue available [9]. Encapsulation has been proposed for the treatment of different diseases, such as dwarfism, hemophilia, kidney and liver failure [10–12], stroke, myocardial infarction [13], and diabetes [9, 14–16]. A commonly used method for immunoisolation is the encapsulation of islets in alginate membranes [17–19]. Alginate is a biocompatible polymer derived from purified algae that allows selective diffusion of molecules

Emmanuel C. Opara (ed.), *Cell Microencapsulation: Methods and Protocols*, Methods in Molecular Biology, vol. 1479,
DOI 10.1007/978-1-4939-6364-5_11,

under a certain molecular weight. Many groups have applied polycations, such as poly-L-lysine (PLL), to control the pore size and provide stability to the capsules [20–22].

Despite it widely being used, many issues remain unresolved with microencapsulation. Among those issues, the best mode of delivery, initial engraftment and tissue biodistribution, and stability of the capsules over time require special attention [23]. Therefore, a method with which one can noninvasively image the fate of the capsules is needed [23–27]. This chapter focuses on several methods that are used in our lab to track microencapsulated cells using MRI, X-ray, CT, and US imaging [24].

2 Materials

2.1 Reagents

Primary cells, cell clusters, cell lines, or spheroids from the tissue of interest.

Cell culture medium (any appropriate culture medium for the cells being used).

10 mM 4-(2-hydroxyethyl)-1-piperazineethanesulfonic acid (HEPES) buffer.

PRONOVA UP LVG ultrapure low-viscosity high-guluronate alginate (NovaMatrix, cat. no. 4200001) (*see* **Note 1**).

PRONOVA UP LVM ultrapure low-viscosity high-mannuronate alginate (NovaMatrix, cat. no. 4200201) (*see* **Note 1**).

Poly-L-lysine (PLL).

Barium sulfate.

Bismuth sulfate.

Superparamagnetic iron oxide (SPIO) particles (*see* **Note 2**)

PFOB ($C_8F_{17}Br$) (Apollo Scientific, cat. no. PC6167).

Perfluoro-15-crown-5-ether ($C_{15}F_{30}O_5$).

Soy lecithin.

Safflower oil USP.

Calcium chloride, anhydrous powder.

Sodium citrate.

0.9 % sodium chloride solution.

Fluorescein diacetate (FDA, Sigma, cat. no. F7378).

Propidium iodide (PI, Invitrogen, cat. no. P3566).

2.2 Equipment

Nano syringe pump (Cole-Parmer, cat. no. EW-74902-00).

Ring stand.

Three-prong extension clamps.

High-voltage power supply (Bertran, model no. 230-30R).

Sonicator.

Electrical tape.

Automotive hookup wire.

Becton-Dickinson tuberculin syringe—1 ml only with slip tip.

Blunt needles (24G × 1/2-in., 20G × 1/2-in.).

Three-way stopcock with swivel male Luer lock (Medex, cat. no. MX5311L).

Millipore filter flask (0.2 μm).

Nylon filter (0.2 μm).

Corning 100 × 20 mm untreated culture dish.

Serological pipettes.

Erlenmeyer screw cap flask.

Ceramic stirring hot plate.

9.4 T horizontal bore magnet (Bruker BioSpin, Karlsruhe, Germany).

40-MHz Vevo 660 scanner (Visual Sonics, Toronto, Ontario, Canada).

2.3 Buffer and Solutions

Primary LVG alginate (2 % w/v): 10 mM HEPES buffer, Pronova UP LVG alginate.

Secondary LVM alginate (0.15 % w/v): 10 mM HEPES buffer, Pronova UP LVM alginate.

Poly-L-lysine solution (0.05 % w/v): 10 mM HEPES buffer, PLL.

Gelation bath: 100 mM calcium chloride, 10 mM HEPES buffer.

Superparamagnetic iron oxide (SPIO): 10 mg Fe/ml, saline or HEPES buffer (*see* **Note 2**).

Perfluorocarbon emulsion. 5 % lecithin (w/v), 2 % safflower oil (v/v), water.

3 Methods

3.1 Solutions

1. Primary LVG alginate (2 % w/v): Prepare a sterile solution of 10 mM HEPES-buffered normal saline (sterilize by filtration through a 0.2 μm filter), and pour the sterilized solution into a flask under sterile conditions. Add a stir bar to the flask and place it on a stirring hot plate with the heater turned off. Slowly add the LVG alginate (*see* **Note 3**). Place a screw cap on the flask and allow it to rotate overnight in a tissue culture hood with the UV light turned off. Alginate aliquots (*see* **Note 4**) can be stored at 2–4 °C for up to 1 month. Remove necessary aliquots from refrigeration the day before encapsulation and allow them to return to room temperature (20–25 °C).

2. Secondary LVM alginate (0.15 % w/v): Prepare secondary alginate in a manner identical to that for the primary alginate using Pronova UP LVM; store aliquots at 2–4 °C for up to 1 month.
3. Poly-L-lysine solution (PLL, 0.05 % w/v): Dissolve PLL in 10 mM HEPES-buffered normal saline and filter-sterilize the solution with a filter flask. Store aliquots at 2–4 °C for up to 1 year.
4. Gelation bath: Prepare a solution of 100 mM calcium chloride in 10 mM HEPES-buffered normal saline and filter-sterilize the solution with a filter flask. Store aliquots at 2–4 °C for up to 1 year.

3.2 Electrostatic Droplet Generator Set-Up

1. Cut two pieces of 5 ft. connection wire and strip 1 in. of insulation tubing off the distal ends of both wires. Attach one end of the wire to a suitable ground (e.g., a ring stand placed on the floor) and place the other end of the wire into a Petri dish containing the cationic gelation bath. Attach the other wire to the ground on the high-voltage generator and connect the other end of the wire to a separate ground (e.g., the inner metal wall of a tissue culture hood).
2. Pass a 20G 1/2-in. blunt needle through the insulation wall of the wire through wire braids and out the opposite insulation wall of the output wire of the high-voltage droplet generator.
3. Gas-sterilize the set-up using ethylene oxide gas or, alternatively, use a UV light assembly for sterilization.
4. Place a culture dish below the needle and ensure that it is properly grounded (e.g., by running a wire from the encapsulation bath out of the tissue culture hood and onto a ring stand on the ground, ensuring that the proximal portion of the wire in the gelation bath and hood remains sterile).
5. Just before encapsulation, add 50 ml of gelation solution to the culture dish (*see* **Note 5**, Fig. 1).

3.3 Cell Encapsulation

1. Affix a blunt-tip needle with the wire from the high-voltage generator to the syringe containing the suspension of alginate, contrast, and cells.
2. Place the syringe with the needle attached in the syringe pump and adjust the height between the needle and the cationic (calcium or barium) solution by adjusting the ring stand. The distance should be approximately on the order of an inch (this results in a voltage differential of ~8 kV).
3. Turn on the high-voltage generator. Adjust the current beforehand, using an alginate solution without cells, to obtain the finest stream (and thus the smallest droplets) possible.

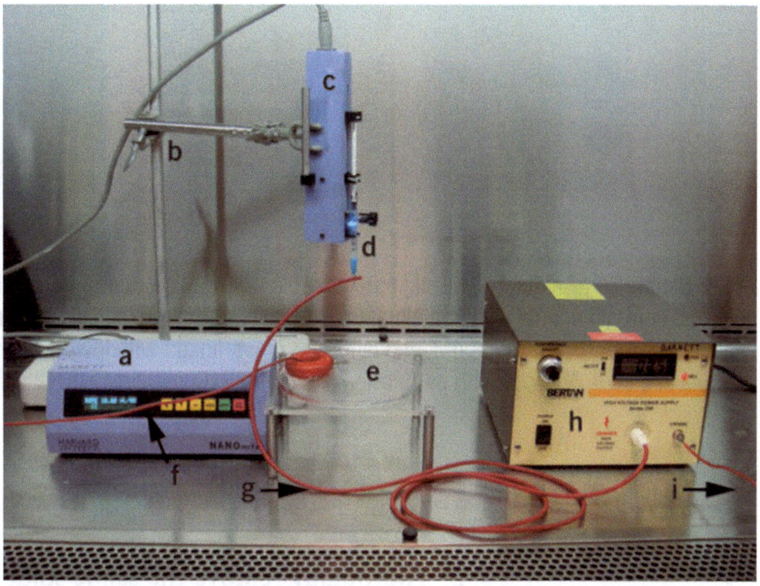

Fig. 1 Essential capsule synthesis equipment. Syringe pump control box (*a*), ring stand and clamp (*b*), syringe pump remote injector (*c*), syringe with blunt-tip needle through the high-voltage output wire (*d*), gelation bath (*e*), gelation bath ground wire (*f*), high-voltage output wire (*g*), high-voltage generator (*h*), and high-voltage generator ground wire (*i*). Reprinted with permission from ref. 24. Copyright © 2011, Nature Publishing Group

4. Preset the microinjector to inject a volume of 1 ml at a rate of 200 μl/min.
5. Ensure that the gelation bath is properly grounded and start the syringe pump (*see* **Notes 6** and 7).
6. Allow microcapsules to remain in the gelation solution for 5 min.
7. Use a 10 ml serological pipette to remove microcapsules from the gelation bath; add them to a 50 ml conical tube.
8. Remove calcium chloride solution by allowing microcapsules to settle to the base of the conical tube, and then, carefully aspirate the entire volume of calcium chloride solution (Fig. 2a).

3.4 Application of Polycation Coat

1. Rinse microcapsules three times with 25 ml of 0.9% (w/v) saline solution. Remove the solution for each change, as described in Subheading 3.3, **step 8**.
2. Add 40–50 ml of the PLL 0.05% solution to microcapsules in the 50 ml conical tube.
3. Seal the conical tube and tape it horizontally onto a platform rocker. Rock at medium speed for exactly 6 min (Fig. 2b).

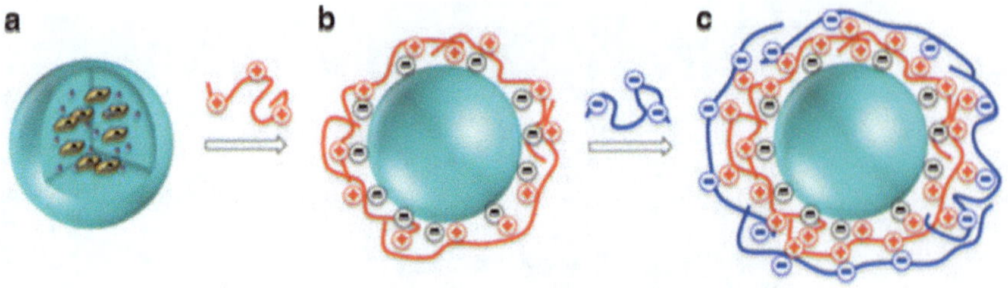

Fig. 2 Synthesis of alginate-PLL-alginate capsules. (**a**) Schematic drawing showing microencapsulation of cells and contrast agents. The first alginate (LVG) layer of the microcapsule is cross-linked with PLL as a polycationic polymer (**b**, *red*) in step 3, and is further cross-linked with low-viscosity mannuronate (LVM) alginate (**c**, *blue*) in step 4. Adapted with permission from ref. 23. Copyright © 2012 John Wiley & Sons, Ltd

3.5 Application of Secondary Alginate Layer

1. Rinse microcapsules three times with 25 ml 0.9% (w/v) saline solution.
2. Thoroughly remove the saline solution and add 40–50 ml of secondary alginate, prepared as described above, to the microcapsules in a 50 ml conical tube.
3. Seal the conical tube and tape horizontally onto a platform rocker. Rock at medium speed for 3 min.
4. Rinse the microcapsules again, three times with 25 ml 0.9% (w/v) saline solution. Immediately transplant or place encapsulated cells in appropriate culture conditions [28] (Fig. 2c).

3.6 Incorporation of Contrast in Primary Alginate

1. Follow the next steps for MR contrast, X-ray contrast, and trimodal (perfluorocarbon) contrast.
2. MR contrast agent: magnetocapsules (*see* **Note 8**) [29–31].
   - (a) To create MR-visible alginate capsules, mix 20% (v/v) SPIO with 80% (v/v) primary alginate.
   - (b) For best results, load alginate and SPIO into two separate syringes and connect the syringes to a three-way stopcock; carefully remove any excess air.
   - (c) Once air is removed, move the stopcock to the closed position and mix the contents of the two syringes back and forth until a homogenous solution is created.
3. X-ray imaging agent (*see* **Note 9**) [32].
   - (a) To create X-ray–visible alginate capsules, first dissolve barium or bismuth sulfate into 0.9% saline at a concentration of 50 mg/ml.
   - (b) Vortex this solution to break down any large aggregates.
   - (c) Mix this barium sulfate solution with primary alginate, as described in Subheading 3.6, **steps 2b** and **2c**.
4. Multimodal contrast agent: fluorocapsules (*see* **Note 10**) [33].

(a) To prepare fluorinated US-visible capsules, first prepare a perfluorocarbon emulsion and filter-sterilize it through a 0.2 μm nylon filter.

(b) Sonicate the lecithin/safflower oil solution at 40% power on ice until the solution is almost transparent.

(c) Add perfluorocabon to sonicated lecithin solution at a concentration ranging from 12% perfluorocarbon (v/v) to 40% perfluorcarbon (v/v).

(d) Sonicate perfluorocarbon lecithin solution for 25 min on ice until a milky homogenous solution is formed. Leave fluorocarbons on ice until incorporation in the primary alginate layer.

(e) Emulsion should be incorporated into the alginate layer within 1 h of emulsion preparation, as described in Subheading 3.6, **steps 2b** and **2c**.

3.7 Examples of X-ray Contrast, ^{1}H MR Contrast, and ^{19}F MR Contrast

1. Fluoroscopic imaging was performed using a Toshiba Infinix VC-i unit, with imaging settings of 64 kV peak, 66 ms exposure time, 112 mA tube current, and 910 mm SID [25] (Fig. 3).

2. ^{1}H MR-trackable capsules (Fig. 4): MR imaging was performed using a 9.4 T horizontal bore magnet with a custom-built animal holder and a whole-body volume transmitting-receiving coil. The microcapsules were imaged by using T1-weighted spin-echo (rapid acquisition with relaxation enhancement) and T2*-weighted gradient-echo (gradient-echo flow compensated) sequences. The T1-weighted imaging parameters for the saline phantoms were 500/14.1 (repetition time msec/echo time msec), a field of view of 1.6×1.6 cm, a matrix of 256×256, a section thickness of 1 mm, and 24 acquired signals. The same parameters were used for T2*-weighted imaging, with the exception of an echo time of 6.7 ms and 12 acquired signals. The T1-weighted imaging parameters for the mice were 500/14.1, a field of view of 2.4×1.8 cm, a matrix of 256×192, a section thickness of 0.8 mm, and 20 acquired signals with fat suppression and zero filling=512×512 (matrix size). Micro-CT of the phantoms and mice was performed by using an X-SPECT Gamma Medica imager with tube settings of 75.16 kV and 240.3 mA, a 64-mm detector setting, and 512 projections [34].

3. ^{19}F-MR imaging (Fig. 5): MR imaging was performed placing 1–50 PFPE capsules in 0.2 ml polymerase chain reaction tubes and covered with 4% gelatin. This phantom was imaged by using a clinical 3.0 T imager. To this end, a custom fluorine phased-array coil with four coil elements was used. Scout proton MR images were obtained by using a surface coil (Body Matrix; Siemens) followed by manual shimming by using the body coil. The custom fluorine coil was then used to obtain fluorine images

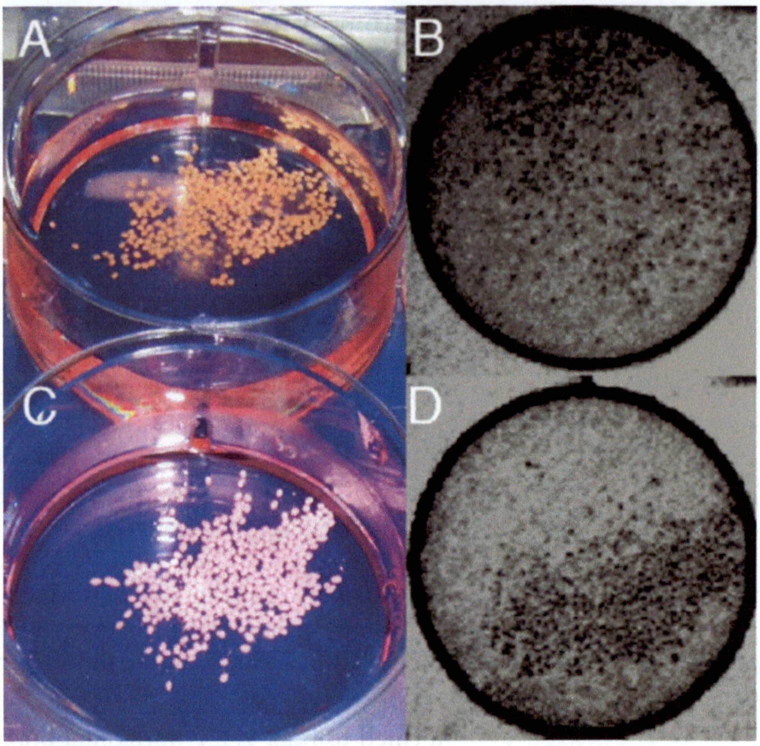

Fig. 3 X-ray-trackable capsules. Macroscopic (**a**, **c**) and fluoroscopic (**b**, **d**) images of bismuth X-caps (**a**, **b**) and barium X-caps (**c**, **d**). Single capsules can be clearly identified. Reprinted with permission from ref. 25. Copyright© 2006, American Chemical Society

by using a three-dimensional true fast imaging with steady-state free precession pulse sequence with the following parameters: 4.1/2.1; number of signals acquired, 32; 1.5 × 1.5 × 2.0 mm voxel size; field of view, 672 × 250; 1002 Hz bandwidth; 70° flip angle; and an acquisition time of about 12 min [35].

3.8 Adding Cells to Primary Alginate/Contrast Solution

1. Pellet cells or allow clusters or spheroids of cells to settle in a 50 ml conical tube and remove all but 100 μl of culture medium. Depending on the expected growth of the cell population, the total concentration of cells should be adjusted to a concentration of 1–5 % (v/v) of that of the primary alginate layer. For growth-arrested cells, a higher number of cells can be encapsulated. For highly proliferative cells, such as immortalized cell lines, cells should be added at lower concentrations (*see* **Note 11**).
2. With the appropriate mixture of contrast and alginate, add the cells at the appropriate concentration to the alginate using a 50 ml conical tube. Use a 1 cc tuberculin syringe to homogenize the alginate/contrast solution with cells (*see* **Note 12**).

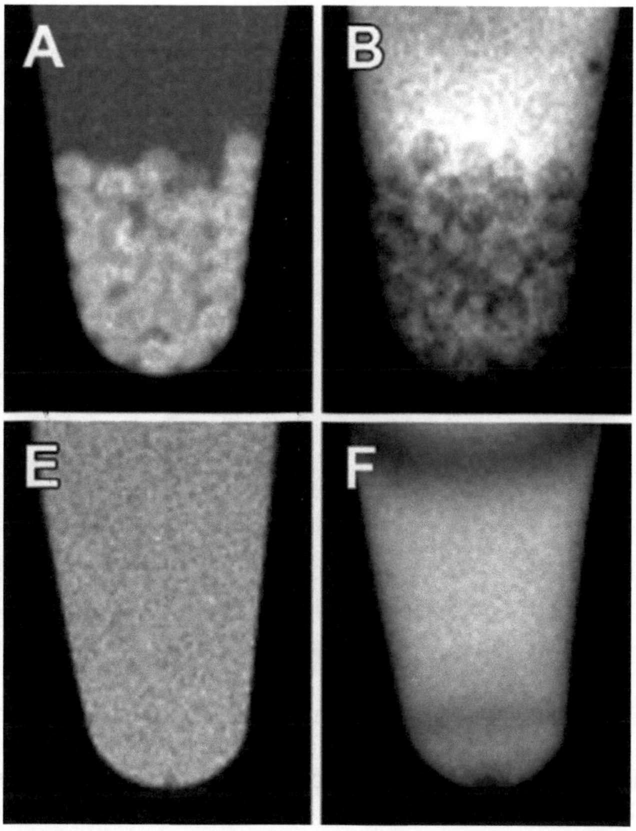

Fig. 4 ^{1}H MR-trackable capsules. (**a**, **e**), T1-weighted and (**b**, **f**) T2*-weighted MR images of alginate-protamine sulfate-alginate (APSA) labeled with gadolinium-gold (GG) (*top*) and APSA without GG (*bottom*) microcapsules in saline. The appearances of the capsules in the MR imaging appear as hyperintense on the T1-weighted and as hypointense on the T2*-weighted images. Reprinted with permission from ref. 34. Copyright © 2015 Radiological Society of North America

4 Notes

1. To avoid endotoxin contamination and potential activation of the cells in the microcapsule, ensure that the source of alginate has low endotoxin levels.
2. The market of commercially available SPIO formulations has seen several products being discontinued (e.g. Feridex®, Endorem®, or Resovist®). In principle, any SPIO formulation with a total hydrodynamic diameter of less than 150 nm should be suitable.
3. To minimize the aggregation of alginate, slowly pour alginate while stirring.
4. As it is difficult to dissolve alginate in small volumes of liquid, it is better to prepare a larger volume and aliquot it into 15 ml conical tubes for later use.
5. Irregularity of the surface of the microcapsule is an indication that swelling has occurred in the saline bath before the cross-linking

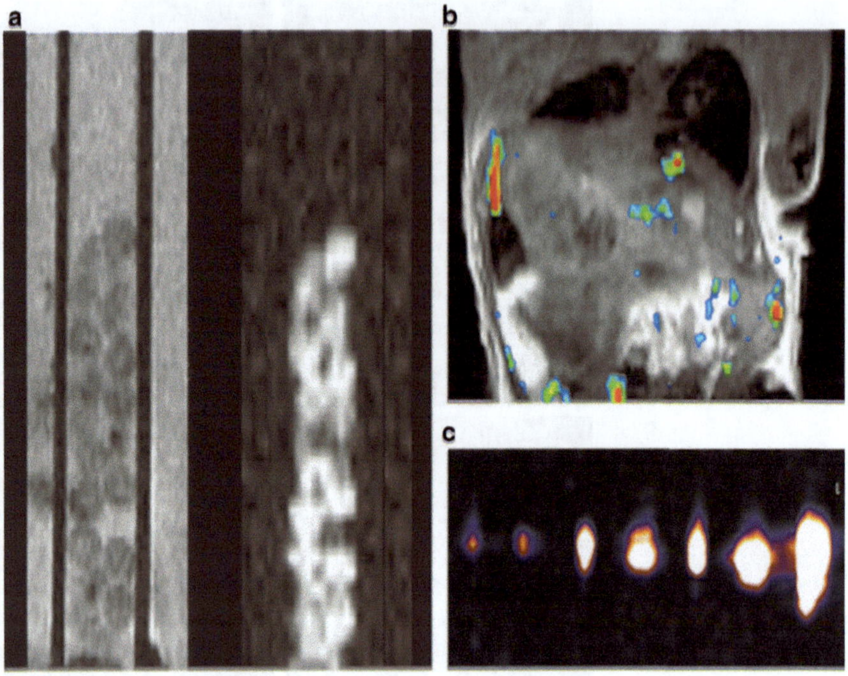

Fig. 5 ^{19}F MR-trackable capsules. (**a**) In vitro proton T2-weighted MR image (*left*) and corresponding ^{19}F MR image (*right*) of fluorocapsules in a standard 5 mm glass tube, obtained with a 11.7 T preclinical system. (The capsules were introduced in a 500 mm capillary tube.) (**b**) In vivo ^{19}F MR image of a mouse after intraperitoneal transplantation of 2000 fluorocapsules. An overlay of the ^{19}F image (*pseudocolor*) on the ^{1}H image (*gray scale*) is shown. (**c**) Fluorine ^{19}MR image of fluorocapsules obtained with a 3.0 T clinical system. From *left to right*, images are of gel phantoms containing 1, 2, 5, 10, 15, 25, and 50 capsules. Reprinted with permission from ref. 35. Copyright © 2015 Radiological Society of North America

with PLL. Thus, cations (i.e., 1 mM $CaCl_2$) should be added to the saline solution to prevent swelling during washes.

6. At this stage, some of the microcapsules might present a tear-drop shape due to a low concentration of alginate, and therefore, the amount of alginate should be increased. If the alginate stream sputters, then the concentration of the primary alginate must be decreased. If the needle becomes frequently clogged during the drip, this is an indication that the cell clusters are too large for the needle and a larger gauge should be used.
7. If microcapsules are too large, as determined using a phase-contrast inverted microscope, increase the distance between the needle tip and the gelling bath or increase the voltage from the generator or lower the concentration of the primary alginate. If microcapsules begin to adhere to each other as they are collecting directly beneath the tip of the needle in the gelling bath, a sterile stir bar can be placed in the gelation bath and placed on a stir plate.

8. If the bloom from the signal created by magnetocapsules is too large, then the concentration of SPIO in the capsules is too high and it should be reduced for field strengths greater than 1.5 T. The magnetic susceptibility induced by these capsules can be quantified using phase map cross-correlation detection and quantification, which automatically identifies the localization of engrafted magnetocapsules, estimating their volume, magnetic susceptibility, and magnetic moment.
9. As barium sulfate is in powder form, it can be added at the necessary concentration for different imaging modalities. For example, for CT imaging, a concentration of 5 % (w/v) is acceptable, whereas for detection on fluoroscopy, a concentration of >10 % (w/v) is required. Single Ba X-caps and Bi X-caps can be clearly identified in vivo in mice and rabbits.
10. Contrast agent leaching from the microcapsules is an indication that the agent is too small or not soluble in water, and therefore, it should be incorporated within a liposome carrier before mixing with alginate.
11. If the cell density is too low or too high, then different ranges of seeding densities should be attempted to find the best range for different cell lines. If low cellular viability is observed, this might be attributable to a lack of critical matrix components that be incorporated into the alginate. To improve the survival of cells, these components should not be in alginate solution too long prior to the formation of beads.
12. Appropriate cell concentrations vary depending on cell types. For human islets, one islet per capsule should be used. For single cells, e.g., mesenchymal stem cells, a maximum of up to 300 single cells can be encapsulated. The duration of this step must be minimized, to avoid large periods of hypoxia to the cells. If the viability of cells is reduced substantially at this step, one could supplement the primary alginate with glucose, amino acids, or other nutrients.

References

1. de Vos P, Spasojevic M, Faas MM (2010) Treatment of diabetes with encapsulated islets. Adv Exp Med Biol 670:38–53
2. Nafea EH, Marson A, Poole-Warren LA, Martens PJ (2011) Immunoisolating semi-permeable membranes for cell encapsulation: focus on hydrogels. J Control Release 154:110–122
3. Lim F, Sun A (1980) Microencapsulated islets as bioartificial endocrine pancreas. Science 210:908–910
4. de Haan BJ, Faas MM, Hamel AF, de Vos P (2006) Experimental approaches for transplantation of islets in the absence of immunosuppression. In: Ford AM (ed) Trends in diabetes mellitus research. Nova Science Publisher, Inc., New York, pp 131–162
5. de Vos P, Spasojevic M, de Haan BJ, Faas MM (2012) The association between in vivo physicochemical changes and inflammatory responses against alginate based microcapsules. Biomaterials 33:5552–5559

6. Paredes-Juarez G, de Haan B, Faas M, de Vos P (2014) A technology platform to test the efficacy of purification of alginate. Materials 7:2087–2103
7. Uludag H, De Vos P, Tresco P (2000) Technology of mammalian cell encapsulation. Adv Drug Deliv Rev 42:29–64
8. Orive G, Santos E, Poncelet D, Hernández RM, Pedraz JL, Wahlberg LU et al (2015) Cell encapsulation: technical and clinical advances. Trends Pharmacol Sci 36:537–546
9. Krishnamurthy NV, Gimi B (2011) Encapsulated cell grafts to treat cellular deficiencies and dysfunction. Crit Rev Biomed Eng 39:473–491
10. Ponce S, Orive G, Hernández R, Gascón AR, Pedraz JL, de Haan BJ et al (2006) Chemistry and the biological response against immunoisolating alginate-polycation capsules of different composition. Biomaterials 27:4831–4839
11. Spasojevic M, Bhujbal S, Paredes G, de Haan BJ, Schouten AJ, de Vos P (2014) Considerations in binding diblock copolymers on hydrophilic alginate beads for providing an immunoprotective membrane. J Biomed Mater Res A 102:1887–1896
12. Spasojevic M, Paredes-Juarez GA, Vorenkamp J, de Haan BJ, Schouten AJ, de Vos P (2014) Reduction of the inflammatory responses against alginate-poly-L-lysine microcapsules by anti-biofouling surfaces of PEG-b-PLL diblock copolymers. PLoS One 9:e109837
13. Huang X, Zhang X, Wang X, Wang C, Tang B (2012) Microenvironment of alginate-based microcapsules for cell culture and tissue engineering. J Biosci Bioeng 114:1–8
14. Calafiore R, Basta G (2014) Clinical application of microencapsulated islets: actual prospectives on progress and challenges. Adv Drug Deliv Rev 67–68:84–92
15. Ngoc PK, Phuc PV, Nhung TH, Thuy DT, Nguyet NTM (2011) Improving the efficacy of type 1 diabetes therapy by transplantation of immunoisolated insulin-producing cells. Hum Cell 24:86–95
16. Steele JAM, Hallé JP, Poncelet D, Neufeld RJ (2014) Therapeutic cell encapsulation techniques and applications in diabetes. Adv Drug Deliv Rev 67–68:74–83
17. Antosiak-Iwańska M, Sitarek E, Sabat M, Godlewska E, Kinasiewicz J, Weryński A (2009) Isolation, banking, encapsulation and transplantation of different types of Langerhans islets. Pol Arch Med Wewn 119:311–317
18. Mallett AG, Korbutt GS (2009) Alginate modification improves long-term survival and function of transplanted encapsulated islets. Tissue Eng Part A 15:1301–1309
19. Scharp DW, Marchetti P (2014) Encapsulated islets for diabetes therapy: history, current progress, and critical issues requiring solution. Adv Drug Deliv Rev 67–68:35–73
20. de Haan BJ, Rossi A, Faas MM, Smelt MJ, Sonvico F, Colombo P et al (2011) Structural surface changes and inflammatory responses against alginate-based microcapsules after exposure to human peritoneal fluid. J Biomed Mater Res A 98:394–403
21. Murua A, Orive G, Hernández RM, Pedraz JL (2009) Xenogeneic transplantation of erythropoietin-secreting cells immobilized in microcapsules using transient immunosuppression. J Control Release 137:174–178
22. Orive G, Tam SK, Pedraz JL, Hallé J-P (2006) Biocompatibility of alginate-poly-L-lysine microcapsules for cell therapy. Biomaterials 27:3691–3700
23. Arifin DR, Kedziorek DA, Fu Y, Chan KWY, McMahon MT, Weiss CR et al (2013) Microencapsulated cell tracking. NMR Biomed 26:850–859
24. Barnett BP, Arepally A, Stuber M, Arifin DR, Kraitchman DL, Bulte JWM (2011) Synthesis of magnetic resonance-, X-ray- and ultrasound-visible alginate microcapsules for immunoisolation and noninvasive imaging of cellular therapeutics. Nat Protoc 6:1142–1151
25. Barnett BP, Kraitchman DL, Lauzon C, Magee C, Walczak P, Gilson WD et al (2006) Radiopaque alginate microcapsules for X-ray visualization and immunoprotection of cellular therapeutics. Mol Pharm 3:531–538
26. Kedziorek DA, Hofmann LV, Fu Y, Gilson WD, Cosby KM, Kohl B et al (2012) X-ray-visible microcapsules containing mesenchymal stem cells improve hind limb perfusion in a rabbit model of peripheral arterial disease. Stem Cells 30:1286–1296
27. Kedziorek DA, Solaiyappan M, Walczak P, Ehtiati T, Fu Y, Bulte JWM et al (2013) Using C-Arm X-ray imaging to guide local reporter probe delivery for tracking stem cell engraftment. Theranostics 3:916–926
28. Chan KW, Liu G, Song X, Kim H, Yu T, Arifin DR et al (2013) MRI-detectable pH nanosensors incorporated into hydrogels for in vivo sensing of transplanted-cell viability. Nat Mater 12:268–275
29. Barnett BP, Arepally A, Karmarkar PV, Qian D, Gilson WD, Walczak P et al (2007) Magnetic resonance-guided, real-time targeted delivery and imaging of magnetocapsules immunopro-

tecting pancreatic islet cells. Nat Med 13:986–991
30. Link TW, Woodrum D, Gilson WD, Pan L, Qian D, Kraitchman DL et al (2011) MR-guided portal vein delivery and monitoring of magnetocapsules: assessment of physiologic effects on the liver. J Vasc Interv Radiol 22:1335–1340
31. Link TW, Arifin DR, Long CM, Walczak P, Muja N, Arepally A et al (2012) Use of magnetocapsules for in vivo visualization and enhanced survival of xenogeneic HepG2 cell transplants. Cell Med 4:77–84
32. Arifin DR, Manek S, Call E, Arepally A, Bulte JW (2012) Microcapsules with intrinsic barium radiopacity for immunoprotection and X-ray/CT imaging of pancreatic islet cells. Biomaterials 33:4681–4689
33. Kim J, Arifin DR, Muja N, Kim T, Gilad AA, Kim H et al (2011) Multifunctional capsule-in-capsules for immunoprotection and trimodal imaging. Angew Chem Int Ed Engl 50:2317–2321
34. Arifin DR, Long CM, Gilad AA, Alric C, Roux S, Tillement O et al (2011) Trimodal gadolinium-gold microcapsules containing pancreatic islet cells restore normoglycemia in diabetic mice and can be tracked by using US, CT, and positive-contrast MR imaging. Radiology 260:790–798
35. Barnett BP, Ruiz-Cabello J, Hota P, Liddell R, Walczak P, Howland V et al (2011) Fluorocapsules for improved function, immunoprotection, and visualization of cellular therapeutics with MR, US, and CT imaging. Radiology 258:182–191

Chapter 12

Retrieval of Microencapsulated Islet Grafts for Post-transplant Evaluation

John Patrick McQuilling, Sivanandane Sittadjody, Rajesh Pareta, Samuel Pendergraft, Clancy J. Clark, Alan C. Farney, and Emmanuel C. Opara

Abstract

Microencapsulation of islets is a procedure used to immunoisolate islets in order to obviate the need for immunosuppression of islet transplant recipients. Although microencapsulated islets have routinely been transplanted in the peritoneal cavity, the ideal site for their engraftment remains to be determined. The omentum, a highly vascularized tissue, has been proposed as an alternative site for microencapsulated islet transplantation. An added benefit to the omentum is that implanted microcapsules can be easily retrieved for post-transplant evaluation. This chapter describes a collagenase-based procedure for the retrieval of microencapsulated islets following the harvest of omentum pouch site of transplantation.

Key words Alginate, Microencapsulation, Islets, Transplantation, Omentum

1 Introduction

The advantages of cell replacement therapy as a treatment option for Type 1 diabetes compared to insulin treatment particularly in controlling diabetic complications are well established [1–3]. The two major β-cell therapy approaches currently in clinical application are pancreas and islet transplantations, albeit islet transplantation remains an experimental procedure [4], whose potential as a treatment option for Type 1 diabetic patients became apparent with the introduction of the Edmonton protocol [5]. However, there is a long-standing interest in islet transplantation because it is a simpler surgical procedure that easily lends itself to routine clinical application, hence the continued development of different strategies to improve the outcomes of islet transplantation [6]. Still, the limited supply of human pancreas and the need for risky immunosuppressive drugs to prevent transplant rejection remain two major barriers to routine clinical islet transplantation. The

Emmanuel C. Opara (ed.), *Cell Microencapsulation: Methods and Protocols*, Methods in Molecular Biology, vol. 1479,
DOI 10.1007/978-1-4939-6364-5_12,

procedure to immunoisolate islets by encapsulation prior to transplantation emerged as a strategy to overcome these two barriers [3, 7–9]. In turn, the clinical application of the encapsulation technology has been slow because of a variety of issues that remain unresolved, as has been recently reviewed [3, 9].

Two general approaches to encapsulation have been examined, microencapsulation which involves the enclosure of one or two islets in a microcapsule measuring <1 mm in diameter [10] and macroencapsulation that entails the enclosure of large numbers of islets in one chamber often with sizes measuring up to several centimeters [11, 12]. Although several biopolymers, including agarose, polyethylene glycol (PEG) and amalgams of PEG with alginate have been examined for cell encapsulation, alginate has been the most investigated of these biomaterials for this purpose [7–10, 13–18] primarily because of its biocompatibility and physicochemical characteristics [19], which have advanced its use to the level of clinical trials [20–24].

One of the critical factors that remains to be resolved in encapsulated islet treatment is determination of the optimal site of transplantation. Encapsulated islets have been routinely transplanted in the peritoneal cavity, which lacks a vascular supply resulting in the need to use large quantities of microencapsulated islets to achieve variable results [20–27]. In attempts to transplant these encapsulated islets closer to the blood supply alternative sites such as the kidney capsule and the omentum have been examined [28, 29]. In addition, these alternative sites are attractive because of the retrievability of the tissue site of implant. However, there is still a scarcity of validated procedures to retrieve the encapsulated islets from a resected implant site for detailed post-transplant analyses. The purpose of the present study was therefore to develop a procedure for retrieval of intact encapsulated islets after a prolonged period of transplantation in the omentum of diabetic animals.

To develop this procedure, rat islets were isolated using standard collagenase digestion techniques [30, 31] then encapsulated into alginate-polyornithine-alginate (APA) microcapsules [32, 33] and were then implanted into STZ-induced diabetic Lewis rats [34]. The animals were maintained for 90 days and then the procedure described below was used to retrieve and evaluate the encapsulated islets.

In this study, we have shown that islets encapsulated in the perm-selective APA microcapsules can be transplanted in the omentum where they remain viable over a long time even in the face of chronic hyperglycemia, consistent with previous observations in mice transplanted with islets encapsulated in agarose microcapsules by Kobayashi et al. [29]. In the study by Kobayashi et al., extensive histological assessment was performed on the encapsulated islets in-situ in the pouch; however, retrieval of the encapsulated islets from the pouch was not performed. In the present study, we have designed a reliable procedure for successful isolation of encapsulated islets after long-term implantation in the omentum pouch. The retrieved islets were shown to be viable as assessed by histological techniques as well as by their appropriate responsiveness to changes in glucose concentrations in the dynamic perifusion system.

Although the present study was performed in an isograft model, we have recently reported that encapsulated allografts have long-term function in immunocompetent diabetic rats, thus showing that the omentum pouch is a viable alternative site to the peritoneum [35]. An inability to retrieve encapsulated islet grafts after extended periods of transplantation is a significant drawback of the use of the peritoneal cavity as a site of encapsulated islet transplantation. Another major obstacle to clinical application of the microencapsulated islet technology is the death of large proportions of encapsulated islet grafts owing to prolonged hypoxia after transplantation in the avascular peritoneal cavity, resulting in the need to use large quantities of islets to achieve normoglycemia in experimental diabetic animals. Although islets constitute approximately 1 % of the pancreas, they receive about 6–10 % of the blood flow to this gland [36, 37], indicating a disproportionate level of perfusion in which islets receive and consume lots of oxygen. The usual high oxygen requirement of islets is interrupted during the period preceding isolated islet transplant revascularization, and studies have shown that hypoxia has significant deleterious effects on the survival and function of islets [38–40]. In the immediate post-transplant period, isolated islet transplants are forced to depend upon diffusion of oxygen and nutrients through peripheral perfusion from the surrounding tissue within the site of transplantation [39], until the islet transplants are revascularized by angiogenesis, a process that requires 7–10 days [41]. However, when isolated islets are microencapsulated and transplanted in the avascular peritoneal cavity, no revascularization takes place, thus subjecting the islet grafts to extended periods of hypoxia and eventual death.

It has also been shown that the function of encapsulated islet grafts in the peritoneal cavity is limited even when inflammatory responses against purified alginate microcapsules amount to less than 10 %. The graft dysfunction is accompanied by a gradual decrease in glucose-stimulated insulin secretion, and gradual cell necrosis, which appears to be triggered by inadequate nutrition [42]. To address the problem of inadequate supply of nutrients and oxygen to encapsulated islets transplants, a variety of approaches are being assessed by different investigators. In a recent study, investigators examined the effect of a combination of growth hormone releasing (GHRH) agonist and controlled oxygen supply on the function of a bioartificial macrochamber. In this study islets were encapsulated and maintained within an alginate slab configuration adjacent to an oxygen-permeable membrane to create an immune barrier and allow for oxygenation of the islet graft. The minimally invasive implantable chamber was shown to normalize blood glucose in streptozotocin-induced diabetic rodents for up to 3 months after subcutaneous transplantation [43]. In another study, investigators showed that encapsulation of solid calcium peroxide within hydrophobic polydimethylsiloxane resulted in sustained oxygen generation that lasted for more than 6 weeks and was enough to prevent hypoxia-induced cell dysfunction and death in insulin-producing cells [44].

We conclude from the present study that the omentum represents a viable alternative site of transplantation to the peritoneal cavity not only because it is a highly vascularized tissue loaded with angiogenic factors [45] to promote revascularization and enhance the delivery of oxygen and nutrients to the islet grafts, but also because of the technical ease in which encapsulated islets can be removed for post-transplant analyses, as shown in our study.

2 Materials

2.1 Materials for Islet Microencapsulation

1. 1.5% Alginate (Pronova UP LVM and UP LVG, Novamatrix, Sandvika Norway).
2. 0.1% Poly-L-ornithine (PLO) (P5061, Sigma-Aldrich).
3. 100 mM $CaCl_2$ solution (C614-10, Fischer Scientific, Waltham, MA, USA).
4. 55 mM Sodium citrate solution (S467-3, Fischer Scientific).
5. 0.9% Sodium chloride solution (normal saline) (71376-5KG, Sigma-Aldrich).

2.2 Materials for Microcapsule Implantation

1. General Surgery Basic Kit (World Precision Instruments, Sarasota, FL, USA).
2. 50 mL Conical tubes.
3. 10% Betadine solution.
4. 70% Ethanol.
5. Sterile gauze.
6. 0.9% Sodium chloride solution (injectable saline).
7. 4-0 Vicryl sutures.

2.3 Materials for Omentum Retrieval

1. General Surgery Basic Kit (World Precision Instruments, Sarasota, FL, USA).
2. 50 mL Conical tubes.
3. 10% Betadine solution.
4. 70% Ethanol.
5. Sterile gauze.
6. Collagenase digestion solution: 1 mg/mL Collagenase P from *Clostridium histolyticum* (Roche, Indianapolis, IN, USA). Add

25 mg of collagenase to 25 mL Hanks' Balanced Salt Solution (HBSS) (Mediatech Inc., Manassas, VA, USA). Sterile filter and store at −20 °C and bring to 4 °C prior to use.

7. Wash solution: 10% Fetal bovine serum (Gibco, Waltham, MA, USA) with 50 U/mL penicillin and 50 μg/mL of streptomycin in HBSS. Add 50 mL FBS and 5 mL Penn/Strep stock to 445 mL of HBSS, adjust pH to 7.4, sterile filter and store at 4 °C.
8. Islet culture media: 10% Fetal bovine serum (Gibco, Waltham, MA, USA), 50 U/mL penicillin and 50 μg/mL of streptomycin, 3.3 mM glucose, 2 g/L sodium bicarbonate, 1 mM HEPES in RPMI 1640 (Sigma). To a bottle of dry RPMI powder add 595 mg glucose, 2 g sodium bicarbonate, 238 mg HEPES, then add 100 mL FBS, 10 mL Penn/Strep stock and 890 mL ultrapure water. Adjust pH to 7.4 with either NaOH or HCL as needed, sterile filter and store at 4 °C (*see* **Note 1**).

2.4 Live/Dead Staining Reagents

1. 1 mg/mL Propidium iodide working solution: To 20 mL HBSS add 20 mg propidium iodide (PI) (Sigma). Protect from light and store at 4 °C.
2. Vybrant CFDA SE Cell Tracer Kit (CFDA) (Invitrogen, Waltham, MA, USA; Cat. No. V12883).
3. HBSS.
4. 4% Paraformaldehyde (PFA) Solution: Add 1 mL 16% PFA (Fisher) to 3 mL of phosphate-buffered Saline (Sigma). Store at 4 °C.
5. Olympus IX71 inverted microscope equipped with standard filters.
6. 10 × 35 mm petri dish with 2 mm grid marks.

2.5 Dithizone Staining Reagents

1. 10 mL Syringe.
2. Sterile syringe filter (pore size 0.45 μm) (Fisher).
3. Hanks' Balanced Salt Solution (HBSS).
4. 5 mL Dimethyl sulfoxide (DMSO) (Sigma).
5. 50 mg Dithizone (Diphenylthiocarbazone) (Sigma).
6. 10 × 35 mm Petri dish with 2 mm grid marks.

2.6 Dynamic Perifusion Reagents

1. Perifusion solution A: Add 70 g NaCl, 3.76 g KCl, and 21 g $NaHCO_3$ to 1 L ultrapure water and store at 4 °C.
2. Perifusion solution B:
   - (a) Prepare a 1 M solution of $MgCl_2$ by adding 4.76 g $MgCl_2$ to 50 mL ultrapure water.
   - (b) Prepare a 1 M solution of $CaCl_2·2H_2O$ by adding 7.35 g of $CaCl_2$·2H2O to 50 mL ultrapure water.

(c) Prepare perifusion solution B by adding 25.6 mL of the 1 M $CaCl_2{\cdot}2H_2O$ solution and 11 mL of the 1 M $MgCl_2$ solution to 963.4 mL of ultrapure water and store at 4 °C.

3. Glucose (Sigma).
4. Bovine serum albumin (BSA) (Sigma).
5. 10 × 35 mm Petri dish with 2 mm grid marks.
6. Peristaltic Pump.
7. Silicone tubing.
8. Perifusion chambers.
9. Water bath.
10. Collection tubes.

3 Methods

3.1 Islet Microencapsulation

Below is a summary of the encapsulation and implantation process; a more detailed description of the process has been described in Opara et al. [10].

1. Suspend islets in sodium alginate.
2. Using microfluidic device generate microspheres and allow spheres to cross-link in $CaCl_2$ for 15 min.
3. Wash with normal saline, and incubate microspheres in 0.1 % PLO for 30 min.
4. Liquefaction of the alginate core of the microcapsules is achieved by a brief (2 min) incubation in 55 mM sodium citrate solution.
5. Wash with normal saline.
6. Coat outer layer of capsule with 1.25 % LVM for 5 min.
7. Separate excess alginate from microcapsules and cross-link outer layer.
8. Place microcapsules in 50 mL conical on ice until transplantation.

3.2 Microcapsule Implantation

1. After animal is appropriately anesthetized, shave and prep the dorsal area of the rodent with 10 % betadine and ethanol.
2. Create a small midline incision over the greater omentum.
3. Place sterile 2 × 2 cm sterile gauze around the midline incision and moisten with saline.
4. Identify and place the greater omentum on the gauze.
5. Orient the omentum flat along the gauze and moisten with saline.
6. Using 4-0 vicryl sutures create a purse string suture along the periphery of the omentum.
7. Place the microcapsules in the center of the omentum and gently tighten the sutures to close the omentum around the capsules.

8. Secure the purse string pouch and place in the peritoneal cavity.
9. Close the midline incision.

3.3 Retrieval of Omentum Pouch and Isolation of Microcapsules

1. Immediately following euthanasia (*see* **Note 2**) shave and disinfect the abdomen with 10% betadine and then with 70% ethanol.
2. Make a small midline incision and expose the greater omentum onto sterile gauze.
3. Identity the omentum pouch (as seen in Fig. 1a) and resect.
4. Place the resected omentum pouch (as seen in Fig. 1b, c) into a 50 mL conical tube with 25 mL of the Collagenase digestion solution (*see* **Note 3**).
5. Place the 50 mL conical tube in a shaking water bath for 20 min at 37 °C.
   (a) Every 5 min during the digestion invert the conical tube.
   (b) Refer to Fig. 2 for a visual representation of the omental pouch over the 20 min digestion period.
6. After the 20-min digestion period move the 50 mL conical tube into a biosafety cabinet and add 10 mL of the cold wash solution.
7. Shake the tube gently for 30 s.
8. Place the tube on ice and allow the capsules to settle to the bottom of the conical tube for approximately 2 min.
9. Aspirate the digestion solution and added wash solution.
10. Add 50 mL wash solution.
11. Again allow the capsules to resettle, aspirate the wash solution and repeat the final wash step. At this point the capsules should be free of omental tissue as seen in Fig. 2c.
12. Culture the retrieved capsules in the islet culture media and culture under standard conditions until islet viability and functionality testing (*see* **Note 4**).

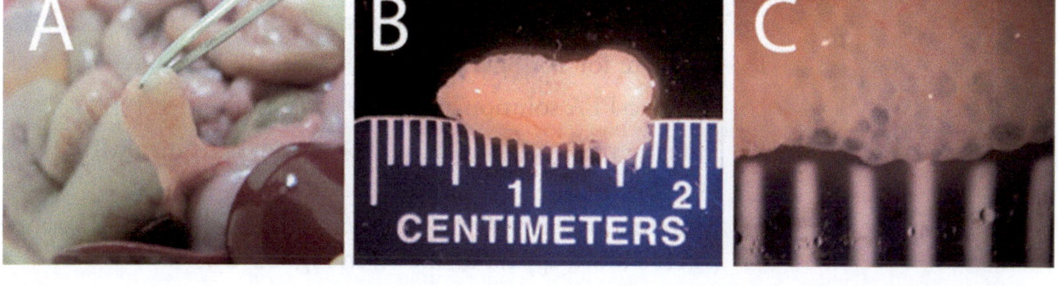

Fig. 1 Omentum pouch retrieved after 90 days in vivo (**a**) Omentum immediately before surgical removal of omentum pouch with the microcapsules. (**b**) Omentum pouch immediately following surgical removal. (**c**) Omentum pouch with microcapsules magnified, spacing between the *white lines* is 1 cm

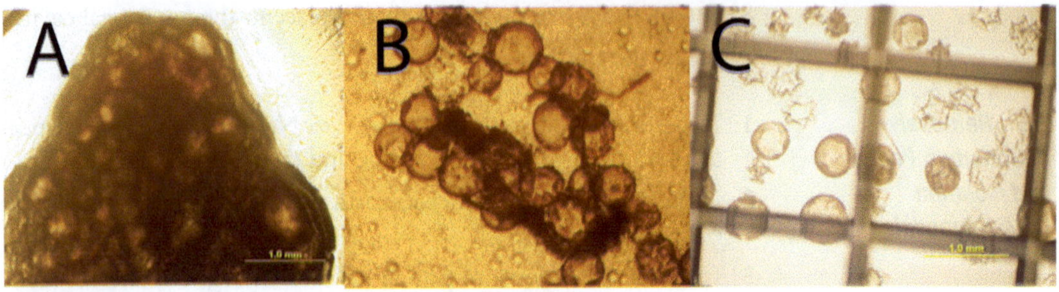

Fig. 2 Digestion of omentum pouch. (**c**) Omentum pouch and microcapsules prior to collagenase P digestion. (**b**) Dispersed omental tissue and microcapsules following 10 min of digestion at 37 °C. (**c**) Free microcapsules after 20-min digestion and wash in HBSS with 10% FBS

3.4 Live/Dead Staining of Retrieved Islets

1. Prepare the CFDA working solution.
   (a) Bring components A and B of the Vybrant CFDA SE Cell Tracer Kit to room temperature.
   (b) Dissolve the contents of one vial (component A) in 90 μL of the DMSO provided within the kit (Component B).
   (c) Wrap the vial in aluminum foil and initial and date, the stock solution is to be stored at −20 °C, and is good for up to 6 months.
2. Add 2.5 μL of CFDA stock solution to 1 mL of HBSS.
3. Vortex to mix.
4. Collect approximately 50–100 retrieved encapsulated islets.
5. Pipette supernatant and discard.
6. Add 200 μL of CFDA working solution.
7. Resuspend capsules and incubate for 15 min at 37 °C.
8. Following 15 min, pipette off CFDA working solution and discard.
9. Add 200 μL of wash solution and incubate for 30 min at 37 °C.
10. Pipette supernatant and discard.
11. Add 200 μL of PI working solution.
12. Incubate for 2 min at room temperature.
13. Remove PI solution and wash three times with HBSS.
14. Add 200 μL of 4% PFA working solution.
15. Incubate for 10 min on ice.
16. Remove PFA solution and wash three times with HBSS.
17. Transfer retrieved islets to a 10 × 35 mm petri dish with 2 mm grid marks.
18. Examine at least 50 islets under the fluorescent microscope (islets should appear as shown in Fig. 3a–d).

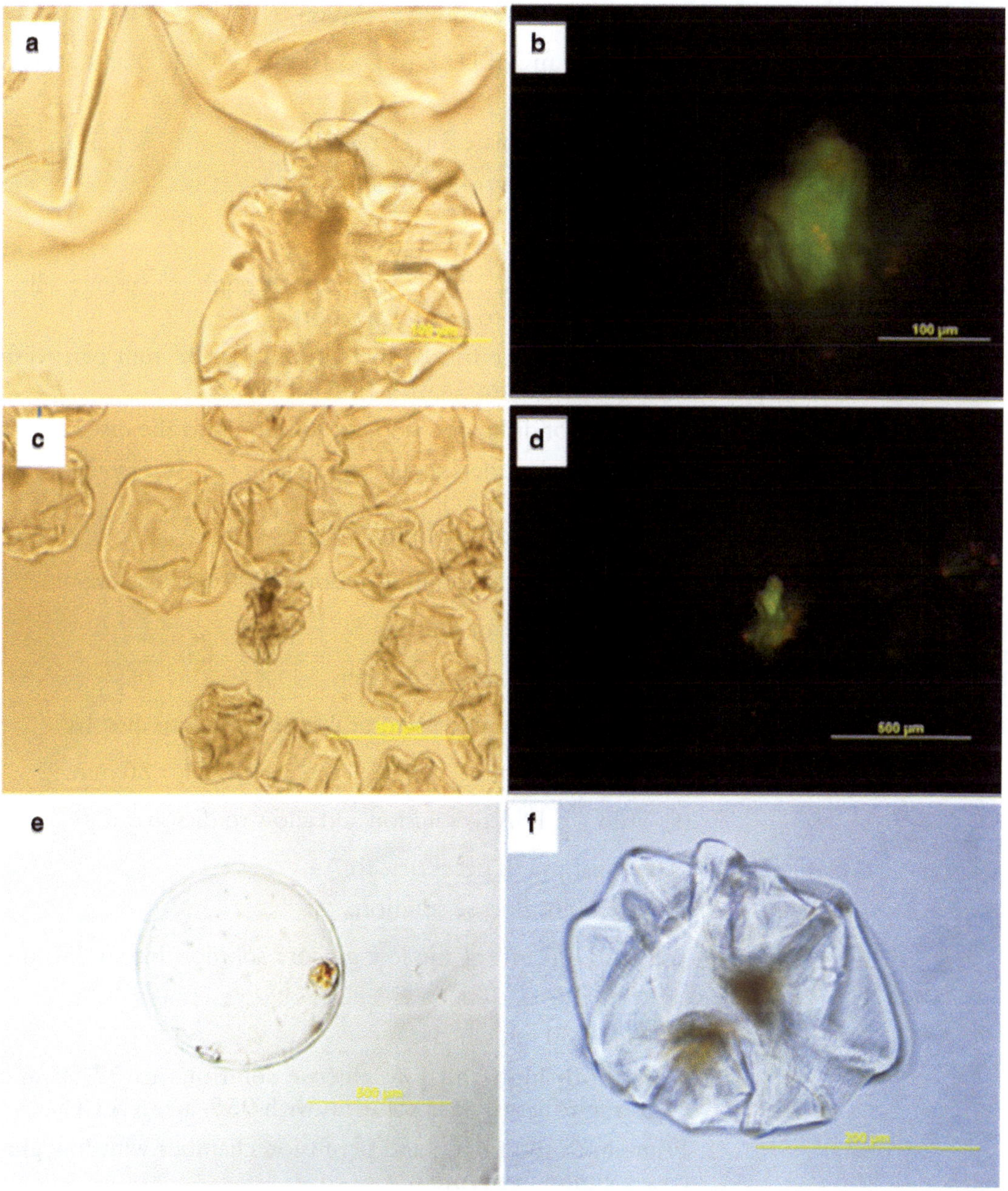

Fig. 3 Live/dead and dithizone staining of retrieved encapsulated islets. While a portion of retrieved microcapsules retained their spherical shape over time, throughout the isolation process the majority of retrieved capsules collapsed as shown above. However as indicated by the live/dead staining majority of the encapsulated islets were viable. (**a**) Phase contrast image of encapsulated islets (10× objective) (**b**) 10× objective of image of encapsulated islets stained with CFDA (*green*) and Propidium Iodide (*red*). (**c**) Phase contrast image of encapsulated islets (4× objective) (**d**) Encapsulated islets stained with CFDA (*green*) and Propidium Iodide (*red*) (4× objective). Additionally, after retrieving islets from the omentum islets were stained with a 2 mg/mL Dithizone solution in HBSS and 25 % DMSO for 2 min, and washed in PBS buffer prior to imaging. (**e**) 4× objective (**f**) 10× objective

3.5 Dithizone Staining of Retrieved Islets

1. Prepare dithizone working solution.
   (a) Add 50 mg of DTZ to a to 10 mL of DMSO and protect from light.
   (b) Mix on stir plate for 15 min.
   (c) Add 50 mL HBSS.
   (d) Mix on stir plate for 15 min.
   (e) Prior to use, filter solution through a syringe filter (0.45 μm) into a new glass bottle.
2. Place sample of retrieved islets into a 10 × 35 mm petri dish with 2 mm grid marks with 1 mL of HBSS.
3. Add 1 mL of dithizone working solution into the dish.
4. Inspect capsules; islets should appear as in Fig. 3e, f (*see* **Note 5**).

3.6 Dynamic Perifusion of Islets and Measurement of Insulin Secretion

1. Prepare low-glucose solution.
   (a) Place 100 mL of solution A in a 1000 mL beaker.
   (b) Add 100 mL of solution B.
   (c) Add 800 mL distilled water.
   (d) Add 600 mg glucose to beaker and allow to dissolve.
   (e) Gas the solution with 95 % air/5 % CO2 for 30 min.
   (f) Add 2 g BSA to solution and allow to dissolve.
   (g) Adjust pH, if necessary, to 7.4.
2. Prepare high-glucose solution.
   (a) Add 200 mL of the low glucose solution into a 250 mL beaker.
   (b) Add 480 mg glucose.
3. Bring both high and low glucose solutions to 37 °C and continue to gently gas solution with 95 % air/5 % CO_2.
4. Prime both the tubing and perifusion chamber with low glucose solution (*see* **Note 6**).
5. Place retrieved capsules into a 10 × 35 mm petri dish with 2 mm grid marks and using either a p1000 or p200 pipetter, handpick 60 islets into each perifusion chamber.

Fig. 4 (continued) (**c**) Glucose stimulation index (GSI) for encapsulated islets prior to transplantation and following retrieval. The GSI is defined as a ratio of the maximum amount of insulin secreted during the high glucose period over the average amount of insulin secreted over low glucose period. The GSI for retrieved islets was not significantly different from that for the pre-transplant group (2.75 ± 0.68 vs. 4.11 ± 0.48 respectively, $p = 0.13$). Error bars represent mean ± SEM, $N = 5$

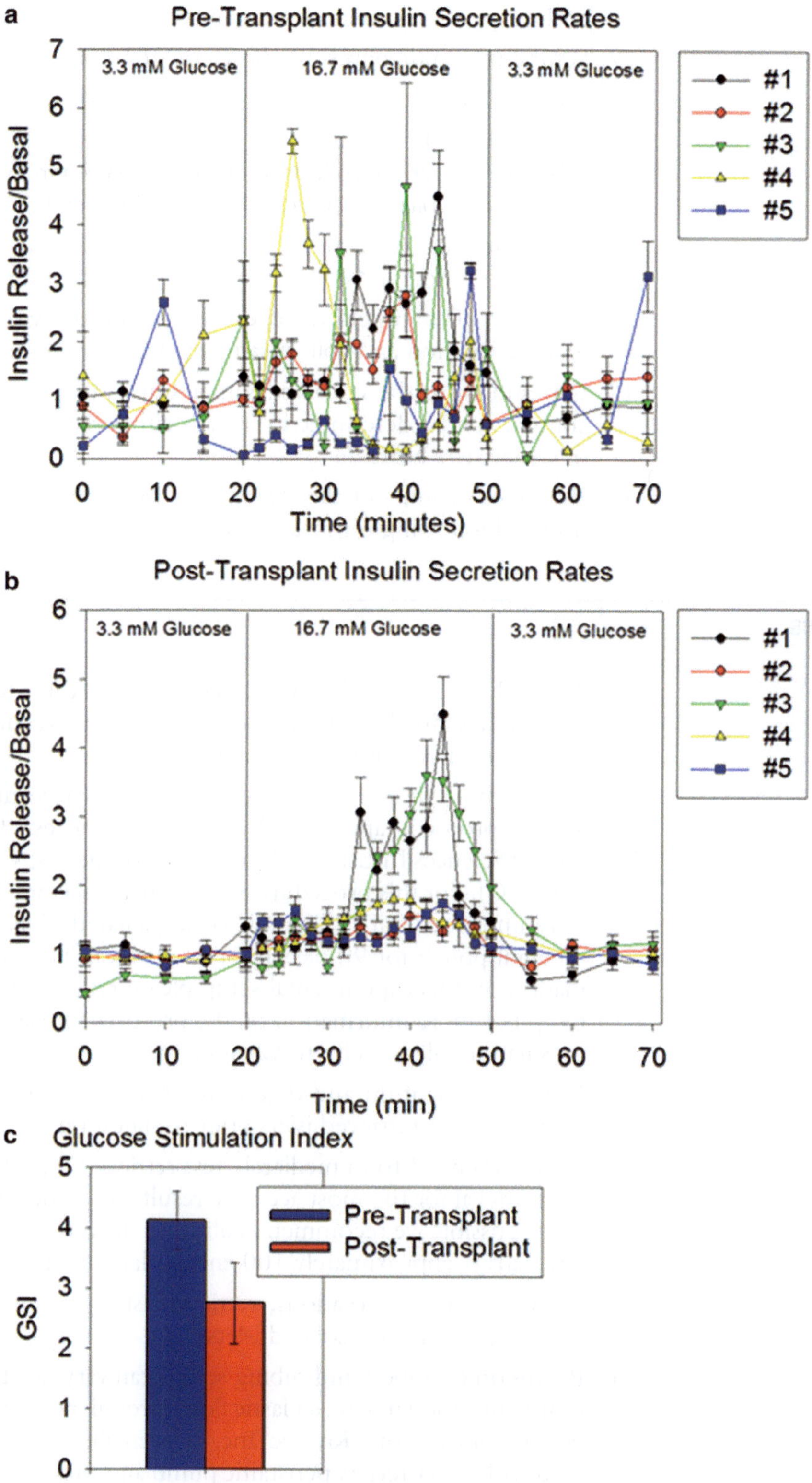

Fig. 4 Islet Function before and after transplantation. (**a**) Insulin secretion by encapsulated islets prior to implantation. (**b**) Insulin secretion by retrieved encapsulated islets 90 days after implantation.

6. Seal the perifusion chambers and place in a water bath at 37 °C.
7. Set the peristaltic pump at a flow rate of 1.4 mL/min.
8. Pre-perifuse islets with the low glucose solution for 1 h prior to collecting samples.
9. After the 1-h period, collect the effluent every 5 min for a 20-min period, keep collected effluent on ice until the end of the assay.
10. After the 20-min period, begin perifusion with the high-glucose solution and collect samples every 2 min for 30 min.
11. After the 30-min period, repeat the perifusion with the low-glucose solution and collect samples every 5 min for 20 min.
12. Cap and store the collected samples at –20 °C until insulin content can be determined through either radioimmunoassay or ELISA (*see* **Note 7**).
13. Examples of functionality testing for islets prior to implantation and following retrieval are shown in Fig. 4.

4 Notes

1. The culture media described here is specifically for rodent islets, as described in Pareta et al. [37] and is not recommended for either human or porcine islets.
2. The methods described here were performed according to the animal protocol approved by the Wake Forest University Health Sciences Institutional Animals Care and Use Committee (IACUC). For this procedure STZ-induced diabetic Lewis rats were implanted with 1000 microencapsulated islets within an omental pouch for 90 days prior to retrieval. For more information on this experimental setup please refer to Pareta et al. [37]. For more information on the protocol for islet microencapsulation, please refer to Chapter 16.
3. Keep the omentum and digestion solution on ice to prevent damage to the retrieved islets prior to digestion.
4. We recommend to immediately test retrieved capsules following retrieval for the most accurate results. If immediate testing is not possible we recommend culturing the retrieved islets at a low density approximately 100 encapsulated islets/cm^2.
5. Based on our observations, retrieved islets may not stain as deeply red as freshly isolated islets.
6. Perifusion chambers and tubing setups can vary; for the perifusion results shown here, a plastic flow-through perifusion chambers obtained from Kiyatec Inc. (Greenville, SC, USA) was used with a Masterflex peristaltic pump and Masterflex L/S 16 tubing available from Cole-Parmer (Vernon Hills, IL, USA).

We also recommend using commercially available perifusion chambers from Biorep Technologies Inc. (Miami, FL, USA).

7. For the examples provided for this protocol, perifusate samples were analyzed via radioimmunoassay as previously described [46].

Acknowledgements

The authors would like to acknowledge financial support from the National Institutes of Health (RO1 DK080897) and the Vila Rosenfeld Estate, Greenville NC for the work in Dr. Opara's laboratory at the Wake Forest Institute for Regenerative Medicine.

Also, research reported in this publication was supported by the National Institute of Biomedical Imaging and Bioengineering (NIBIB) of the National Institutes of Health under award # T32EB014836. The content is solely the responsibility of the authors and does not necessarily represent the official views of the National Institutes of Health.

References

1. White SA, Shaw JA, Sutherland DER (2009) Pancreas transplantation. Lancet 373(9677):1808–1817
2. Hills CE, Brunskill NJ (2009) Cellular and physiological effects of C-peptide. Clin Sci (Lond) 116(7):565–574
3. Opara EC, Mirmalek-Sani SH, Khanna O et al (2010) Design of a bioartificial pancreas. J Investig Med 58(7):831–837
4. Berney T, Ricordi C (1999) Islet transplantation. Cell Transplant 8:461–464
5. Shapiro AMJ, Lakey JR, Ryan EA et al (2000) Islet transplantation in seven patients with type 1 diabetes mellitus using a glucocorticoid-free immunosuppressive regimen. N Engl J Med 343:230–238
6. Barton FB, Rickels MR, Alejandro R et al (2012) Improvements in outcomes of clinical islet transplantation: 1999-2010. Diabetes Care 35(7):1436–1445
7. Lim F, Sun A (1980) Microencapsulated islets as bioartificial pancreas. Science 210(4472):908–910
8. Prokop A (2001) Bioartificial pancreas: materials, devices function, and limitations. Diabetes Technol Ther 3(3):431–449
9. O'Sullivan ES, Vegas A, Anderson DG et al (2011) Islets transplanted in immunoisolation devices: a review of the progress and the challenges that remain. Endocr Rev 32(6):827–844
10. Opara EC, McQuilling JP, Farney AC (2013) Microencapsulation of islets for use in a bioartificial pancreas. In: Basu J, Ludlow JW (eds) Organ regeneration: methods and protocols. Humana Press, New York, pp 261–266
11. Jain K, Yang H, Cai BR et al (1995) Retrievable, replaceable, macroencaspulated pancreatic islet xenografts. Long-term engraftment without immunosuppression. Transplantation 59(3):319–324
12. Muthyala S, Raj VRR, Mohanty M et al (2011) The reversal of diabetes in rat model using mouse insulin producing cells – a combination approach of tissue engineering and macroencapsulation. Acta Biomater 7(5):2153–2162
13. Opara EC, Kendall WF (2002) Immunoisolation techniques for islet cell transplantation. Expert Opin Biol Ther 2:503–511
14. Weber LM, He J, Bradley B et al (2006) PEG-based hydrogels as an in vitro encapsulation platform for testing controlled beta-cell microenvironments. Acta Biomater 2(1):1–8
15. Capretto L, Mazzitelli S, Luca G et al (2010) Preparation and characterization of polysaccharidic microbeads by a microfluidic technique: application to the encapsulation of Sertoli cells. Acta Biomater 6(2):429–435
16. Rokstad AM, Brekke OL, Steinkjer B et al (2011) Alginate microbeads are complement compatible, in contrast to polycation containing

microcapsules, as revealed in a human whole blood model. Acta Biomater 7(6):2566–2578
17. Hall KK, Gattas-Asfura KM, Stabler CL (2011) Microencapsulation of islets within alginate/poly(ethylene glycol) gels cross-linked via Staudinger ligation. Acta Biomater 7(2):614–624
18. Lee KY, Mooney DJ (2012) Alginate: properties and biomedical applications. Prog Polym Sci 37(1):106–126
19. Tam SK, Bilodeau S, Dusseault J et al (2011) Biocompatibility and physicochemical characteristics of alginate-polycation microcapsules. Acta Biomater 7(4):1683–1692
20. Soon-Shiong P, Heintz RE, Merideth N et al (1994) Insulin independence in a type 1 diabetic patient after encapsulated islet transplantation. Lancet 343(8903):950–951
21. Calafiore R, Basta G, Luca G et al (2006) Microencapsulated pancreatic islet allograft into non-immunosuppressed patients with Type 1 diabetes. Diabetes Care 29(1):137–138
22. Elliott RB, Escobar L, Tan PL et al (2007) Live encapsulated porcine islets from type 1 diabetic patient 9.5 yr. after xenotransplantation. Xenotransplantation 14(2):157–161
23. Tuch BE, Keogh GW, Williams LJ et al (2009) Safety and viability of microencapsulated human islets transplanted into diabetic humans. Diabetes Care 32(10):1887–1889
24. Living Cell Technologies. Clinical Trials 2012 Update. http://www.lctglobal.com/
25. Sun Y, Ma X, Zhou D et al (1996) Normalization of diabetes in spontaneously diabetic Cynomolgus monkeys by xenografts of microencapsulated porcine islets without immunosuppression. J Clin Invest 98(6):1417–1422
26. Wang T, Adcock J, Kuhtreiber W et al (2008) Successful allotransplantation of encapsulated islets in pancreatectomized canines for diabetic management without the use of immunosuppression. Transplantation 85(3):331–337
27. Cui H, Tucker-Burden C, Cauffiel SM et al (2009) Long-term metabolic control of autoimmune diabetes in spontaneously diabetic non-obese diabetic mice by nonvascularized microencapsulated adult porcine islets. Transplantation 88(2):160–169
28. Dufrane D, Goebbels RM, Saliez A et al (2006) Six month survival of microencapsulated pig islets and alginate biocompatibility in primates: proof of concept. Transplantation 81(9):1345–1353
29. Kobayashi T, Aomatsu Y, Iwata H et al (2006) Survival of microencapsulated islets at 400 days post transplantation in the omental pouch of NOD mice. Cell Transplant 15(4):359–365
30. Lacy PE, Kostianovsky M (1967) Method for the isolation of intact islets from the pancreas. Diabetes 16(1):35–39
31. Field J, Farney A, Sutherland DE (1996) Improved islet isolation from rat pancreas using 35% bovine serum albumin in combination with Dextran gradient separation. Transplantation 61(10):1554–1556
32. Tendulkar S, Mirmalek-Sani SH, Childers C et al (2012) A three-dimensional microfluidic approach to scaling up microencapsulation of cells. Biomed Microdevices 14(3):461–469
33. Darrabie MD, Kendall WF, Opara EC (2005) Characteristics of poly-L-ornithine-coated alginate microcapsules. Biomaterials 26(34):6846–6852
34. McQuilling JP, Arenas-Herrera J, Childers C et al (2011) New alginate microcapsule system for angiogenic protein delivery and immunoisolation of islets for transplantation in the rat omentum pouch. Transplant Proc 43(9):3262–3264
35. Pareta R, McQuilling JP, Sivanandane S et al (2014) Long-term function of islets encapsulated in a redesigned alginate microcapsule construct in omentum pouches of immune-competent diabetic rats. Pancreas 43:605–613
36. Lifson N, Lassa CV, Dixit PK (1985) Relation between blood flow and morphology in islet organ of rat pancreas. Am J Physiol 249(1):E43–E48
37. Jansson L, Hellerström C (1983) Stimulation by glucose of the blood flow to the pancreatic islets of the rat. Diabetologia 25(1):45–50
38. Dionne KE, Colton CK, Yarmush ML (1993) Effect of hypoxia on insulin secretion by isolated rat and canine islets of Langerhans. Diabetes 42(1):12–21
39. Davalli AM, Scaglia L, Zangen DH et al (1996) Vulnerability of transplanted islets in the immediate post transplantation period; dynamic changes in structure and function. Diabetes 45(9):1161–1167
40. Mendoza V, Klein D, Ichii H et al (2005) Protection of islets in culture by delivery of oxygen binding neuroglobin via protein transduction. Transplant Proc 37(1):237–240
41. Jones GL, Juszczak MT, Hughes SJ et al (2007) Time course and quantification of pancreatic islet revascularization following intraportal transplantation. Cell Transplant 16(5):505–516
42. De Vos P, Van Straaten JFM, Nieuwenhuizen AG et al (1999) Why do microencapsulated

islet grafts fail in the absence of fibrotic overgrowth? Diabetes 48(7):1381–1388

43. Ludwig B, Rotem A, Schmid J et al (2012) Improvement of islet function in a bioartificial pancreas by enhanced oxygen supply and growth hormone agonist. Proc Natl Acad Sci U S A 109(13):5022–5027
44. Pedraza E, Coronel MM, Fraker CA et al (2012) Preventing hypoxia-induced cell death in beta cells and islets via hydrolytically activated oxygen generating biomaterials. Proc Natl Acad Sci U S A 109(11):4245–4250
45. Goldsmith HS, Griffith AL, Kupferman A et al (1984) Lipid angiogenic factor from omentum. JAMA 252(15):2034–2036
46. Herbert V, Lau KS, Gottlieb CW et al (1965) Coated charcoal immunoassay of insulin. J Clin Endocrinol Metab 25(10):1375–1384

Part IV

Isolation of Cells for Microencapsulation

Chapter 13

A Method of Porcine Pancreatic Islet Isolation for Microencapsulation

William F. Kendall Jr. and Emmanuel C. Opara

Abstract

Since the discovery of insulin by Banting and Best in 1921, the prognosis and treatment options for individuals with diabetes have improved. The development of various insulin types, various oral agents, and insulin pumps have improved the available medical options for individuals afflicted with diabetes. The current need for frequent blood glucose monitoring imposed by multiple daily insulin injections, result in significant life-style challenges for in individuals afflicted with Type 1 diabetes (T1D). In contrast the use of surgical interventions, such as whole organ pancreas transplantation (PT) requires less-intensive glucose monitoring while the organ is viable. Also, isolated human pancreatic islet transplantation (IT) holds similar promise as PT; however, the limited availability of human pancreata exacerbated by, the need for multiple pancreata per individual IT recipient, and issues with prolonged viability, still hamper widespread successful, and routine use of IT. The use of porcine pancreata holds promise as a viable alternative to human pancreas to significantly increase the volume of islets available to meet the needs of millions of patients afflicted with T1D. This chapter outlines our protocol utilized to reliably isolate and microencapsulate porcine islets.

Key words Diabetes, Porcine, Pig, Transplantation, Islets, Xenotransplantation

1 Introduction

In 2014, the global prevalence of diabetes among individuals 18 years of age and older was estimated to be about 9% [1]. In 2012, approximately 1.5 million deaths were directly related to diabetes, with more than 80% of the deaths occurring in low to middle income countries [1]. It is projected that by 2030 diabetes will be the seventh leading cause of death [1]. The development of various treatment modalities such as different formulations of insulin, various oral agents, better blood glucose monitoring devices and insulin pumps have provided improved nonsurgical management for diabetes since insulin was discovered in 1921. Overall, however, these treatment options are still cumbersome and require careful glucose monitoring and often the need for multiple insulin injections each

Emmanuel C. Opara (ed.), *Cell Microencapsulation: Methods and Protocols*, Methods in Molecular Biology, vol. 1479, DOI 10.1007/978-1-4939-6364-5_13, © Springer Science+Business Media New York 2017

day [2]. Whole organ pancreatic transplant has emerged as a viable treatment option with much less need for numerous invasive daily blood glucose checks, while the organ is viable. Islet transplantation has also been vigorously pursued as a treatment option, but has had limited success compared to pancreas transplantation. The need for multiple human pancreata in order to obtain adequate functional islet mass to attain adequate glucose control in recipient patients along with the need for islet re-transplantation have plagued this option clinically making it less attractive than whole organ transplant, especially as it leads to increased consumption of an already very scarce resource. The use of porcine islet cells is thus vigorously being explored as a potential alternative unlimited source of pancreatic islets. As we work through issues such as overcoming the concerns regarding zoonosis and xenotransplant rejection [3], one technical challenge has been the development of a reliable technique for isolation of viable porcine islets for transplantation. This chapter discusses a protocol that was developed by our research team to accomplish this goal [4] and the potential to use such islets in the fabrication of a bio-artificial pancreas.

2 Materials and Methods

2.1 Equipment

1. COBE 2991.
2. Inverted light microscope (Olympus CK40), linked to the UTHSCA image tool computer program (University of Texas, Austin).
3. Nylon mesh with a pore size of 100 μm.

2.2 Chemicals/Biochemicals

1. 2% Isoflurane.
2. Liberase PI enzyme blend (0.5 mg/mL), obtained from Roche Molecular biochemical (Indianapolis, IN)).
3. Deoxyribonuclease I, 10,000 U.
4. 1 mM/L Trolox.
5. (400 mg/200 mL) Dithizone.
6. 0.5% (w/v) Trypan Blue.

2.3 Solutions

1. University of Wisconsin solution.
2. Hanks balanced salt solution, from GIBCO BRL (Grand Island, NY).
3. Solution A (10X)—Perfusion solution.

(a) NaCl	70 g/L
(b) KCl	3.76 g/L
(c) $NaHCO_3$	21 g/L

4. Solution B (10×)—Perfusion solution

(a) Stock $MgCl_2$ (1 M)	95.2 g/L
(b) Stock $CaCl_2$, $2H_2O$ (1 M)	147 g/L

To prepare stock solution B, add 25.6 mL $CaCl_2$ to 11 mL $MgCl_2$, fill volumetric flask to 1 L with distilled water

5. Hepes buffer solution.
   (a) 25 mM HEPES (651 mg = 0.651 g/100 mL).
   (b) 118 mM NaCl (690 mg = 0.690 g/100 mL).
   (c) 5.6 mM KCl 41.7 mg = 0.042 g/100 mL).
   (d) 2.5 mM $MgCl_2$ (23.8 mg = 0.024 g/100 mL).
   (e) 100 mL of distilled water.
   (f) Adjust to pH 7.4.
6. Dithizone Solution (400 mg/200 mL) (*see* **Note 1**).
   Store in sterile 200 mL centrifuge tube. Pour into 30 mL syringe when ready to use. (*see* **Note 2**).
   (a) 400 mg DTZ (Dithizone).
   (b) 40 mL DMSO.
   (c) 160 mL Hanks solution.
7. Porcine Culture Medium.
   For a total of 500 mL, add the following using sterile technique:
   (a) RPMI (may substitute with Hams as directed): 384 mL.
   (b) Porcine Serum (makes 20%): 100 mL.
   (c) 2-Mercaptoethanol 5.5 x10^{-2}M (makes 50 μM): 500 μL.
   (d) Sodium Pyruvate 100 mM (makes 1 mM) 5 mL
   (e) HEPES buffer 1 M (makes 10 mM) 5 mL.
   (f) Penicillin/Streptomycin (makes 1%) 5 mL.
   (g) Trolox 0.25025 g (makes 2 mM).
   (h) If RPMI without glutamine, ON DAY OF USE Add 0.0292 g glutamine/100 mL culture medium.
8. Cocktail for Porcine Culture Medium.
   To make 100 mL of cocktail/culture (Add to 100 mL of culture medium).
   (a) Dimethylthiourea (make 10 mM) 0.1048 g.
   (b) Citiolone (makes 5 mM) 0.0796 g.
   (c) LNMMA (makes 2 mM) 0.0496 g.
   (d) Glutathione [GSH] (makes 10 mM) 0.3073 g (*see* **Note 3**).
9. To prepare 200 mL of Porcine Pancreas Digestion Medium.

   (a) 1.5 mg/mL Collagenase P (300 mg).
   (b) 20,000 U DNase I.
   (c) 1 mM Trolox (0.050058 g).
   (d) 200 mL Hanks.
10. To prepare 200 mL of 0.75 M Sucrose.
   (a) 51.345 g Sucrose.
   (b) 200 mL RPMI 164.

2.4 Animals

Young (25–30 kg) adult male Yorkshire pigs, from Shenise and sons. Henderson, NC.

2.5 Methods

1. Pig Pancreatectomy (pig placed under anesthesia per protocol with isoflurane).
   (a) Positioning: Dorsal Recumbency.
   (b) Approach: Ventral Midline Celiotomy.
   (c) Left Pancreatic Limb Dissection.
      - Open omental bursa.
      - Incise deep leaf of omentum near the extremity of the left pancreatic limb (tail), working medially towards the head.
         – Stay extracapsular at this stage.
      - Isolate the splenic artery (*see* **Note 4**).
      - Place two sutures for subsequent ligation (will decrease spasm during cannulation if artery is isolated early in the procedure).
   (d) Celiac Artery/Cranial (Superior) Mesenteric Artery (SMA) Isolation (*see* **Note 4**).
      - Reflect mesoduodenum to work in the right paravertebral gutter; Deliver the bowel from the abdomen and incise right crus of diaphragm to improve exposure.
   (e) Isolate left renal vein-ligate/divide: this improves exposure to the aorta.
   (f) Dissect through lymphatics to expose aorta cranial to celiac and caudal to the SMA.
   (g) Isolate the hepatic artery via dissection in the lesser omentum: ligate/divide.
   (h) Isolate the SMA: ligate/divide.
   (i) Free up as much of the pancreas as possible without injury to vascular supply.
   (j) Leave mesoduodenal attachment intact at this stage.

2. Cannulate the splenic artery.
   (a) Heparin: 5000 U IV.
   (b) Cannulate splenic artery with large bore catheter, tying off the artery distally.
   (c) Secure the catheter to the artery.
   (d) Tie/ clamp aorta cranial to celiac and caudal to SMA.
   (e) Drain venous blood from pancreas.
   (f) Working in right paravertebral gutter, make portal venotomy, just caudal to hepatic branching.
   (g) Pour cold saline on pancreas, and infuse (300 mL chilled UW solution) via splenic artery.
   (h) Remove capsule from tail during dissection to allow for maximum expansion during infusion. The pancreas should become pale in color and cold.
3. Pancreatic harvest.
   (a) Resect pancreas, leaving only a small portion of the uncinate process.
   (b) Obtain incisional biopsy.
   (c) Weigh the gland.
   (d) Process only the splenic portion if the gland is more than 90 g.
   (e) Cannulate accessory duct (transverse incision across head [right lobe extremity] exposes cut surface of duct) with 20 G angiocath;
   (f) Slowly inject up to 2 mL of collagenase solution (1–1.5 mg collagenase/mL Hank's solution or UWS) per gram tissue. (Temperature of solution should be 24 °C.)
   (g) Clamp areas of leakage during infusion.

   If transport is needed before isolation, solution is U W S , 4 °C.
4. Isolation of pig islets [4].
   (a) Pancreas intraductally cannulated with 14 G catheter and 100 cc of digestion medium (1.5 mg/mL collagenase P, 10,000 U DNASE I, and 1 mM Trolox, injected in slowly.
   (b) Pancreas placed on ice, (approx. 30 min cold ischemic time).
   (c) Pancreas placed in 37° water bath for 20 min.
   (d) Pancreas manually shook for 1 min.
   (e) Pancreas filtered over two layered gauze into blue top centrifuge tubes.

(f) Rinsed with approximately 100 cc cold RPMI 1640.

(g) Centrifuged at speed 3.5 at 4 °C for 3 min.

(h) Supernatant discarded and pellet resuspended in 100 mL RPMI 1640.

(i) Centrifugation/Rinse repeated for a total of three centrifugations.

(j) Resuspend in 50–100 mL of RPMI 1640 (with 10–20% Porcine Serum added).

(k) If doing COUNTS, at this point, after making sure cells are well suspended, measure 25 μL aliquots into 16 different wells, using good sterile technique.

(l) If doing COBE counts, then centrifuge at speed 3.5 at 4 °C for 3 min; discard supernatant and then process pellet using COBE Procedure.

(m) At Completion of COBE processing on machine, repeat rinse procedure for each tube, do counts after third centrifugation.

(n) Place pellet from each tube into different wells containing porcine culture medium.

(o) Overnight 24 h culture (or modify as experiment requires i.e.… encapsulation).

5. COBE 2991 Procedure for Islet Purification.

   (a) Euroficoll Gradient Components.

   - Make up this solution no more than 48 h prior to use.
     - Stock Euroficoll.

       Take 2 L beaker with stir bar.

       Weigh out 500 g Ficoll (Molecular Grade-Sigma F2637).

       Add 8.94 g HEPES (Sigma, Catalog # H-1016).

       Add 1 L Eurocollins Solution.

       Stir overnight

       Next day

       pH to 7.4 with 10 N NaOH.

       Autoclave 15 min at 100 °C.

       Cool, store at 4 °C Density measurement(usually 1.140).

       The following components will be enough for 2 COBE runs

       1.108:

       550 mL Stock + 175 mL Eurocollins solution.

1.096:

320 mL Stock + 175 mL Eurocollins solution.

1.069:

150 mL Stock + 150 mL Eurocollins solution.

1.037:

50 mL Stock + 250 mL Eurocollins solution.

Check density of each gradient ±0.002 (Adjust accordingly)

(prepare, autoclave, store at 4 °C).

6. Set-up COBE tubing.
   (a) Load tubing set bag into place.
   (b) Press tube load, run tubing through valves (except V3 and V1).
   (c) Place main tubing from diaphragm to the rear of steel weighted lid.
   (d) Using hemostats clamp all tubings as proximal as possible, except for V3.
   (e) Attach feed line to V3.
   (f) Set: Speed = 2200 RPM; Super Out rate = 100; Super out vol = max.
   (g) Label series of 50 mL centrifuge tubes from 1 to 20 (more if needed).
7. Tissue loading.
   (a) Resuspend tissue pellet (≤25 g pcv) in 50 mL 1.108 Euroficoll solution.
   (b) Add resuspended pellet to remaining 225 mL of 1.108 Euroficoll solution.
   (c) Pump into COBE at 80 mL/min (or allow to run in by gravity), COBE is not spinning.
   (d) When all solution is in, turn off pump, and release line (unclamp pump head).
   (e) Purge COBE of Air.
   (f) Press start/spin button, allow COBE time to come up to speed.
   (g) Ensure that super out valve is active (light is green-toggle switch collect/SOV).
   (h) Press superout valve.

   Superout rate = 100, allow solution to be pushed up the tube past the rotary seal to end of "feed" line.

   - Clamp line, press stop/reset button.
   - Adjust superout rate = 0.

(i) Establish Discontinuous Gradient.

- Press start/spin button and begin pumping Euroficoll solutions in at 90 mL/min (*see* **Note 5**)
  - Pump 125 mL of 1.096.
  - Pump 100 mL of 1.069.
  - Pump 100 mL of 1.037.
  - Pump 50 mL of Hanks until fluid reaches rotary seal.
- Unclamp pump head.
- Press Super out to relieve excess pressure.

(j) Separation.

- Spin 5 min.

(k) Collection.

- Set: Superout rate = 175; Superout Vol = 600.
- Collect 20–30 mL per centrifuge tube (Record volume in each tube).
- Add 20–30 of RPMI 1640/ 10–20% Porcine Serum/ 10–20% UWS added to each tube.
- Centrifuge.
- Carefully remove supernatant.
- Resuspend in 20-30 mL of RPMI 1640/10–20% Porcine Serum/10–20% UWSMIXTURE.-
- Repeat rinse procedure for each tube (as outlined in PROTOCOL FOR ISOLATION OF PIG ISLETS), do counts after third centrifugation.
- Place pellet from each tube into different wells containing porcine culture medium (*see* Fig. 1).

(l) Overnight 24 h culture (*see* Fig. 2).

(m) Microencapsulation of pig islets (*see* Fig. 3).

- Centrifuge the 15 cm centrifuge tube, containing the handpicked islets, suspended in RPMI solution, for 2 min at 4 °C.
- Carefully, pipette off the supernatant as much as possible, without disturbing the pellet.
- Resuspend the pellet in the residual RPMI solution.
- Add 1 mL of pure Alginate to the centrifuge tube (or modify volume as experiment or pellet size dictates).
- Suspend the islets by hand-vortexing.
- Prepare the microencapsulator.
  - Ensure that the needle, metal tip, syringe, and funnel (be sure filter is in funnel) are clean.

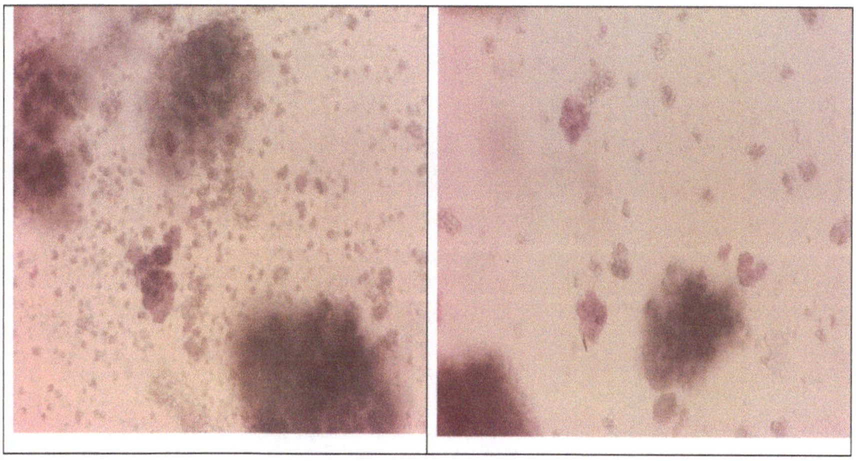

Fig. 1 Samples of Dithizone-stained islets post COBE purification

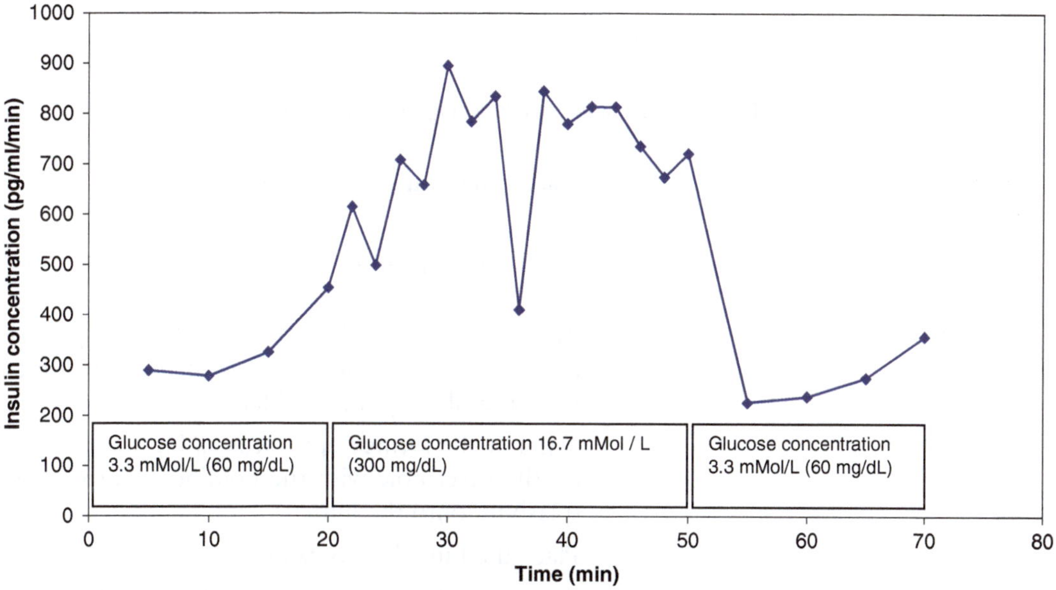

Fig. 2 A representative sample of perifusion runs performed during our various experiments while devising this protocol: Porcine islets were perifused at 37 °C for 1 h in Krebs-Ringer bicarbonate biuffer, pH 7.4, which contained basal glucose at 3.3 mMol/L (60 mg/dL)We then collected basal effluent samples on ice for 20 min. The glucose concentration in the perifusate was then increased to 16.7 mMol/L (300 mg/dL)and effluent perifusate was collected for 30 min. We then decreased the glucose level of the perifusate back to basal levels (3.3 mMol/L) and collected the perifusate for 20 min. The samples were frozen and then assessed by a Radioimmunoassay to assess insulin concentration

- Turn on the air compressor until it hisses.
- Attach needle to syringe.
- Place white hard plastic ring around syringe, with screw beds facing upward.

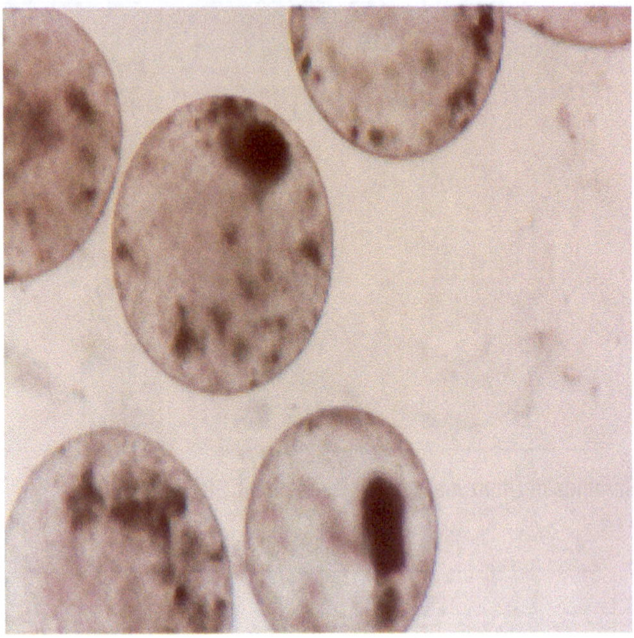

Fig. 3 Samples of encapsulated pig islets after Dithizone-staining

- Place black ring around syringe.
- Connect white ring to silver syringe holder (be sure it is appropriately connected to the silver holding arm).
- Adjust syringe so that tip of needle is approximately 1–2 mm from the end of the exit lumen from the silver syringe holder.
- Tighten the syringe apparatus as high as possible on the silver pole with the holding arm (to avoid splashing onto the lumen).
- Place the funnel onto its holder.
- Turn on and test gauges (Adjust with black knobs).
- Set gauges to appropriate settings
- After testing and adjusting gauges, leave gas on; turn off air jacket and alginate switches.
- Put 5 mL of cross-linking solution into the syringe (be sure gas is on).
- Adjust gas so that slow bubbling is occurring in cross-linking solution (No large bubbling).
- Re-vortex tube containing islets suspended in alginate.
- Using pipettman, place islet/alginate mixture into syringe.

  - Attach white connector on tubing to the syringe.
  - Turn on air jacket switch.
  - Turn on alginate switch.
- Monitor droplet formation.
- Add additional cross-linking solution as needed.
- Turn on air compressor as needed.
- After capsule formation is completed, empty contents of funnel into sterile capsule-washing-container (use sterile funnel); turn gas up slightly to assist with blowing remaining capsules out from bottom of funnel. Use additional cross-linking solution as needed to rinse capsules into the container.
- Turn off switches.
- Open exhaust switch on capsule-washing-container, hook air to input switch on capsule-washing-container.
- Incubate for 15 min, in capsule-washing-container, placed on ice on rotator.
- Close air input switch on capsule-washing-container, then close exhaust switch on capsule-washing-container.
- Let sit for 1 min on ice, then remove supernatant with disposable bulb pipet.
- Add Saline to tube to 30 mL mark (If transplant planned use culture medium), gently invert capsule-washing-container; Let tube sit in ice for 1 min; Remove supernatant carefully with disposable bulb pipet (don't disturb capsules).
- Add Saline to tube to 30 mL mark, gently invert capsule-washing-container; Let tube sit in ice for 1 min; Remove supernatant carefully with disposable bulb pipet (don't disturb capsules).
- Coat the alginate bead with 0.05 % polylysine in saline for 6 min.
- Fill container (containing capsules) with polylysine to 20 mL mark.
- Gently invert.
- Open exhaust switch on capsule-washing-container, hook air to input switch on capsule-washing-container.
- Incubate for 4 min, in capsule-washing-container, placed on ice on rotator.
- Close air input switch on capsule-washing-container, then close exhaust switch on capsule-washing-container.

- Let sit for 2 min on ice, then remove supernatant with disposable bulb pipet.
- Carefully remove supernatant with a disposable bulb pipet.
- Fill tube to 30 mL mark with saline.
- Gently invert.
- Sit upright in ice bucket for 2 min.
- Carefully remove supernatant with a disposable bulb pipet.
- Apply outer alginate layer by suspension of capsules in 0.06% (wt/vol) sodium alginate in saline for 4 min.
- Fill tube to 20 mL with 0.06% sodium alginate.
- Open exhaust switch on capsule-washing-container, hook air to input switch on capsule-washing-container.
- Incubate for 2 min, in capsule-washing-container, placed on ice on rotator.
- Close air input switch on capsule-washing-container, then close exhaust switch on capsule-washing-container.
- Let sit for 2 min on ice, then remove supernatant with disposable bulb pipet.
- Carefully remove supernatant with a disposable bulb pipet.
- Fill tube to 30 mL mark with saline.
- Gently invert.
- Sit upright in ice bucket for 2 min.
- Carefully remove supernatant with a disposable bulb pipet.
- Liquefaction of the inner calcium alginate core by suspension in 50 mM sodium citrate for 7 min.
  - Fill tube to 20 mL mark with sodium citrate.
  - Open exhaust switch on capsule-washing-container, hook air to input switch on capsule-washing-container.
  - Incubate for 5 min, in capsule-washing-container, placed on ice on rotator.
  - Close air input switch on capsule-washing-container, then close exhaust switch on capsule-washing-container.
  - Let sit for 2 min on ice, then remove supernatant with disposable bulb pipet.
  - Carefully remove supernatant with a disposable bulb pipet.

  - Fill tube to 30 mL mark with saline.
  - Gently invert.
  - Sit upright in ice bucket for 2 min.
  - Carefully remove supernatant with a disposable bulb pipet.
  - Fill tube to 30 mL mark with saline.
  - Gently invert.
  - Sit upright in ice bucket for 2 min.
  - Carefully remove supernatant with a disposable bulb pipet (Two washings Total with saline).
- Treat capsule with treatment solution as dictated by experiment being conducted (four subgroups: (1) PLL 6 min coating; (2) PLL 20 min coating; (3) PLO 6 min coating; (4) PLO 20 min coating.)
- Add treatment solution to tube to 20 mL mark (If using Sodium Sulfate incubation for planned Transplantation, add 5 mM Glucose to the Sodium Sulfate Solution).
- Gently invert.
- Open exhaust switch on capsule-washing-container, hook air to input switch on capsule-washing-container.
- Incubate for 30 min (or modify as dictated by experiment), in capsule-washing-container, placed on ice on rotator.
- Close air input switch on capsule-washing-container, then close exhaust switch on capsule-washing-container.
- Let sit for 2 min on ice, then remove supernatant with disposable bulb pipet.
- Add saline to tube to 30 mL mark, gently invert container; Sit upright in ice bucket for 2 min
- Remove supernatant carefully with disposable bulb pipet (don't disturb capsules). Repeat this entire step once (Two washings with saline).

3 Results from Our Studies

During the course of 12 different preparations we obtained the islet shown in Table 1 (reprinted with kind permission from Ref. [4])

A sample of isolated pig islets is shown in Fig. 1, In perifusion experiments with these islets shown in Fig. 2, we found that glucose stimulation of freshly isolated islets, had biphasic insulin secretion

Table 1
Islet yield at different stages of processing

Variable		Measurement
Pancreas weight (grams)		53.6 ± 3.7
Islet yield (yield per gram of tissue)		
	Before automatic cell separator use	3115 ± 325
	Purified	2398 ± 243
Purified islet clusters, μm		644 ± 85

with increase from a mean basal rate of 1415 + 61 to a peak of 3423 + 56 pg per six islets per minute ($n = 8$) ($p < .01$) with prompt return to basal rate with reduction of glucose to basal levels. (reprinted with kind permission from Ref. [4]). In Fig. 3, we have a representative picture of porcine islets encapsulated using the air-syringe pump microbead generator, as we previously described [5].

4 Notes

1. Make up this solution monthly.
2. Place blue filter on the tip of the syringe. Place needle on the tip of the blue filter.
3. Add GSH at time of use. Make up this solution no more than 48 h prior to use.
4. Blood supply to the pancreas in the pig:
   - Celiac artery → Splenic artery → Pancreatic branch → head of spleen, anastomosis with the cranial pancreaticoduodenal artery.
   - Hepatic artery → gastroduodenal artery (distal to hepatic branches → cranial pancreaticoduodenal artery, entering gland at the cranial duodenal flexure; anastomoses with pancreatic branch of splenic artery and caudal pancreaticoduodenal artery.
   - Cranial (superior) mesenteric artery: Caudal pancreaticoduodenal artery branches off the SMA distal to the ileocolic branch, proximal to the jejunal arteries. Enters gland at the distal extremity of the right limb. To further improve isolation of the pancreatic arterial supply from the remainder of the abdominal viscera, consider ligating the gastroduodenal, celiac, SMA, and caudal pancreaticoduodenal arteries. This approach would rely upon the intraglandular

arterial anastomoses for perfusion to the head and right limb from the tail.

5. Actually found it simpler to just allow all of the different solutions to run in by gravity, through a syringe/tubing apparatus prior to turning the machine on.

References

1. Johanesson B, Sui L, Freytes DO, Creusot RJ, Egli D (2015) Toward beta cell replacement for diabetes. EMBO J 34:841–855
2. http://www.who.int/mediacentre/factsheets/fs312/en/
3. Reichart B, Niemann H, Chavakis T, Denner J, Jaeckel E, Ludwig B, Marckmann G, Schnieke A, Schwinzer R, Seissler J, Tonjes RR, Klymiuk N, Wolf E, Bornstein SR (2015) Xenotransplantation of porcine islets as a potential option for the treatment of Type 1 diabetes in the future. Horm Metab Res 47:31–35
4. Ching DC, Harland RC, Collins BH, Kendall W, Hobbs H, Opara EC (2001) A reliable method for isolation of viable porcine islet cells. Arch Surg 136:276–279
5. Darrabie MD, Kendall WF, Opara EC (2005) Characteristics of poly-L-ornithine-coated alginate microcapsules. Biomaterials 26/34: 6846–6852

Chapter 14

Selective Osmotic Shock (SOS)-Based Islet Isolation for Microencapsulation

Kevin Enck, John Patrick McQuilling, Giuseppe Orlando, Riccardo Tamburrini, Sittadjody Sivanandane, and Emmanuel C. Opara

Abstract

Islet transplantation (IT) has recently been shown to be a promising alternative to pancreas transplantation for reversing diabetes. IT requires the isolation of the islets from the pancreas, and these islets can be used to fabricate a bio-artificial pancreas. Enzymatic digestion is the current gold standard procedure for islet isolation but has lingering concerns. One such concern is that it has been shown to damage the islets due to nonselective tissue digestion. This chapter provides a detailed description of a nonenzymatic method that we are exploring in our lab as an alternative to current enzymatic digestion procedures for islet isolation from human and nonhuman pancreatic tissues. This method is based on selective destruction and protection of specific cell types and has been shown to leave the extracellular matrix (ECM) of islets intact, which may thus enhance islet viability and functionality. We also show that these SOS-isolated islets can be microencapsulated for transplantation.

Key words Islets, Nonenzymatic, Digestion, Microencapsulation, Alginate, Diabetes, Transplantation

1 Introduction

Type 1 Diabetes Mellitus (T1DM) is a condition of absolute deficiency of the beta cell component of the pancreas. Currently, beta cell replacement is considered the gold standard for the treatment of T1DM [1]. The two forms of beta cell replacement procedures currently in clinical use are whole pancreas and islet transplantations. Islet transplantation requires isolation of the islets from the pancreas. Currently, this process involves enzymatic digestion of the pancreas which has significant harmful effects to the islets and the extracellular matrix (ECM) surrounding them. Although the architecture and physical interactions within and surrounding islets are complex and not completely understood, there is significant evidence that islets, like other tissues, are also heavily influenced by cell–ECM interactions [2]. In mature, intact islets, interactions with ECM or synthetic matrix materials have been shown to

Emmanuel C. Opara (ed.), *Cell Microencapsulation: Methods and Protocols*, Methods in Molecular Biology, vol. 1479,
DOI 10.1007/978-1-4939-6364-5_14, © Springer Science+Business Media New York 2017

regulate cell survival, proliferation, and insulin secretion, and aid in the preservation and restoration of spherical islet morphology [2, 3]. Similarly, ECM-based materials have been shown to regulate survival, proliferation, and insulin secretion in purified β-cell cultures, while incompletely isolated islets that retain some of their native ECM have been shown to have markedly reduced apoptosis rates and significantly greater function with respect to insulin response in vitro than completely isolated and purified islets. This nonenzymatic method, first described by Atwater et al., uses rapid osmosis to isolate islet cells from the rest of the pancreatic cells [4]. This is possible because the beta cells of islets contain glucose transporters (Glut 2). These transporters allow islet cells to quickly adapt to different levels of glucose, process it, and produce insulin in response to the prevailing glucose concentration [5]. When islets are exposed to high glucose concentrations, islet cells take up the glucose via Glut-2 transporters in the membrane in order to equilibrate the osmotic strength of the external solution, whereas acinar cells rapidly lose water in order to equilibrate internal and external solutions, and this causes the cells to shrink. Acinar cells take up potassium ions to recover cell volume and to adapt to the extracellular osmotic strength. When the glucose solution is removed, the islet cells lose the extra glucose through Glut-2 transporters. However, acinar cells rapidly take up water to dilute the hypertonic intracellular solution, and this causes the cells to swell. The acinar cells swell so fast that they burst [4]. Therefore, we are able to isolate the beta cells from the rest of the pancreatic tissue using this nonenzymatic digestion method.

2 Materials

2.1 Equipment

1. Four sterile disposable 600 mL Beakers [Fisher Scientific; Cat. No. 01-291-6].
2. One Sterile 400 mL glass Beaker [Pyrex; Cat. No. 1000].
3. One Sterile basic Surgical Kit [World Precision Instruments; Cat. No. 503407].
4. One sterile Instrument Tray [Polarware; Ref. # E2306].
5. 2×½ in. Full-Size Heavy Weight Steam Table Pan [Polarware; Ref. # S20122].
6. One 500 μM Sterile Stainless Steel U.S. Standard Test Sieve (4 in. diameter, 2 in. depth) [Cole-Parmer].
7. One Sterile Stainless Steel Bowl 6 5/8 in. × 2½ in. [Polarware; Ref. # 12A].
8. One 150× 25 mm Suspension Culture Dish (Non-Tissue Culture Treated) [Corning; Cat. No. 430597].
9. One 60 mL Syringe [BD; Ref. No. 309653].

10. One 60 mL Piston/Irrigation Syringe with Protective Cap [Covidien Ref. # 68000].
11. One Package of 50 mL Conical Tubes [BD; Ref. # 352070].
12. Scale (Adventure Pro) [Ohaus; Cat. No. AV201c].
13. Densito 30PX Densitometer [Mettler-Toledo; Mfr. No. 51324450].

2.2 Reagents

1. 1 L (2× 500 mL Bottles) Sterile PBS [Hyclone, Cat. No. SH30028.02].
2. 1 M HEPES Buffer Solution [Cellgro; Cat. No. 25-060-C1].
3. Glucose [Sigma; Cat. No. G7021-1KG].
4. RPMI-1640 Medium [Sigma; Cat. No. R1383-1L].
5. 10% Betadine [Fischer; Cat. No. NC0304148].
6. 76 mL Optiprep™ [Sigma Cat. No. D1556].
7. 390 mL CMRL-1066 Medium [Fischer; Cat. No. 15-110-CV].
8. 100 mL Fetal Bovine Serum.
9. 5 mL Penicillin/Streptomycin.
10. 339 mL Waters IGL Cold Storage Solution [Hospital Pharmacy Man. No. Css/1000/US].

3 Methods

3.1 Preparation of Reagents

1. 300 mM RPMI (100–200 mL per pancreas).
   (a) Reconstitute RPMI in 990 mL of ultrapure H_2O.
   (b) Add 10 mL of 1 M HEPES Stock.
   (c) Add 54 g Glucose and allow to dissolve on stir plate.
   (d) Adjust pH to 7.4.
   (e) Filter Sterilize and store at 4 °C.
2. 0 mM RPMI (about 1 L per pancreas).
   (a) Reconstitute RPMI in 990 mL of ultrapure H2O.
   (b) Add 10 mL of 1 M HEPES Stock.
   (c) Adjust pH to 7.4.
   (d) Filter Sterilize and store at 4 °C.
3. CMRL Culture Media.
   (a) Start with 390 mL of CMRL-1066 Medium.
   (b) Add 5 mL of 1 M HEPES Stock.
   (c) Add 5 mL of Penn/Strep.

(d) Add 100 mL of FBS.
(e) Adjust pH to 7.4.
(f) Filter Sterilize and store at 4 °C.

4. 265 mL Optiprep (1.10 g/mL).
    (a) Calibrate densitometer according to user's manual.
    (b) Add 214 mL of IGL (1.045 g/mL) and 51 mL of Optiprep (1.32 g/mL) to a plastic beaker.
    (c) Mix on shaker for 30 min.
    (d) Take a 1.5 mL sample for density measurements.
    (e) Measure density of sample then adjust concentration by adding IGL or Optiprep until it is 1.10 g/mL.
5. 150 mL Optiprep (1.09 g/mL).
    (a) Calibrate densitometer according to user's manual.
    (b) Add 125 mL of IGL (1.045 g/mL) and 25 mL of Optiprep (1.32 g/mL) to a plastic beaker.
    (c) Mix on shaker for 30 min.
    (d) Take a 1.5 mL sample for density measurements.
    (e) Measure density of sample then adjust concentration by adding IGL or Optiprep until it is 1.09 g/mL.
    (f) Filter Sterilize and store at 4 °C.

3.2 Isolation of Islets

1. Harvest pancreas from animal/donor into container of cold PBS.
2. Remove as much viscera as possible from organs on small surgical plate (on ice or placed on ice packs) using blunt dissection with forceps and shears.
3. Weigh tissue in sterile stainless steel bowl.
4. Rinse with 500 mL of ice cold PBS.
5. Incubate in 200 mL of 10% ice cold betadine for 5 min in 500 mL beaker.
6. Rinse with 500 mL ice cold PBS X3 into 500 mL beakers.
7. In large surgical plates, finely mince tissue (at least <0.25 cm^2) and place in 50 mL conical tubes (*see* **Notes 1** and **2**).
8. Expose to 300–600 mM glucose RPMI medium for 20 min on ice.
    (a) Use a 1:1 ratio of tissue to high glucose RPMI.
    (b) Fill up to 50 mL mark on conical tubes.
9. Centrifuge for 2 min at 200 × *g* 4 °C.
10. Quickly decant supernatant and add 0 mM glucose RPMI.
11. Gently invert tubes to expose cells to low glucose media.
12. Centrifuge for 2 min at 200 × *g* 4 °C.

13. Repeat **steps 10–12** twice and decant 0 mM RPMI Solution.
14. Remove tissue from 50 mL conical tubes and place in a sterile 500 mL beaker (*see* **Note 3**).
15. Add 50 mL of 0 mM RPMI.
16. Using an irrigation syringe, gently disrupt tissue by suctioning up and down for approximately 5–10 min (*see* **Note 4**).
17. If possible maintain the 500 mL beaker on ice or frozen aluminum beads while dissociating.
18. Once the tissue has been dispersed enough to easily pass through the irrigation syringe switch to a 60 mL syringe with a leur lock and continue disrupting tissue for an additional 5–10 min. Maintain temperature at ~4 °C.
19. Rinse tissue through 500 μM strainer into a sterile 500 mL glass beaker.
20. Take the remaining tissue which did not pass through the 500 μM filter and place back in the 500 mL beaker with 50 mL of additional 0 mM RPMI.
21. Repeat dissociation for an additional 5 min.
22. Rinse tissue through 500 μM strainer into 500 mL glass beaker from **step 19**.
23. Repeat **steps 20–22**.
24. Take the filtered tissue located in the glass beaker and pipette into 50 mL conical tubes.
25. Centrifuge for 3 min at $100 \times g$ (accel/decel = 4, 4 °C) (*see* **Note 5**).
26. Aspirate supernatant and fill conical tubes with fresh 0 mM RPMI.
27. Repeat **steps 25** and **26** twice, on the last wash, do not fill with fresh RPMI (*see* **Note 6**).
28. Layer a 50 mL conical tube with 30 mL 1.10 g/mL Optiprep, 8 mL 1.09 g/mL Optiprep, then 10 mL 1.045 g/mL IGL (*see* **Note 7**).
29. Syringe tissue pellet gently on top of layered Optiprep.
30. Centrifuge for 3 min at 375 g (accel/decel = 1, 4 °C).
31. Purified islets will be located in the middle layer containing 1.09 g/mL Optiprep (*see* **Note 8**).
32. Carefully use a syringe with a large opening to collect purified islets (*see* Figs. 1 and 2) (*see* **Note 9**).
33. Resuspend purified tissue pellet in 25 mL of RPMI Culture Media and plate into one large non-tissue culture treated petri dish.
34. Place in 37 °C, 5 % CO_2 incubator and culture overnight.
35. Change media in the morning.

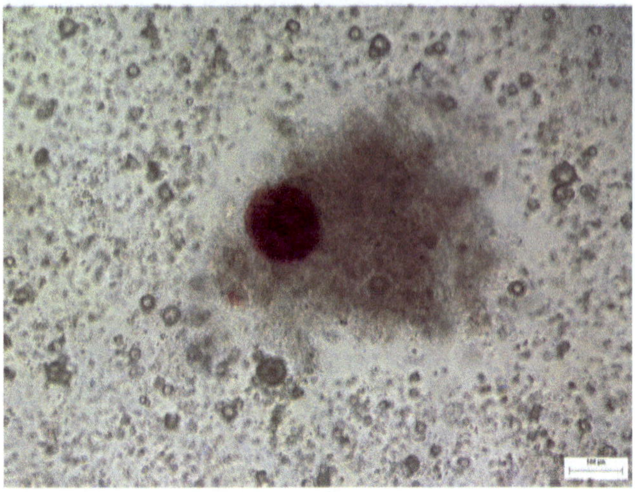

Fig. 1 Rat islet isolated using SOS protocol with 300 mM. Islet is stained *red* with dithizone to measure viability

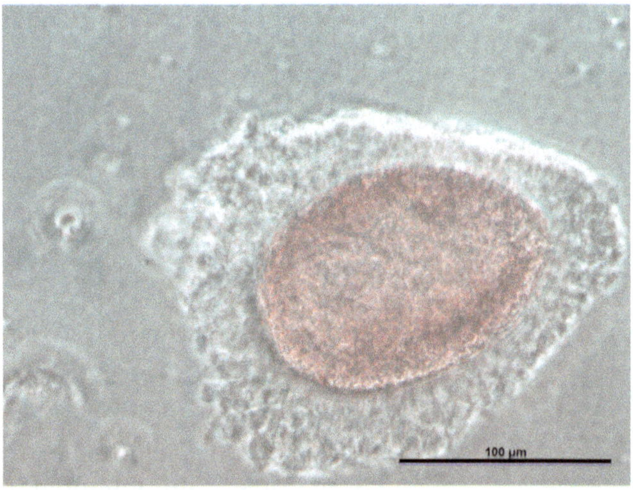

Fig. 2 Human islet isolated using SOS method with 300 mM glucose. The islet surrounded by pancreatic ECM is stained *red* using Dithizone to assess viability

3.3 Microencapsulation of Islets

Encapsulation of islets is performed using an alginate-derived method we developed in our lab [6].

1. The islets are suspended in the alginate solution then loaded into the droplet generator.
2. The droplets form while falling into a $CaCl_2$ bath for cross-linking for 15 min.
3. The encapsulated islets are collected and washed with normal saline (*see* Fig. 3) (*see* **Note 10**).

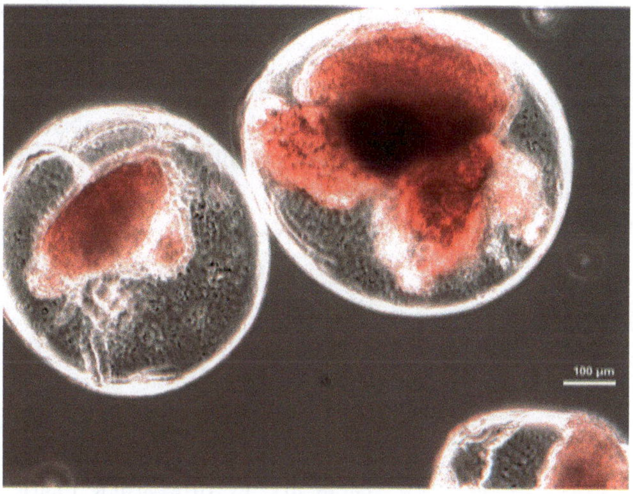

Fig 3 Rat islet isolated with SOS method. Islets are purified, encapsulated in alginate, then stained *red* with Dithizone to assess viability

4. Following washing, the microbeads are incubated in PLO for 10–20 min.
5. Repeat **step 38**.
6. The alginate core is liquefied by incubating in 55 mM Sodium Citrate for 2 min.
7. Repeat **step 38**.
8. The microbeads are incubated in alginate solution for 30–45 min followed by Calcium-supplemented saline washes to cross-link the outer layer.
9. Encapsulated islets can then be implanted for in vivo studies.

4 Notes

1. Taking extra time in this step will make mechanical dissociation with syringe much easier and quicker.
2. It is also possible to place minced tissue all in one 500 mL beaker with 300 mM glucose RPMI. If using 50 mL conical tubes, only place about 10–15 mL of tissue per tube.
3. If pancreas is small (rat or mouse), placing in a smaller container such as the stainless steel bowl used for weighing (2.1.7) may allow for quicker dissociation.
4. Slowly push out tissue. High pressures can occur if nozzle is filled with large pieces of tissue and can quickly spray tissue out of beaker if pushed too aggressively. Holding the syringe still and twisting the pump while pushing tissue out can control flow better. This is also true for 60 mL syringe.

5. Acceleration/Deceleration numbers may differ between centrifuge equipment. We used a machine with a scale from 1 to 9 with 1 being the slowest and 9 being the fastest.
6. Tissue pellet should be about 2 mL, if larger, adjust Optiprep volumes to allow for more tissue. Ratios of layers should remain the same.
7. The Optiprep density gradients are very small meaning that they will mix easily if not added slowly. Add Optiprep at a rate of <2 mL/min so the layers are well defined.
8. If a larger yield is required, collect from the top of the 1.045 gradient as well.
9. 25 mL or 50 mL syringes have larger openings which allow for islets to be collected without getting stuck on the tip. If many islets are getting stuck inside the syringe chamber, collect smaller volumes of islets at a time. Again, the density gradient is very sensitive, move slowly so the layers are not disturbed.
10. Do not centrifuge microbeads, as they are fragile. Allow microbeads to settle via gravity then aspirate.

Disclosure

Giuseppe Orlando is member of the BioCaPan Consortium which has received funding from the European Union's Horizon2020 research and innovation program under grant agreement No 646272. Kevin Enck and Riccardo Tamburrini are funded through BioCaPan.

References

1. Barton FB, Rickels MR, Alejandro R et al (2012) Improvement in outcomes of clinical islet transplantation: 1999–2010. Diabetes Care 35(7):1436–1445
2. Stendahl JC, Kaufman DB, Stupp SI et al (2009) Extracellular matrix in pancreatic islets: relevance to scaffold design and transplantation. Cell Transplant 18(1):1–12
3. Mirmalek-Sani SH, Orlando G, McQuilling JP et al (2013) Porcine pancreas extracellular matrix as a platform for endocrine pancreas bioengineering. Biomaterials 34(22):5488–5495
4. Atwater I, Guajardo M, Caviedes P et al (2010) Isolation of viable porcine islets by selective osmotic shock without enzymatic digestion. Transplant Proc 42(1):381–386
5. Kohnert KD, Wohlrab F, Hahn HJ, Cossel L (1997) Glucose transporter isoform (GLUT) 2 expression in beta-cells of long-term syngeneic islet grafts. Acta Diabetol 34(4):301–304
6. Opara EC, McQuilling JP, Farney AC (2013) Microencapsulation of pancreatic islets for use in a bioartificial pancreas. Methods Mol Biol 1001:261–266

Chapter 15

Preparation and Characterization of Alginate–Chitosan Microcapsule for Hepatocyte Culture

Lanjuan Li, Yanhong Zhang, and Xiaoping Pan

Abstract

This method is used to prepare microencapsulation of hepatocytes. The key steps in the protocol are (a) isolation of porcine hepatocytes, followed by (b) production of hepatocytes encapsulation, and (c) evaluation of encapsulations and hepatocytes. This method of collecting porcine hepatocytes was first used in 2003 (Maruyama et al. Cell Transplant 12:593–598, 2003). Cell encapsulation technology was first introduced by Chang (Science 146:524–525, 1964) and later by Lim and Sun in 1980 (Science 210:908–910, 1980). Since then, it has been well adopted in tissue engineering and controlled delivery of biological therapeutics such as bioartificial liver.

Key words Hepatocytes, Microcapsule, Alginate, Chitosan, Encapsulation

1 Introduction

Bioartificial liver is a novel method for the treatment of liver failure. Primary hepatocytes are the ideal cell source for bioartificial liver. However, primary hepatocytes are always unavailable because of the loss of function and their inability to significantly proliferate and expand in vitro. Microencapsulation is a type of nontoxic polymer membrane material in liver cells forming tens of microns to hundreds of microns diameter microcapsule technology [1, 2]. Microencapsulation serves several functions: immunoisolation; protection of hepatocytes from shear forces; and provision of an anchorage support for maintaining hepatocyte functionality. Simultaneously, the large surface area and possible coating with ultrathin semi-permeable membranes allow permanent substrates and products to diffuse rapidly. Thus, hepatocyte microencapsulation could be carried out. Hepatocyte microencapsulation could provide a new technology for large-scale preparation of liver cells, highly active liver cells in vitro, and long-term cryopreservation.

Up to now, the alginate-poly-L-lysine-alginate (APA) membranes have been used in hepatocyte microencapsulation. However,

Emmanuel C. Opara (ed.), *Cell Microencapsulation: Methods and Protocols*, Methods in Molecular Biology, vol. 1479,
DOI 10.1007/978-1-4939-6364-5_15, © Springer Science+Business Media New York 2017

APA capsules may cause fibrotic reactions and exhibit cytotoxic effects in some cases [3, 4]. Therefore, finding a biocompatible polymer enabling long-term entrapment and growth of hepatocytes without causing adverse host immune responses is still an ongoing area of research.

Alginate–chitosan (AC) microcapsules have been studied as an alternative to microcapsule membranes. Chitosan is used instead of poly-L-lysin (PLL) to form an external membrane around the alginate core and to entrap the cells. The use of chitosan instead of PLL was suggested to give higher mechanical strength and stability because of the stronger binding of chitosan to alginate gel [3]. Moreover, the use of chitosan in hepatocytes encapsulation has been reported previously [5, 6].

In this chapter, we describe in detail the preparation and characterization of alginate–chitosan microcapsule for hepatocyte culture. The main content in the protocol is followed: (a) isolation of primary porcine hepatocytes; (b) preparation and production of hepatocytes encapsulation; and (c) evaluation of microencapsulation and hepatocytes.

2 Materials

2.1 Buffers and Solutions

0.9% NaCl (pH 7.4): 9 g/L NaCl. Dissolve 9 g of NaCl in 800 mL of distilled H_2O. Adjust the pH to 7.4 and add H_2O to 1 L.

0.5% chitosan (pH 6.0): 12.764 g/L $CaCl_2$, 5 g/L chitosan, 9 g/L NaCl, 2.38 g/L Hepes. Sterilize the solution by passing it through a 0.22 μm filter; store the filtrate in aliquots at 4 °C (*see* **Note 1**).

115 mM calcium chloride (pH 7.4): 12.764 g/L $CaCl_2$, 9 g/L NaCl, 2.38 g/L Hepes. Sterilize them by autoclaving for 20 min at 15 psi (1.05 kg/cm^2) on liquid cycle.

2% sodium alginate (pH 7.4): 10 g/L sodium alginate, 9 g/L NaCl, 2.38 g/L Hepes. Sterilize the solution by passing it through a 0.22 μm filter; store the filtrate in aliquots at 4 °C (*see* **Note 2**).

Albumin (ALB) (2 mg/mL): Dissolve ALB in 2% sodium alginate solution.

Dulbecco's modified Eagle's medium complete (DMEM medium): DMEM, 100 units/mL penicillin, 100 μg/mL streptomycin, 2 mM glutamine, 10% bovine calf serum.

FITC-IgG: Dissolve FITC-IgG 1:10,000 in 2% sodium alginate solution just before test.

Krebs-Henseleit buffer: 2 g/L D-Glucose, 0.141 g/L $MgSO_4$, 0.16 g/L KH_2PO_4, 0.35 g/L KCl, 6.9 g/L NaCl. Sterilize the solution by passing it through a 0.22 μm filter; store the filtrate in aliquots at 4 °C.

Lysis buffer: 50 mM Tris (pH 7.4), 150 mM NaCl, 1% Triton X-100, 1% sodium deoxycholate, 0.1% SDS, 1 mM sodium orthovanadate, 1 mM EGTA, 1 μg/mL leupeptin.

Microbeads broken buffer (pH 7.4): 0.2 mM $NaHCO_3$, 0.06 mM $Na_3C_6H_5O_7.2H_2O$, 9 g/L NaCl. Sterilize the solution by passing it through a 0.22 μm filter; store the filtrate in aliquots at 4 °C.

MTT (5 mg/mL): Dissolve 500 mg MTT into 100 mL PBS and sterilize the solution by passing it through a 0.22 μm filter; store the filtrate in aliquots at −20 °C.

Perfusion buffer (PB): 9 g/L NaCl, 0.42 g/L KCl, 2.1 g/L $NaHCO_3$, 0.9 g/L glucose, 4.78 g/L Hepes, 0.37 g/L of EDTA. Adjust the pH to 7.4 and sterilize the solution by passing it through a 0.22 μm filter; store the filtrate into sterile containers at 4 °C.

Perfusion buffer with collagenase (PBC): 0.5 g/L collagenase, 0.55 g/L $CaCl_2.2H_2O$, 1000 mL PB. Adjust the pH to 7.4 and sterilize the solution by passing it through a 0.22 μm filter; store the filtrate into sterile containers at 4 °C.

Perfusion buffer with dispase (PBD): 8.4 g/L dispase, 1000 mL PB. Adjust the pH to 7.4 and sterilize the solution by passing it through a 0.22 μm filter; store the filtrate into sterile containers at 4 °C.

Perfusion buffer with EDTA (PBE): 0.37 g/L EDTA, 1000 mL PB. Adjust the pH to 7.4 and sterilize the solution by passing it through a 0.22 μm filter; store the filtrate into sterile containers at 4 °C.

Phosphate-buffered saline (PBS): 137 mM NaCl, 2.7 mM KCl, 10 mM Na_2HPO_4, 2 mM KH_2PO_4. Dissolve 8 g of NaCl, 0.2 g of KCl, 1.44 g of Na_2HPO_4, and 0.24 g of KH_2PO_4 in 800 mL of distilled H_2O. Adjust the pH to 7.4 and add H_2O to 1 L. Sterilize solution by autoclaving for 20 min at 15 (1.05 kg/cm^2) on liquid cycle or by filter sterilization. Store the buffer at room temperature.

Trypan blue dye: 4 g/L trypan blue. Dissolve 0.4 g of trypan blue in 80 mL of PBS, and add PBS to 100 mL. Sterilize the solution by passing it through a 0.22 μm filter; store the filtrate in aliquots at room temperature.

Washing buffer (WB): 7.0 g/L NaCl, 0.461 g/L KCl, 0.13 g/L $CaCl_2.2H_2O$, 2.38 g/L Hepes, 1.0 g/L BSA. Adjust the pH to 7.4 and sterilize the solution by passing it through a 0.22 μm filter; store the filtrate into sterile containers at 4 °C.

Washing buffer with DNase I (WBD): 0.1 g/L $MgCl_2.6H_2O$, 0.1 g/L $MgSO_4.7H_2O$, 0.1 g/L DNase I, 1000 mL WB. Adjust the pH to 7.4 and sterilize the solution by passing it through a 0.22 μm filter; store the filtrate into sterile containers at 4 °C.

Cells and Tissues: Liver lobes, cultured hepatocytes.

2.2 Kits

BCA Assay kit.

Human Albumin ELISA Quantitation Set.

P450-Glo CYP1A2 assay and CYP3A4 assay.

Reverse transcription kit.

Rneasy plus mini kit.

Urea Assay Kit.

2.3 Special Equipment

Electrostatic microencapsulator unit.

Forma orbital shaker.

Fluorescent microscopy.

Microplate reader.

3 Methods

3.1 Isolation of Porcine Hepatocytes

Hepatocytes are isolated from Chinese experimental miniature pigs using a modified four-step collagenase perfusion method [7, 8], as illustrated in Fig. 1.

(a) Excised liver lobes are perfused at 37 °C with 300 mL and then at 39 °C with 500 mL of PB.

(b) The liver lobes are perfused at 39 °C with 500 mL of PBS and 500 mL of PBD.

(c) The liver lobes are perfused with 500 mL of PBC. The capsule of the liver lobes is then opened and digested. Parenchyma is collected on ice after filtration through 80-mesh filter.

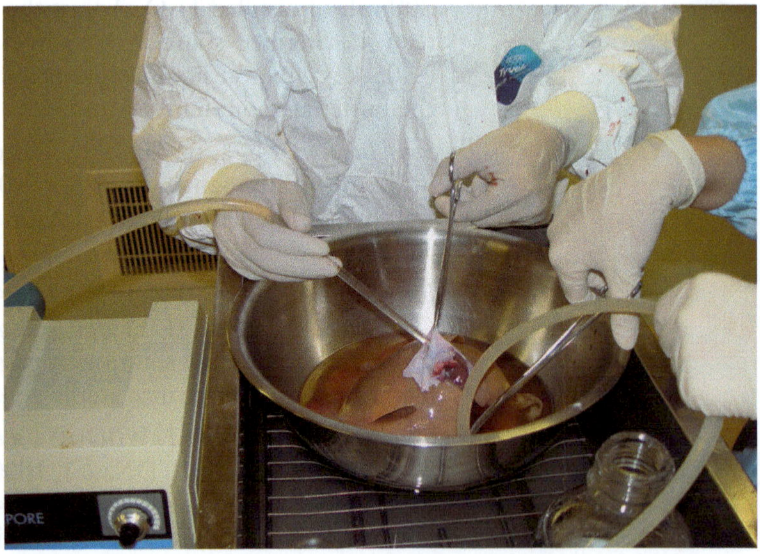

Fig. 1 Isolation of primary porcine hepatocytes

(d) The cells are washed three times with ice-cold WB through centrifugation at 4 °C and 50×*g* for 2 min, followed by washing two times with WBD through centrifugation at 4 °C and 50×*g* for 75 s. The cells are resuspended in DMEM medium and stored in ice-cool space.

(e) The yield varied from 1.2 to 1.5×10^7 hepatocytes per gram of wet liver weight, with viability ranging from 90 to 99%, as determined by trypan blue exclusion test. The viabilities of the hepatocytes ranged from 90 to 99%, as determined by trypan blue exclusion.

3.2 Production of Hepatocytes Encapsulation

A single-stage procedure is adopted for alginate–chitosan (AC) encapsulation [4].

(a) Hepatocytes are counted and suspended in 2% sodium alginate solution to a final ratio of approximately 3×10^6 per 1 mL of alginate.

(b) The mixture is sprayed through an electrostatic microencapsulator unit (Fig. 2; 300 mm nozzle was used in this experiment) (*see* **Note 3**). The vibration frequency of the nozzle is kept at 0.5 kHz, and the amplitude of frequency at 100% into a 0.5% chitosan solution where they are allowed to gel for 30 min (*see* **Note 4**).

Fig. 2 Preparation of alginate–chitosan microcapsule for primary porcine hepatocytes

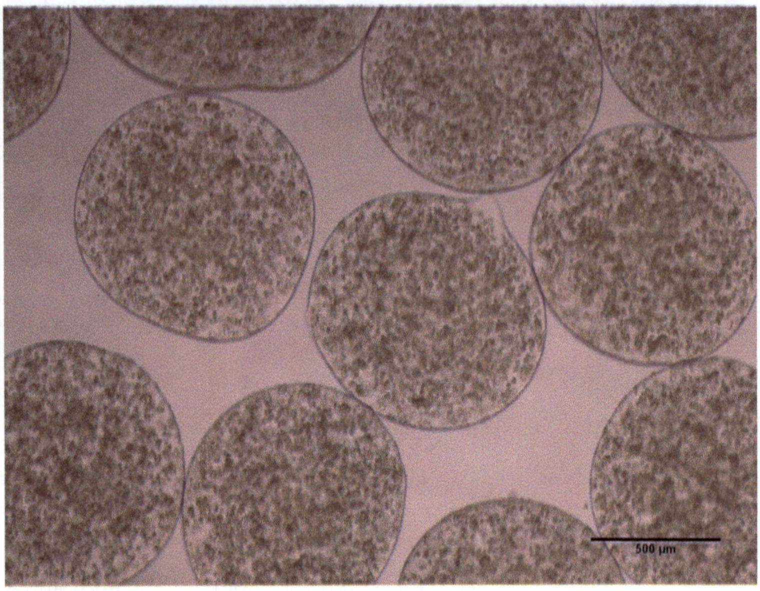

Fig. 3 Hepatocytes microencapsulation with alginate–chitosan

(c) Alginate–chitosan microcapsules with a diameter of 800–1000 μm are produced and then washed three times with 0.9 % NaCl solution.

(d) Microencapsulated hepatocytes are cultured as 10 % suspension in DMEM medium containing 10 % FBS. The medium is refreshed every other day, and the growth morphology of the encapsulated hepatocyte is observed under microscopy Fig. 3.

3.3 Evaluation of Encapsulations and Hepatocytes

3.3.1 Microcapsule Mechanical Stability

The mechanical stability of the microcapsules is determined in a shaking test.

(a) About 100 microcapsules are added to a 15 mL centrifuge tube.

(b) The microcapsules are equilibrated in 5 mL of 0.9 % NaCl solution, placed on a Forma orbital shaker, and shaken at 180 rpm at 37 °C for 24 h.

(c) The percentage of intact microcapsules is assessed by count under an optical microscope and is recorded at various time points.

3.3.2 Microcapsule Solute Permeability

(a) Two mg/mL human serum albumin (ALB) or fluorescein isothiocyanate–immunoglobulin G (FITC-IgG; 1:10,000) is suspended in a 2 % sodium alginate solution that is used to prepare the microcapsules with an outer shell containing chitosan by means of the usual procedure.

(b) During preparation, the microcapsules are incubated at 37 °C for 30 min. After incubation, the microcapsules are transferred to a solution containing an equal concentration of ALB or FITC-IgG and equilibrated for 2 h.

(c) The microcapsules are then washed with 0.9% NaCl solution, and the ALB release profile is obtained at a 3–30 min interval within 2 h.

(d) ALB concentration is determined by using the BCA Assay kit. The percentage of released ALB in the culture medium per total amount of ALB encapsulated is plotted over time.

(e) The microcapsules containing FITC-IgG (FITC, 390 Da; IgG, 150,000 Da) are observed by fluorescent microscopy over time.

3.3.3 MTT Assay for Cell Viability [9]

(a) Microcapsules are placed in a 96-well plate with 200 μL DMEM medium per well.

(b) Twenty microliter MTT solution (5 mg/mL) is added into each well and incubated at 37 °C in a humidified atmosphere (95% air and 5% CO2) for 4 h.

(c) The medium is removed and replaced with 150 μL dimethyl sulfoxide to solubilize the MTT tetrazolium dye. The plates are swirled until the dye appears fully dissolved.

(d) About 200 μL of the solution is collected to read the absorbance (A) at 570 nm by using a plate reader.

3.3.4 Albumin Assay and Urea Production Assay

(a) The release of albumin synthesized by the encapsulated hepatocytes into DMEM medium at different time points during the culture period is measured by Human Albumin ELISA Quantitation Set and following the conditions recommended by the supplier.

(b) Urea concentration of encapsulated hepatocytes in microcapsules is assayed using Urea Assay Kit and following the conditions recommended by the supplier.

(c) All the results are analyzed by CurveExpert 1.3 software and fitted with logistic model with $r^2 > 0.99$.

3.3.5 Cytochrome P450 1A2 and 3A4 Metabolic Activity Assay

Cytochrome P450 metabolic activities are assessed by measuring luciferase activity using the P450-Glo CYP1A2 assay and CYP3A4 assay.

(a) The encapsulated hepatocytes are incubated at 37 °C for 1 h in fresh DMEM medium containing Luciferin-IPA or Krebs–Henseleit buffer containing Luciferin-1A2.

(b) Approximately 50 μL solution from each well is taken to a 96-well opaque white plate and mixed with 50 μL luciferin detection reagent to initiate a luminescent reaction.

(c) After 20 min at room temperature, luminescence is measured using a plate reader.

3.3.6 Real-Time PCR

(a) Microcapsules are dissolved with two bed volumes of microbead broken solution.

(b) The cells are collected by centrifugation at $50 \times g$ for 5 min and then washed with PBS three times.

(c) RNAs are extracted via Rneasy plus mini kit.

(d) The cDNAs are synthesized using reverse transcription kit.

(e) Real-time quantitative PCR analyses are performed using a commercial packaging kit and following the conditions recommended by the supplier.

3.3.7 Western Blot

(a) Hepatocytes are suspended in lysis buffer supplemented with phosphatase inhibitors and protease inhibitors and stored on ice for 30 min.

(b) The suspension is centrifuged at 14,000 × g for 10 min at 4 °C.

(c) The protein concentrations are measured by BCA assay.

(d) The key proteins, such as protein kinases, in the lysates are determined by Western blot.

4 Notes

1. Normal chitosan is only dissolved in water containing acetic acid, so water-soluble chitosan should be chosen.
2. Sodium alginate with low viscosity could be better carrier for cells. Do not attempt to sterilize sodium alginate solution by autoclaving as it affects the structure and reduces solution viscosity.
3. Microcapsule size can be controlled by adjusting the nozzle aperture, nozzle flux, and viscosity of alginate.
4. Microcapsules can easily adhere in chitosan solution. Therefore, magnetic stirring is required in the preparation process.

References

1. Chang TM (1964) Semipermeable microcapsules. Science 146:524–525
2. Lim F, Sun AM (1980) Microencapsulated islets as bioartificial endocrine pancreas. Science 210:908–910
3. Gaserod O, Sannes A, Skjak-Braek G (1999) Microcapsules of alginate-chitosan. II. A study of capsule stability and permeability. Biomaterials 20:773–783
4. Rokstad AM, Holtan S, Strand B et al (2002) Microencapsulation of cells producing therapeutic proteins: optimizing cell growth and secretion. Cell Transplant 11:313–324
5. Yu CB, Lv GL, Pan XP et al (2009) In vitro large-scale cultivation and evaluation of microencapsulated immortalized human hepatocytes (HepLL) in roller bottles. Int J Artif Organs 32(5):272–281
6. Lv G, Zhao L, Zhang A et al (2011) Bioartificial liver system based on choanoid fluidized bed bioreactor improve the survival time of fulminant hepatic failure pigs. Biotechnol Bioeng 108(9):2229–2236
7. Maruyama M, Totsugawa T, Kunieda T et al (2003) Hepatocyte isolation and transplantation in the pig. Cell Transplant 12:593–598
8. Li LJ, Du WB, Zhang YM et al (2006) Evaluation of a bioartificial liver based on a nonwoven fabric bioreactor with porcine hepatocytes in pigs. J Hepatol 44(2):317–324
9. Yang Y, Li J, Pan X et al (2013) Co-culture with mesenchymal stem cells enhances metabolic functions of liver cells in bioartificial liver system. Biotechnol Bioeng 110(3):958–968

Chapter 16

Use of Flow Focusing Technique for Microencapsulation of Myoblasts

J. Ciriza, L. Saenz del Burgo, R.M. Hernández, G. Orive, and J.L. Pedraz

Abstract

Alginate cell microencapsulation implies the immobilization of cells within a polymeric membrane that allows the bidirectional diffusion of nutrients and oxygen inside the microcapsules and the release of waste and therapeutic molecules outside them. This technology has been applied to several cell types and it has been extensively described with pancreatic islets. However, other cells such as myoblasts are being currently studied and showing high interest. Moreover, different systems and approaches have been developed for cell encapsulation such as electrostatic extrusion and Flow focusing technology. When Flow focusing technology is applied for myoblast encapsulation, several factors should be considered, such as the pressure, the flow of the system, or the diameter size of the nebulizer, which will determine the final diameter size and shape of the microcapsules containing the myoblasts. Finally, viability of encapsulated myoblasts needs to be assessed before further studies are performed.

Key words Flow focusing, Alginate, Cell microencapsulation, Myoblasts, Erythropoietin, Drug delivery, Nanomedicine

1 Introduction

Cell microencapsulation is extensively used in research as a delivery system for therapeutic drugs. The structure of the capsule enables the flow of nutrients and gases (O_2 and CO_2) between its outer and inner side, as well as the waste outlet and release of therapeutic molecules produced *de novo* by encapsulated cells. The aim is to create a suitable microenvironment that promotes and controls cell viability and proliferation. In vivo, microcapsules protect the cells from the immune system while allowing the release of therapeutic molecules by the entrapped cells [1].

Several biomaterials have been used for microencapsulation including agarose [2], chitosan [3], and hyaluronic acid [4], but alginate is by far the most frequently employed biomaterial due to its excellent biocompatibility and suitable mechanical properties [5, 6].

J. Ciriza and L. Saenz del Burgo contributed equally with all other contributors.

Emmanuel C. Opara (ed.), *Cell Microencapsulation: Methods and Protocols*, Methods in Molecular Biology, vol. 1479, DOI 10.1007/978-1-4939-6364-5_16, © Springer Science+Business Media New York 2017

The encapsulation of pancreatic islets in alginate matrices is the field in which cell microencapsulation more progress has made, with promising results starting in the early 1990s [5, 7, 8]. But alginate cell encapsulation has also been applied to other diseases and other cell types. Our group has previously studied the microencapsulation of genetically engineered C_2C_{12} myoblasts to secrete erythropoietin (EPO) as an ex vivo gene therapy approach without implementation of immunosuppressive protocols. In vitro, microcapsules containing genetically engineered C_2C_{12} myoblasts are able to release EPO into the culture medium at rates up to 15,490 ± 600 IU EPO/24 h. After implantation in the peritoneum and subcutaneous tissue of syngeneic and allogeneic mice, encapsulated C_2C_{12} myoblasts secreting EPO are able to maintain high and constant hematocrit levels, showing that encapsulation allows long-term expression after allogeneic transplantation, up to 210 days, as well as in xenogeneic transplantation, up to 98 days [9–11]. The results showed by our research group highlighted the feasibility of cell encapsulation technology for the long-term delivery of EPO independent of the method of administration and the mouse strain. Following our previous experience, we will describe in this chapter the microencapsulation of these myoblasts genetically modified to secrete EPO.

Currently there are two main techniques for cell microencapsulation: Electrostatic extrusion and Flow focusing. Electrostatic extrusion allows the immobilization of cells and it has been described for various cell types such as brewing yeast cells, insect cells, and plant cells [12]. Electrostatic droplet generation is based on the use of electrostatic forces to disrupt a liquid filament at the capillary–needle tip and form a charged stream of small droplets. This is reproducible, controllable, with uniform bead sizes, and easy to use under sterile conditions. During the production process it is not required organic solvents that can inhibit cell activity and cause serious damaging effects and although a high electric field is involved, the applied potential does not damage cells without loss in cell activity or viability.

Another technology for cell microencapsulation is Flow Focusing. The aim of this technology is the production of drops or bubbles by straightforward hydrodynamic means, with an output of a dispersed liquid or gas, frequently in the form of a fine aerosol or an emulsion. The basic principle consists of a continuous phase fluid (focusing or sheath fluid) to give rise to droplet or bubble break-off in the vicinity of an orifice through which the fluid is extruded. No other driving force is required, apart from traditional pumping. A flow focusing device consists of a pressure chamber pressurized. Inside, one fluid is injected through a capillary feed tube whose extremity opens up in front of a small orifice linking the pressure chamber with the exterior ambient. Capillary instability breaks up the steady jet into homogeneous droplets or bubbles. Flow focusing ensures an extremely fast as well as controlled production of up to millions of droplets per second as the jet breaks up. Currently, our group is working with the Cellena® Flow Focusing technology that will be described in this book chapter. Cellena® is a bioencapsulation portable platform able to encapsulate high molecular weight compounds, microorganisms, and cells in homogeneous particles of predictable and controllable size. Its performance is based on Flow Focusing® technology. By means of this technology, the elements to be encapsulated are immobilized within a semipermeable membrane that protects them from mechanical stress and the host immune system, allowing the bidirectional diffusion of nutrients. Here we will describe the procedure to encapsulate myoblasts genetically modified to secrete EPO within 200 μm diameter alginate microcapsules by means of Cellena® Flow Focusing technology and their subsequent viability evaluation.

2 Materials

For cell encapsulation all solutions should be sterile prepared by means of filtration through a 0.22 μm filter, while sterilization is not necessary for microcapsules performance without cells. However, under our criteria we always prepare microcapsules with sterile solutions regardless of microcapsules will contain cells within or not. All solutions for cell encapsulation and further covering should be prepared using 1 % mannitol.

2.1 Myoblast Culture

1. Myoblast culture solution: Dulbecco's modified Eagle's medium supplemented with 10% inactivated fetal bovine serum, 2 mM L-glutamine, and 1 % antibiotic/antimycotic solution.
2. Trypsin solution: we usually work with trypsin-EDTA 0.25 %.

2.2 Cell Encapsulation

1. 1 % Mannitol: Dissolve 1 g of mannitol into 100 ml of distillated water or scale up to the required volume. Stir until mannitol gets dissolved and filter through a Millipore 0.22 μm filter Stericup.
2. 55 mM calcium chloride: For 400 ml solution, dissolve 2.44 g of calcium chloride (≥96 % purity) into 400 ml of 1 % mannitol solution. Scale up the amounts of calcium chloride to higher volumes. Stir until calcium chloride is dissolved and filter through a Millipore 0.22 μm filter Stericup.
3. 1.5 % alginate LVG (NovaMatrix): Ultra pure low-viscosity and high glucuronic acid alginate (LVG) should be purchased since it gives good outcomes on cell encapsulation. For a 1.5 % alginate solution, dissolve 1.5 g alginate into 100 ml of 1 % mannitol and stir (*see* **Note 1**). For higher volumes of alginate solution, scale up as required. Lower or higher alginate concentrations will require modifying the weight of alginate dissolved into 1 % mannitol. When alginate solution is transparent, filter through a 0.20 μm syringe filter. We recommend to filter

2 ml each time since higher alginate volumes require too strength to get filtered due to its high viscosity.

2.3 Microcapsules Coatings

1. 0.05% poly-L-lysine solution (PLL solution): Dissolve 0.05 g of poly-L-lysine (molecular weight: 15,000–30,000 Da) into 100 ml of distillated water or scale up to the required volume. Stir until poly-L-lysine gets dissolved and filter through a Millipore 0.22 μm filter Stericup.
2. 0.1% alginate LVG: Dissolve 0.1 g alginate LVG into 100 ml of 1% mannitol. For higher volumes of alginate solution, scale up as required. When alginate solution is transparent, filter through a Millipore 0.22 μm filter Stericup.

2.4 Viability Assessment of Encapsulated Cells

1. Calcein/ethidium staining solution for microscopy: Dilute 0.25 μl of ethidium and 0.125 μl of calcein provided in the LIVE/DEAD® Viability/Cytotoxicity Kit for mammalian cells (Life Technologies) into 1 ml of DPBS (*see* **Note 2**).
2. Solution for viability quantification: all reagents are provided in the Cell Counting Kit-8 (Sigma-Aldrich) (*see* **Note 3**).

3 Methods

All procedures are carried out at room temperature under aseptic conditions in a flow laminar cabinet unless otherwise specified.

3.1 Myoblast Culture

1. Myoblasts are stored at liquid nitrogen and they should be thawed quickly at 37 °C. To remove DMSO used as cryoprotectant, cell suspension should be added quickly to a tenfold volume culture medium.
2. Spin cell suspension and resuspend it into 20 ml of culture medium. Place the solution onto a T-175 flask. Incubate at 37 °C in humidified 5% CO_2/95% air atmosphere.
3. Cells should be passaged every 2–3 days by detaching them with trypsin-EDTA 0.25% and plating them with a 1:3 dilution.

3.2 Cell Suspension in Alginate Preparation

1. Remove culture supernatant from T-175 flask and rinse with 20 ml of PBS. Add 10 ml of trypsin-EDTA 0.25% to the T-175 flask (Fig. 1a) and incubate at 37 °C for 5 min. Next, inactivate trypsin by adding 10 ml of cell medium and collect the cell suspension into a 50 ml conical tube.
2. Filter the cell suspension through a 40 μm filter in order to remove cell aggregations and count the cell density under a microscope or with an automatic counter.
3. Collect the volume of cell suspension in order to get the number of cells required for cell encapsulation. We usually encapsulate at a density of 5×10^6 cell/ml of alginate, although this amount should

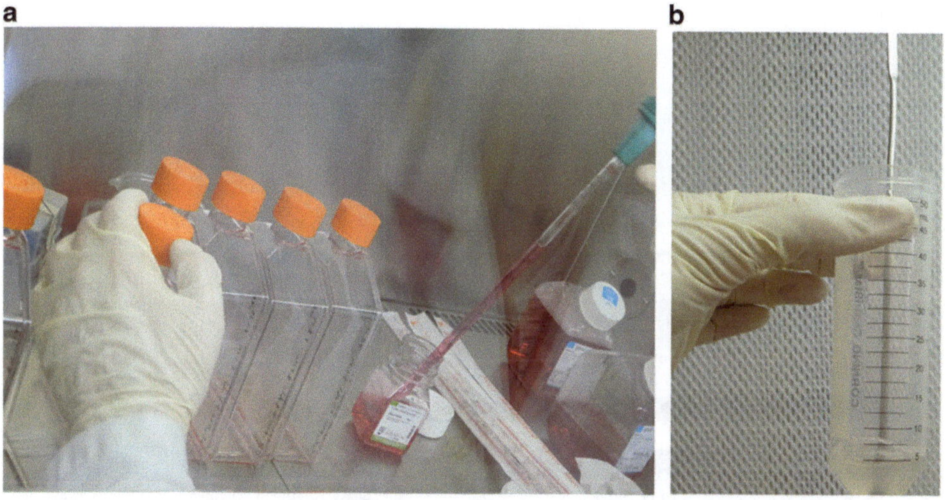

Fig. 1 Cell suspension in alginate preparation: (**a**) Image of myoblast trypsinization. (**b**) Myoblast suspension in alginate

be modified if different cell suspension is desired such as 10^7 cells/ml alginate or 2.5×10^6 cells/ml of alginate (*see* **Note 4**).

4. Spin the cell suspension collected for encapsulation and carefully discard supernatant.
5. Add the volume of 1.5% alginate LVG required to get the desired cell/alginate concentration, for example 5×10^6 cells/ml of alginate. Carefully resuspend cells into alginate solution with the help of a sterile spatula (*see* **Note 5**) (Fig. 1b).

3.3 Cell Encapsulation

1. Transfer the alginate cell suspension into a syringe avoiding the formation of bubbles. Place the syringe into the peristaltic pump included in Cellena® (Fig. 2a).
2. Connect the syringe to the disposable nebulizer (*see* **Note 6**) through the provided connector (*see* **Note 7**) and place it at a height of 4.5 cm (Fig. 2b).
3. Connect Cellena® portable platform to a vacuum pump and set up the conditions of pressure and flow. For 200 μm diameter alginate microcapsules we set up the conditions at 53 mbar and 4 ml/h for pressure and flow respectively (*see* **Note 8**).
4. Push the syringe containing alginate until the alginate suspension reaches the tip of the nebulizer. With a 24-well plate containing calcium chloride solution below the nebulizer, turn on the flow to asses that alginate microcapsules show the appropriate shape and diameter. Alginate will start to leave the nebulizer as droplets and fall down into the calcium chloride solution. Observe under an inverted microscope the 24-well plate containing the alginate microcapsules (*see* **Note 9**).

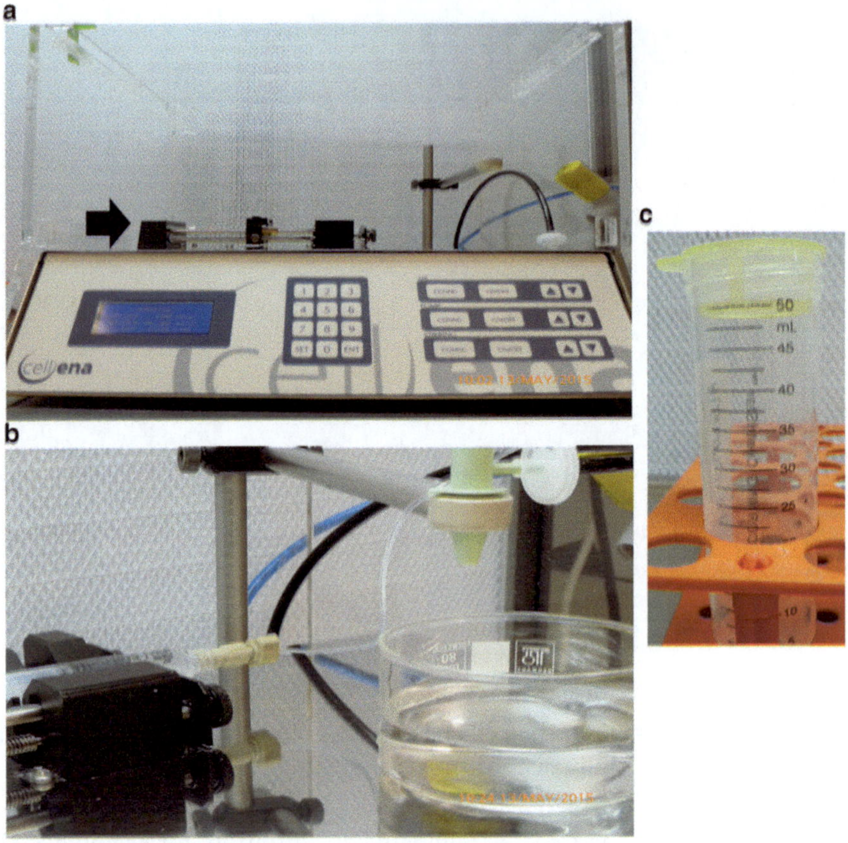

Fig. 2 Cell encapsulation: (**a**) Cellena® Flow Focusing portable platform. *Black arrow* indicates the peristaltic pump. (**b**) Set up of the cell encapsulation system. (**c**) Filtering the solution through a 100 μm strainer

5. Fill a 80 mm height sterile glass containing magnet stirrer with 80 ml of a 55 mM calcium chloride solution and place it under the disposable nebulizer. Turn on agitation at 300 rpm and let the alginate solution push into the glass through the nebulizer until syringe is empty (*see* **Note 10**) (Fig. 2).
6. Once the syringe is empty, stop the flow of the system and stir the alginate microcapsules containing myoblasts in the calcium chloride solution for at least 7 min.

3.4 Micro-encapsulated Cells Coating

1. Collect microcapsules from calcium chloride solution by filtering the solution through a 100 μm strainer (Fig. 2c).
2. Invert the 100 μm strainer and collect alginate microcapsules into a conical tube with 10 ml of 0.05 % PLL solution.
3. Place the conical tube with alginate microcapsules into 0.05 % PLL solution on a roller and incubate for 5 min at room temperature.
4. Collect microcapsules from 0.05 % PLL solution by filtering the solution through a 100 μm strainer.

5. Rinse microcapsules in the 100 μm strainer with 10 ml of 1% mannitol solution.
6. Invert the 100 μm strainer and collect alginate microcapsules into a conical tube with 10 ml of 0.1% alginate LVG solution.
7. Place the conical tube with alginate microcapsules into 0.1% alginate LVG solution on a roller and incubate for 5 min at room temperature.
8. Collect microcapsules from 0.1% alginate LVG solution by filtering the solution through a 100 μm strainer.
9. Rinse microcapsules in the 100 μm strainer with 10 ml of 1% mannitol solution.
10. Invert the 100 μm strainer and collect alginate microcapsules into a conical tube with 10 ml of myoblast culture solution. Collect alginate microcapsules in myoblast culture solution with a pipette and place the solution into a T-75 flask. Observe the microcapsules under an inverted microscope and keep the flask in the incubator for further studies (Fig. 3a).

3.5 Viability Assessment of Encapsulated Cells Under Fluorescent Microscope

1. Collect suspension with encapsulated cells from the T-75 flask and place it into a tube. Let it sediment by gravity and remove medium. Repeat this process until a volume of 20 μl of encapsulated cells is reached.
2. Rinse there times with DPBS containing Ca^{2+} and Mg^{2+}.
3. Dilute rinsed microcapsules with 300 μl of Calcein/ethidium staining solution for microcopy.
4. Dispense this staining solution containing the microcapsules into three wells of a 96-well plate (100 μl solution/well).
5. Incubate at room temperature for 45 min protected from light.
6. Assess under a fluorescent microscope the viability of encapsulated cells. Calcein and ethidium can be viewed simultaneously with a conventional fluorescein longpass filter. The fluorescence from these dyes may also be observed separately; calcein can be viewed with a standard fluorescein bandpass filter and ethidium can be viewed with filters for propidium iodide or Texas Red® dye (*see* **Note 11**) (*see* Fig. 3b).

3.6 Viability Quantification of Encapsulated Cells

1. Collect suspension with encapsulated cells from the T-75 flask and place it into a tube. Let it sediment by gravity and remove medium. Repeat this process until a volume of 25 μl of encapsulated cells is reached.
2. Dilute the encapsulated cells (25 μl) into 500 μl of myoblast culture solution.
3. Dispense this solution containing the microcapsules into five wells of a 96-well plate (100 μl solution/well). Also dispense 100 μl of myoblast culture solution into two wells (negative control).

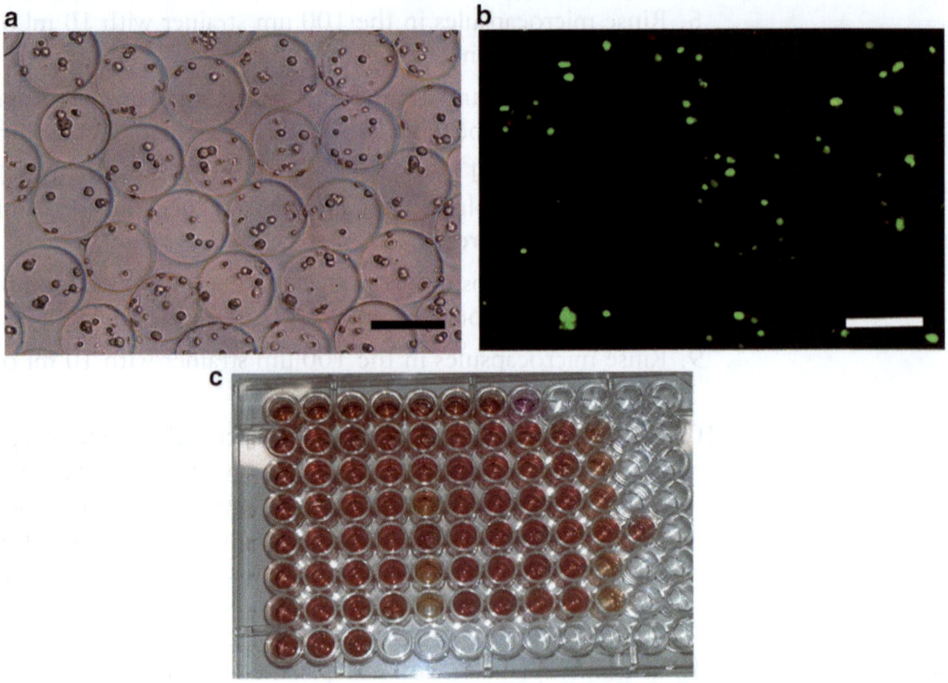

Fig. 3 Viability assessment of encapsulated cells: (**a**) Myoblasts encapsulated in 200 μm diameter size alginate capsules under inverted microscope. (**b**) Viability assessment of encapsulated cells under fluorescent microscope: *Green color* indicates calcein staining (live cells) while *red color* indicates ethidium staining (dead cells). (**c**) Viability quantification of encapsulated cells: 96-well plate with a CCK8 quantification assay. Scale bar: 200 μm

4. Add 10 μl of solution for viability quantification (CCK-8) into each well.
5. Incubate the 96-well plate inside a wet chamber on an incubator for 4 h at 37 °C.
6. Measure absorbance at 450 nm with reference wavelength at 650 nm using a microplate reader (*see* **Note 12**) (Fig. 3c).

4 Notes

1. Alginate needs several hours stirring until it gets dissolved. Sometimes alginate gets stuck at the bottom of the flask and requires to be taken off using a spatula. Other times alginate gets agglomerated on the top of the solutions and again the agglomerate requires to be broken using a spatula.
2. Calcein/ethidium staining solution for microcopy should be freshly prepared and protected from light.
3. For the viability quantification with Cell Counting Kit-8 a wet chamber should be prepared.

4. At this step it is important to avoid the collection of foam from the cell suspension.
5. This is a critical step for the success of the encapsulation. Cell suspension in alginate should be completely homogeneous without presence of bubbles. Try to remove any bubble in the new solution with a micropipette. The presence of bubbles and the presence of cell aggregation imply the failure of cell encapsulation.
6. For 200 μm diameter alginate microcapsules, we use the nebulizer 350 μm but for other diameter alginate microcapsules the recommended nebulizer should be used:

Nebulizer (μm)	Diameter alginate microcapsules (μm)
200	95–120
350	160–190
500	260–350
700	360–460

7. We usually clean the nebulizer and the connector pushing through 1 % mannitol by means of a syringe.
8. With the described conditions we obtain 160 μm diameter alginate microcapsules that will slightly increase in diameter after coating.
9. If alginate microcapsules containing myoblasts do not show a spherical shape wait until more alginate gets out from the nebulizer as the flow might need a few minutes before it stabilizes. Asses again with a 24-well plate until alginate microcapsules show a spherical shape with the expected diameter. If after several tests alginate microcapsules do not show the appropriate spherical shape and diameter, modify pressure conditions up or down until the desired shape and diameter is obtained.
10. The correct performance of alginate nebulization can be tested by moving a glass slide under the nebulizer and observe if a perfect alginate droplets line is deposited on the glass slide. If the nebulizer gets plugged, all the system should be cleaned with 1 % mannitol by pushing the solution with a syringe.
11. We usually observe samples under a Nikon TMS confocal microscope with the excitation/emission settings for calcein AM (excitation 495 nm, emission 515 nm) and ethidium homodimer (excitation 495 nm, emission 635 nm). Next, we analyze images with Eclipse Net software, version 1.20.0.
12. To measure the absorbance later, add 10 μl of 1 % w/v SDS to each well, cover the plate, and store it with protection from light at room temperature. No absorbance change should be observed for 48 h.

References

1. Orive G, Santos E, Pedraz JL, Hernandez RM (2013) Application of cell encapsulation for controlled delivery of biological therapeutics. Adv Drug Deliv Rev 67–68:3–14
2. Dang SM, Gerecht-Nir S, Chen J, Itskovitz-Eldor J, Zandstra PW (2004) Controlled, scalable embryonic stem cell differentiation culture. Stem Cells 22(3):275–282
3. Li Y, Du Q, Liu T, Sun J, Wang Y, Wu S, Wang Z, Xia Y, Xia L (2013) Methylene blue adsorption on graphene oxide/calcium alginate composites. Carbohydr Polym 95(1):501–507
4. Gerecht S, Burdick JA, Ferreira LS, Townsend SA, Langer R, Vunjak-Novakovic G (2007) Hyaluronic acid hydrogel for controlled self-renewal and differentiation of human embryonic stem cells. Proc Natl Acad Sci U S A 104(27):11298–11303
5. de Vos P, Faas MM, Strand B, Calafiore R (2006) Alginate-based microcapsules for immunoisolation of pancreatic islets. Biomaterials 27(32):5603–5617
6. Siti-Ismail N, Bishop AE, Polak JM, Mantalaris A (2008) The benefit of human embryonic stem cell encapsulation for prolonged feeder-free maintenance. Biomaterials 29(29):3946–3952
7. Lim F, Sun AM (1980) Microencapsulated islets as bioartificial endocrine pancreas. Science 210(4472):908–910
8. Soon-Shiong P, Heintz RE, Merideth N, Yao QX, Yao Z, Zheng T, Murphy M, Moloney MK, Schmehl M, Harris M et al (1994) Insulin independence in a type 1 diabetic patient after encapsulated islet transplantation. Lancet 343(8903):950–951
9. Murua A, Orive G, Hernandez RM, Pedraz JL (2009) Xenogeneic transplantation of erythropoietin-secreting cells immobilized in microcapsules using transient immunosuppression. J Control Release 137(3):174–178
10. Orive G, De Castro M, Ponce S, Hernandez RM, Gascon AR, Bosch M, Alberch J, Pedraz JL (2005) Long-term expression of erythropoietin from myoblasts immobilized in biocompatible and neovascularized microcapsules. Mol Ther 12(2):283–289
11. Ponce S, Orive G, Gascon AR, Hernandez RM, Pedraz JL (2005) Microcapsules prepared with different biomaterials to immobilize GDNF secreting 3T3 fibroblasts. Int J Pharm 293(1–2):1–10
12. Manojlovic V, Djonlagic J, Obradovic B, Nedovic V, Bugarski B (2006) Immobilization of cells by electrostatic droplet generation: a model system for potential application in medicine. Int J Nanomedicine 1(2):163–171

Chapter 17

Alginate Microbeads for Cell and Protein Delivery

Sami I. Somo, Omaditya Khanna, and Eric M. Brey

Abstract

Alginate hydrogels have been used for a broad variety of medical applications. The ability to assemble alginate gels at neutral pH and mild temperatures makes alginate a promising choice for the encapsulation and delivery of cells and proteins. This chapter covers the basics of cell encapsulation and protein delivery using two different variations of alginate microbeads, single layered and multilayer systems. The first section describes a method for encapsulating cells within alginate microbeads coated with a permselective polymer layer. The second section describes a multilayer alginate microbead system that allows simultaneous encapsulation of cells and delivery of growth factors. The primary goal of the systems described is for encapsulation of islets as a treatment for type I diabetes. However, these microbeads can be used for a broad variety of applications in tissue engineering, cell encapsulation, and regenerative medicine.

Key words Alginate, Encapsulation, Growth factor delivery, Multilayered alginate beads

1 Introduction

1.1 Encapsulation of Cells Within the Alginate Beads

The main goal of cell encapsulation is to provide an environment that maintains cell viability and function while regulating nutrient exchange with host tissue. Protection of the encapsulated cells from factors contributing to rejection may reduce the need for immunosuppressive drugs [1]. Encapsulating cells within a hydrogel network coated by a permselective polymer layer provides a suitable environment that permits the diffusion of oxygen and nutrients essential for cell survival, while preventing the host immune system from recognizing or destroying the encapsulated cells. Microencapsulation of cells within alginate-based hydrogels coated with a permselective layer has been investigated clinically as a treatment for type 1 diabetes [2–5].

Alginate is a naturally occurring polysaccharide that is capable of undergoing ionic crosslinking in the presence of divalent cations. Cells can be immobilized within the alginate when ionic bridges are formed between neighboring mannuronic and guluronic groups trapping the cells within the crosslinked polymer mesh [6]. Prior to

Emmanuel C. Opara (ed.), *Cell Microencapsulation: Methods and Protocols*, Methods in Molecular Biology, vol. 1479,
DOI 10.1007/978-1-4939-6364-5_17, © Springer Science+Business Media New York 2017

gelation, the cells are suspended within dissolved alginate, and upon exposure to the cation, a hydrogel network forms around the cells. Unlike many other polymerization procedures, this crosslinking can occur at pH, temperature, and salt conditions with little effect on cell viability or protein activity. When combined with the relatively straightforward techniques for crosslinking, alginate has become a popular material for cell-based therapies [7, 8].

Coating the alginate core with a permselective layer polymer may reduce or possibly completely eliminate the requirements of immunosuppressive drugs when delivering allogeneic or xenogeneic cells. Alginate microcapsules are typically coated with a polycation, such as poly-L-ornithine (PLO) or poly-L-lysine (PLL). The positive cationic chains interact with the negative alginate core to form a polycation-polyanion complex that coats the entire alginate surface. This layer influences the rate of diffusion of solutes and is typically designed to limit the immune system from recognizing the cells contained within the core.

PLO and PLL play an important role in transport but their positive charge results in a robust inflammatory response following implantation. This may resolve over time but even under the best conditions the final result is a thick layer of encapsulation tissue that serves as a substantial transport barrier. This barrier hinders delivery of essential nutrients to the cells and ultimately contributes to failure of the cells. In clinical and preclinical studies, the surface is coated with an outer monolayer of alginate to prevent direct contact of cells from the host with the polycation, effectively reducing the host inflammatory response.

1.2 Multilayered Alginate Beads

The typical alginate coating is a thin monolayer. However, alginate capsules have also been generated with a thicker layer of alginate [9]. An advantage of these multilayered alginate beads lies in their ability to allow simultaneous encapsulation of cells within the alginate core and delivery of other factors/cells from the outer layer of alginate. Growth factors can be encapsulated and delivered from the outer layer with release kinetics determined by the composition of the outer layer [9]. Delivery of fibroblast growth factor-1 (FGF-1) from multilayered alginate beads has been shown to stimulate blood vessel formation around encapsulated cells [10]. In order for islet transplantation to succeed, the local tissue response, include vascularization and modulation of the foreign body response are essential. These multilayer structures allow for delivery of factors directly around the encapsulated cells in order to provide both therapeutic cells and local control of tissue development.

2 Materials

2.1 Encapsulation of Cells Within Alginate Beads

1. Ultrapure Alginate (Pronova LVM or LVG, Novamatrix).
2. Alginate solution: 25 mM HEPES, 118 mM NaCl, 5.6 mM KCl, 2.5 mM $MgCl_2$.

3. Crosslinking solution: 100 mM $CaCl_2$, 10 mM HEPES.
4. 0.9 % w/v saline solution.
5. Min6 cells (*see* **Note 1**).
6. Min6 culture medium: DMEM medium supplemented with 15 % FBS, 1 % Antibiotic/Antimycotic, 0.05 mM 2-mercaptoethanol (*see* **Note 2**).

2.2 Encapsulation of Proteins in Multilayered Alginate Beads

1. Ultrapure Alginate (Pronova LVM or LVG, Novamatrix).
2. Poly-L-Ornithine (PLO) (MW 15,000–30,000).
3. Alginate solution: 25 mM HEPES, 118 mM NaCl, 5.6 mM KCl, 2.5 mM $MgCl_2$.
4. Crosslinking solution: 100 mM $CaCl_2$, 10 mM HEPES.
5. 0.9 % w/v saline solution.
6. Wash solution: 0.2 % $CaCl_2$ in 0.9 % saline solution.

3 Methods

3.1 Encapsulation of Cells Within Alginate Beads (Fig. 1)

1. Dissolve alginate 1.5 % (w/v) in alginate solution (*see* **Note 3**).
2. Cells cultured should be washed twice with 10 mL 1× DPBS (*see* **Note 4**).
3. Trypsinize cells with 2 mL Trypsin per T75 flask (*see* **Note 5**).

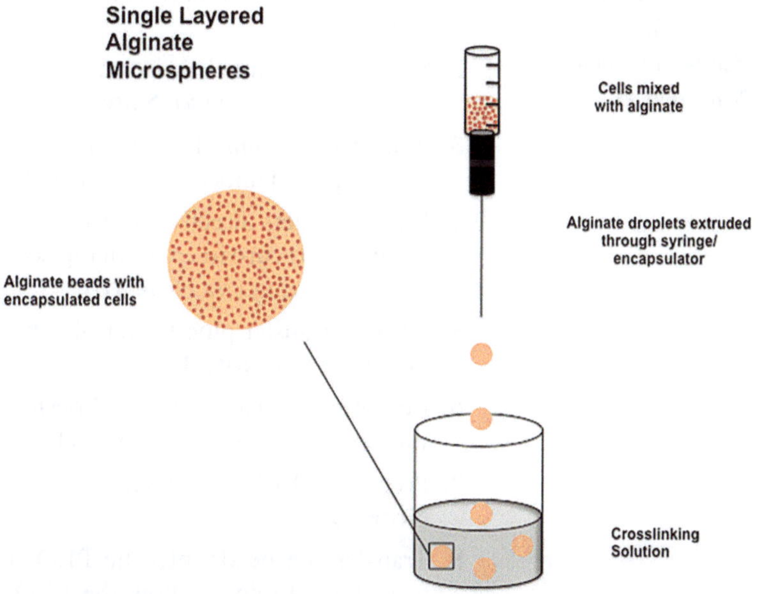

Fig. 1 Single layered alginate microspheres with cells suspended homogenously throughout the bead. Cell-alginate solution is extruded through a syringe into a $CaCl_2$ bath to allow crosslinking of alginate

4. Resuspend cells in 8 mL of Min6 medium to have a total volume of 10 mL.
5. Count cells using a hemocytometer.
6. Aliquot two million cells into individual Eppendorf tubes (*see* **Note 6**).
7. Centrifuge the cells for 5 min at 150 × *g*.
8. Aspirate the supernatant from the centrifuged Eppendorf tubes using pipette (*see* **Note 7**).
9. Resuspend the cell pellet with the dissolved alginate solution to have a cell density of two million cells/1 mL of dissolved alginate solution (*see* **Note 8**).
10. Collect the cell-alginate solution using syringe and extrude through 25-gauge needle into 15 mL crosslinking solution, while maintaining constant stirring (*see* **Note 9**).
11. Beads should be left to crosslinking solution for 15 min.
12. Collect cell-loaded spheres using transfer pipette and place in a petri dish (*see* **Note 10**).
13. Wash cell-loaded spheres with saline solution two times (*see* **Note 11**).
14. Using a spatula, place the cell-loaded beads in a six-well plate containing 4 mL Min6 medium and maintain at 37 °C in a humidified 5 % CO_2–95 % air atmosphere (*see* **Note 12**).

3.2 Encapsulation of Proteins in Multilayered Alginate Beads (Fig. 2)

1. Dissolve LVM alginate (1.5 % w/v) in alginate solution (*see* **Note 13**).
2. Set up two-channel air jacket alginate microencapsulator using a 25-gauge needle (*see* **Note 14**).
3. A flask containing 10 mL of crosslinking solution should be directly placed underneath the needle with stir bar (*see* **Note 15**).
4. Inject droplets directly into the crosslinking solution and allow the beads to cure in crosslinking solution for 15 min with continuous stirring (*see* **Note 16**).
5. Using a transfer pipette transfer the beads to a 15 mL centrifuge tube (*see* **Note 17**).
6. Remove residual volume and perform three washes with wash solution, each lasting 2 min each (*see* **Note 18**).
7. Prepare a 0.1 % (w/v) solution of PLO in normal saline (*see* **Note 19**).
8. Transfer the beads into the PLO solution and place on vortexer for 30 min to allow the PLO adequate time to bind to alginate (*see* **Note 20**).
9. Remove the PLO solution and wash with wash solution three times for 2 min each.

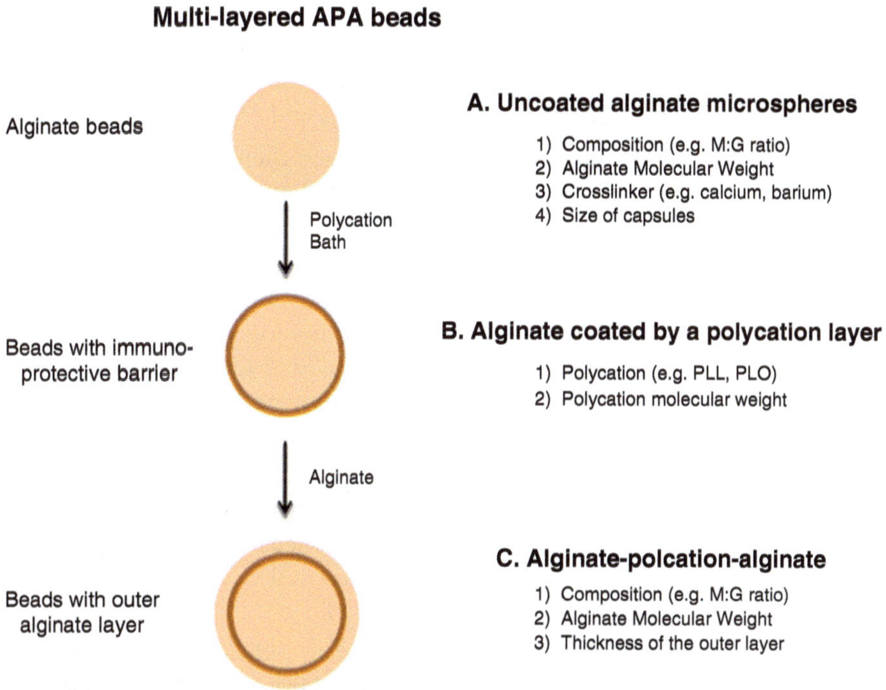

Fig. 2 Multilayered APA bead consists of a permselective membrane around the alginate core followed by an outer layer alginate. The outer layer alginate allows for encapsulation of proteins and growth factors while also maintaining biocompatibility

10. Prepare an alginate solution of desired concentration that is used to create the outer layer (*see* **Note 21**).
11. Transfer the alginate beads to a cell strainer to allow solution to be removed (*see* **Note 22**).
12. Transfer the microbeads to a parafilm surface (*see* **Note 23**).
13. Transfer the alginate solution created from **step 10** onto the alginate microbeads and incubate for 45 min to allow enough time for alginate to interact with PLO layer (*see* **Note 24**).
14. Remove excess alginate solution not bound to the microbeads using a pipette.
15. Transfer the beads into the crosslinking solution to allow formation of outer layer.
16. Perform three washes with wash solution for 2 min each (*see* **Note 25**).

4 Notes

Encapsulation of cells within alginate

1. Subheading 3 describes encapsulation of Min6 cells cultured in T75 flasks. This approach can be modified depending on cell type and culture conditions.

2. All materials should be sterilized prior to use or if bought sterile should be kept in sterile conditions. For crosslinking solution, alginate solution, and wash solution they can be prepared under nonsterile conditions and filtered using 0.22 μm filter.
3. If alginate powder is sterilized then it can be dissolved in sterile alginate solution. If the powder is not sterile it can be dissolved in alginate solution and sterilized with syringe filter using 0.22 μm filter.
4. The remainder of the methods for encapsulation of cells will be based on cells entrapped homogenously within the alginate. Encapsulation of cell aggregates within the alginate is also possible. The cell aggregate can be dispersed within the alginate solution and extruded through a syringe into CaCl2 bath. The number of aggregates per bead will vary depending on size of alginate beads desired and number of initial aggregates in alginate solution.
5. To speed up process of cells separating from flask, flask can be placed in incubator maintained at 37 °C in a humidified 5 % CO_2–95 % air atmosphere for 5 min.
6. It is recommended that each individual group being encapsulated be completed followed prior to start the following to prevent the cells from being in pellet form for long periods of time.
7. The pellet should not be touched with the pipette so some supernatant will remain in tube.
8. Polymer solution should be directly added to the pellet to allow detachment. Minimize bubble formation by slowly pipetting the solution up and down and also this allows for cells to distribute evenly within the solution. Ten cycles is sufficient to allow cells to distribute cells.
9. Different syringes sizes can be used depending on polymer volume. It should be noted the diameter of syringe and diameter of tube to allow for aspiration of cell-polymer solution.
10. A transfer pipette can be cut to allow for suction of beads from crosslinking solution.
11. The pipette tip can be directly placed onto the bottom of petri dish to remove residual solution.
12. The beads should be split into different wells, no more than 25 beads per well. For 1 mL of polymer solution, roughly 60 beads are formed.

Encapsulation of alginate within the outer core

13. Mix on vortex until the alginate powder has fully dissolved.
14. Different gauge needles can be used depending on the size of the alginate microbeads desired. This step can also be performed with the use of a syringe should a microencapsulator not be available. The dissolved alginate should be added into

the syringe and a needle gauge based on the size of microbeads desired. The use of a syringe allows for larger beads to be formed (greater than 2 mm).

15. Appropriate size stir bar should be used to allow mixing of crosslinking solution, without deforming beads themselves.
16. If injecting droplets with syringe, droplets should be injected one after another to prevent from crosslinking to each other.
17. The tip of the transfer pipette should be cut to allow beads to be easily transferred.
18. To completely remove the residual volume, a 200 μL pipette should be used. The pipette tip should be placed completely at the end of the centrifuge tube to prevent beads from obstructing the tip. This should be repeated until all residual volume is removed.
19. Enough PLO should be dissolved to the alginate microbeads completely submerged; used roughly 3 mL of PLO solution per mL of alginate solution. A different polycation (poly-L-lysine) can be used to replace the PLO if desired.
20. The PLO solution can be added to the test tube with the beads inside, the wash solution should be completely removed beforehand. At the end of the 30 min, there should be a visible white coating around the alginate beads.
21. The size of the outer layer is dependent on the composition and concentration of alginate used. For protein encapsulation, it should be added to the outer layer alginate solution prior to addition of the PLO-coated alginate microbeads.
22. A Kimwipe can be used to absorb the excess solution in order to dry the microbeads as much as possible.
23. Other smooth surfaces can be substituted for the parafilm surface such as glass or plastic.
24. The volume of alginate solution used should completely cover the alginate microbeads.
25. Viewing the alginate beads under a microscope should allow visualization of a distinct outer alginate layer.

References

1. Kollmer M, Appel AA, Somo SI et al (2015) Long-term function of alginate-encapsulated islets. Tissue Eng Part B Rev [Epub ahead of print]
2. Calafiore R, Basta G, Luca G et al (2005) Microencapsulated pancreatic islet allografts into nonimmunosuppressed patients with type 1 diabetes: first two cases. Diabetes Care 29:137–138
3. Basta G, Montanucci P, Luca G et al (2011) Long-term metabolic and immunological follow-up of nonimmunosuppressed patients with type 1 diabetes treated with microencapsulated islet allografts: four cases. Diabetes Care 34:2406–2409
4. Elliott RB, Escobar L, Tan PL et al (2007) Live encapsulated porcine islets from a type 1 dia-

betic patient 9.5 yr after xenotransplantation. Xenotransplantation 14:157–161

5. Tuch BE, Keogh GW, Williams LJ et al (2009) Safety and viability of microencapsulated human islets transplanted into diabetic humans. Diabetes Care 32:1887–1889
6. Moya ML, Morley M, Khanna O et al (2012) Stability of alginate microbead properties in vitro. J Mater Sci Mater Med 23:903–912
7. McQuilling JP, Arenas-Herrera J, Childers C et al (2011) New alginate microcapsule system for angiogenic protein delivery and immunoisolation of islets for transplantation in the rat omentum pouch. Transplant Proc 43:3262–3264
8. Richardson T, Kumta PN, Banerjee I (2014) Alginate encapsulation of human embryonic stem cells to enhance directed differentiation to pancreatic islet-like cells. Tissue Eng Part A 20:3198–3211
9. Khanna O, Moya ML, Opara EC et al (2010) Synthesis of multilayered alginate microcapsules for the sustained release of fibroblast growth factor-1. J Biomed Mater Res A 95:632–640
10. Moya ML, Cheng M-H, Huang J-J et al (2010) The effect of FGF-1 loaded alginate microbeads on neovascularization and adipogenesis in a vascular pedicle model of adipose tissue engineering. Biomaterials 31:2816–2826

Chapter 18

Compartmentalization of Two Cell Types in Multilayered Alginate Microcapsules

Sivanandane Sittadjody, Justin M. Saul, and Emmanuel C. Opara

Abstract

Two-dimensional (2D) culture systems do not represent the native microenvironment of the cells which is known to be three dimensional (3D), and surrounded by other cells from all directions. There exist interactions with other cell types in the same vicinity and this also cannot be replicated in a 2D culture. To study the cell–cell interactions between two or more cell types and their biological functions, a few 3D models have been used by different investigators. We have designed a 3D model to investigate the cell–cell interactions between various types of ovarian cells. The same model was also used to study the interactions between prostate cancer epithelial cells and stromal cells. This model uses hydrogel as the anchor matrix to fabricate the constructs and microencapsulation techniques to design multilayered microcapsules. In these multilayer microcapsules the different types of cells are compartmentalized by a sequential encapsulation process. In this chapter, we provide the protocol to compartmentalize two cell types in the same multilayer microcapsules. Although this chapter describes the fabrication of multilayer microcapsules with ovarian cells, the same approach could be applied to other multi-cell tissue-engineered constructs that require cell–cell interactions.

Key words Compartmentalization, 3D culture, Multilayer microcapsules, Ovarian constructs

1 Introduction

Our current knowledge of a variety of biological phenomena including cellular proliferation, differentiation, migration, and other cell-specific functions is based largely on studies from two-dimensional (2D) cell culture systems on glass or plastic surfaces. However, in vivo, the cells naturally exist in a three-dimensional (3D) microenvironment and interact with other neighboring cell types. Indeed, many cell types when isolated from tissues and cultured on planar surface become progressively flatter, divide aberrantly, and lose their differentiated phenotype [1, 2]. Interestingly, some of these cell types can regain their physiological form and function once they are embedded in 3D culture environment [2–5]. More recently, the field of cell biology has come to appreciate the dissimilarity between the 2D surfaces and topographically complex 3D

Emmanuel C. Opara (ed.), *Cell Microencapsulation: Methods and Protocols*, Methods in Molecular Biology, vol. 1479,
DOI 10.1007/978-1-4939-6364-5_18, © Springer Science+Business Media New York 2017

microenvironment in which the cells primarily exist in vivo. The difficulty in replicating not only the 3D microenvironment but also the interactions with other cell types has spurred into the development of various 3D models in order to assess the native functions of various cells and tissues. One such 3D system has been developed in our laboratory for the purpose of studying the biological functions of ovarian cells and their potential application in cell-based hormone replacement therapy for ovarian failure conditions. The 3D ovarian cell construct fabrication protocol described in this chapter involves sequential multilayer microencapsulation.

The fabrication of multilayered microcapsules with two different ovarian cell types (granulosa and theca cells, referred as GC and TC, respectively) compartmentalizes these two cell types in different layers of alginate microcapsules. This design helps in studying the significance of paracrine interactions between these two cell types in a 3D microenvironment [6]. The compartmentalization process is carried out with the help of encapsulation techniques using alginate hydrogels and poly-L-Ornithine (PLO) as perm-selective material that separates the two compartments housing the different cell types. The advantage of this system is that it not only harbors the two cell types in different compartments of the same microcapsule but also provides protection against host immune system (immune-isolation) as would be required for the use of cells from allogenic and xenogenic sources for in vivo studies.

We have also applied the same approach in our laboratory to study interactions between prostate cancer epithelial cells with stromal cells in a 3D microenvironment [7]. However, in this chapter, we specifically describe the fabrication of ovarian tissue construct with compartmentalized granulosa and theca cells. To demonstrate the compartmentalization of individual cell types, the cells were stained with fluorescent Cell Tracker dyes prior to encapsulation. These are vital dyes which are harmless to the cells and do not affect any cellular functions.

2 Materials

2.1 General Equipments

1. Rocker-mixer (Easy-mixer, SKSC).
2. Centrifuge (Beckman Coulter).
3. Confocal microscope (Zeiss LSM510).
4. Microfluidic encapsulator device (Dept. of Mechanical Engineering, Clemson University, SC).
5. Surgical tools.

2.2 Chemicals and Materials

1. Medium 199 (GIBCO 11043-023).
2. McCoy's 5A (GIBCO 16600-082).

3. HEPES (Sigma H4034).
4. Penicillin/Streptomycin (Sigma P4333).
5. Amphotericin-B (Sigma A2942).
6. Bovine serum albumin—BSA (Sigma A2153).
7. Collagenase type 1 (Worthington 4196).
8. Deoxyribonuclease—DNase (Sigma DN25).
9. Percoll (Sigma P4937).
10. Follicle Stimulating Hormone—FSH (Sigma F8174).
11. Luteinizing Hormone—LH (Sigma L5269).
12. 17β-estradiol—E_2 (Sigma E2758).
13. Insulin-like growth factor-I—IGF-I (Pepro Tech 100-11).
14. Insulin-Transferrin-Selenium—ITS (Sigma I2521).
15. Dulbecco's phosphate buffered saline—DPBS (Sigma D5652).
16. Dulbecco's phosphate buffered saline—DPBS (Hyclone SH30028.03).
17. Hanks buffered salt solution—HBSS, without calcium (Ca^{2+}) or magnesium (Mg^{2+}) (Hyclone SH30588.02).
18. HBSS, with Ca^{2+} and Mg^{2+} (Hyclone SH 30030.03).
19. Cell Tracker™ green CMFDA (Invitrogen C2925).
20. Cell Tracker™ Orange CMRA (Invitrogen C34551).
21. Dimethyl sulfoxide—DMSO (Sigma D2650).
22. Ultra-pure low viscous high mannuronic acid, up-LVM (Novamatrix 4200206).
23. Ultra-pure low viscous high guluronic acid, up-LVG (Novamatrix 4200006).
24. Poly-L-Ornithine—PLO (Sigma P2533).
25. Hydrochloric acid—HCl (Fisher A144).
26. Sodium hydroxide—NaOH (Amresco 0583).
27. Calcium Chloride—$CaCl_2$ (Fisher C77).
28. Sodium Chloride—NaCl (Fisher S671).
29. Cell strainer—100 μm (BD Falcon 352360).
30. Syringes—5 cc (BD Falcon 309646).
31. Syringe filter—0.2 μm (Fisherbrand 09-719C).
32. 15 ml centrifuge tubes (BD Falcon 14-959-53A).
33. 50 ml centrifuge tubes (BD Falcon 14-432-22).

2.3 Cells

1. Ovarian cells: granulosa (GC) and theca cells (TC) isolated from 3-week-old fisher 344 female rats.

2.4 Reagents and Solutions

2.4.1 Cell Culture

1. Isolation medium: Medium 199 supplemented with 25 mM HEPES, 0.1 % BSA, penicillin (10,000 IU/ml), streptomycin (10, 000 μg/ml), and amphotericin B (25 μg/ml).
2. Culture medium: McCoy's 5A supplemented with 10,000 IU/ml penicillin, 10,000 μg/ml streptomycin, 25 μg/ml amphotericin B, ITS and 10 nM IGF-I, plus supplementation of different hormones to maintain their respective phenotypes of GC and TC.
3. Cell-specific culture media: Supplement culture medium with 50 ng/ml FSH and 100 nM E_2 for GC culture. Similarly supplement culture medium with 50 ng/ml LH for TC culture.
4. Collagenase type 1 (100 mg/ml): dissolve 1 g in 10 ml of isolation medium, filter with syringe filter, aliquot into 0.5 ml, and store in –20 °C freezer until used (*see* **Note 1**).
5. DNase (500 μg/ml): dissolve 100 mg in 10 ml of isolation medium to obtain 10 mg/ml, filter with syringe filter, label it as stock A, aliquot into 1 ml, and store in –20 °C freezer until used (*see* **Note 1**). Take 1 ml from stock A and add 19 ml of isolation medium to obtain 500 μg/ml (Stock B). Aliquot into 0.5 ml and store in –20 °C freezer until used (*see* **Note 1**).
6. Collagenase/DNase mixture: Add 0.5 ml of 100 mg/ml collagenase, 0.5 ml of 500 μg/ml DNase (Stock B), and 24 ml of isolation medium to obtain a final concentration of 2 mg/ml collagenase and 10 μg/ml DNase (prepare freshly).
7. 10× DPBS: dissolve the entire content from the pack (Sigma D5652), which is 9.6 g of powder in 100 ml (instead of 1 l) of diH_2O to obtain 10× DPBS.
8. Percoll solutions (44, 20% and $d = 1.055$): As per the Pertoft [8], make 100% (w/v) solution of percoll solution by mixing nine parts of manufacturer's percoll (Sigma P4937, which is 1.13 g/ml) with one part of 10× DPBS. From this 100% percoll, take 22 ml in a 50 ml conical tube and make up to 50 ml with 28 ml of culture medium to obtain 44% percoll solution. Similarly, take 10 ml of 100% percoll in a 50 ml conical tube and make up to 50 ml with 40 ml of culture medium to obtain 20% percoll solution. However in order to make $d = 1.055$, mix approximately 19.3 ml of manufacturer supplied percoll (Sigma P4937) with 30.7 ml of isolation medium and check the density of the solution using a densitometer. Adjust the density with either percoll or medium to obtain the desired density of 1.055.
9. Discontinuous Percoll gradient: In a 15 ml centrifuge tube, add 2 ml of 44% percoll solution to the bottom of the tube. On top of 44% percoll, carefully add 2–3 ml d = 1.055 percoll solution without disturbing the bottom layer. Similarly, on top of $d = 1.055$ percoll layer, carefully add 2 ml of 20% percoll solution without disturbing the bottom two layers (*see* **Note 2**).

2.4.2 Fluorescent Staining

1. Cell Tracker™ green CMFDA: Dissolve 1 mg of Cell Tracker green in 215.12 μl DMSO to obtain 10 mM stock. Aliquot and store in −20 °C freezer until used (*see* **Note 1**). From the 10 mM stock, make 10 μM working solution by diluting 1:1000 with serum-free culture medium prior to use (*see* **Note 3**).
2. Cell Tracker™ orange CMRA: Dissolve 1 mg of Cell Tracker green in 181.67 μl DMSO to obtain 10 mM stock. Aliquot and store in −20 °C freezer until used (*see* **Note 1**). From the 10 mM stock, make 10 μM working solution by diluting 1:1000 with serum-free culture medium prior to use (*see* **Note 3**).

2.4.3 Encapsulation

1. 1.5 % (w/v) ultra-pure low viscous mannuronic acid (up-LVM): In a 50 ml centrifuge tube take 0.5 g of up-LVM and add 33.33 ml of HBSS without Ca^{2+} or Mg^2. Mix the content overnight at 4 °C in order to dissolve the alginate completely in HBSS since it takes a while for the alginate to get dissolved and hydrated (*see* **Note 4**).
2. 0.5 % (w/v) ultra-pure low viscous guluronic acid, up-LVG: 0.1 g of up-LVG: In a 50 ml centrifuge tube take 0.1 g of up-LVG and add 20 ml of HBSS without Ca^{2+} or Mg^2. Mix the content overnight at 4 °C in order to dissolve the alginate completely in HBSS since it takes a while for the alginate to get dissolved and hydrated (*see* **Note 4**).
3. 1 N HCl: Take 8.26 ml of 12.1 N and make it to 100 ml with diH_2O.
4. 1 N NaOH: Dissolve 4 g of NaOH in 100 ml diH_2O.
5. Cross-linking solution (100 mM $CaCl_2$ with 10 mM HEPES): Dissolve 11.1 g of $CaCl_2$ and 2.383 of HEPES in approximately 800 ml of diH_2O. Adjust the pH to 7.4 using 1 N HCl and 1 N NaOH diH_2O. Sterile filter the solution (*see* **Note 5**).
6. 0.1 % (w/v) PLO solution: Dissolve 100 mg of PLO in 100 ml of HBSS with Ca^{2+} and Mg^2 and sterile filter it.
7. Saline solution (0.9 % NaCl): Dissolve 9 g of NaCl in approximately 1 l of diH_2O and sterile filter it.
8. Storage solution (25 mM $CaCl_2$; 0.9 % NaCl; 10 mM HEPES): Dissolve 2.75 g of $CaCl_2$; 9 g of NaCl and 2.383 of HEPES in approximately 800 ml of diH_2O. Adjust the pH to 7.4 using 1 N HCl and 1 N NaOH diH_2O. Sterile filter the solution.

3 Methods

3.1 Cell Isolation and Purification

The isolation of endocrine cells from ovaries of immature rats as shown in Fig. 1 was described briefly in Sittadjody et al. [6].

1. Collect the ovaries in ice-cold isolation medium.

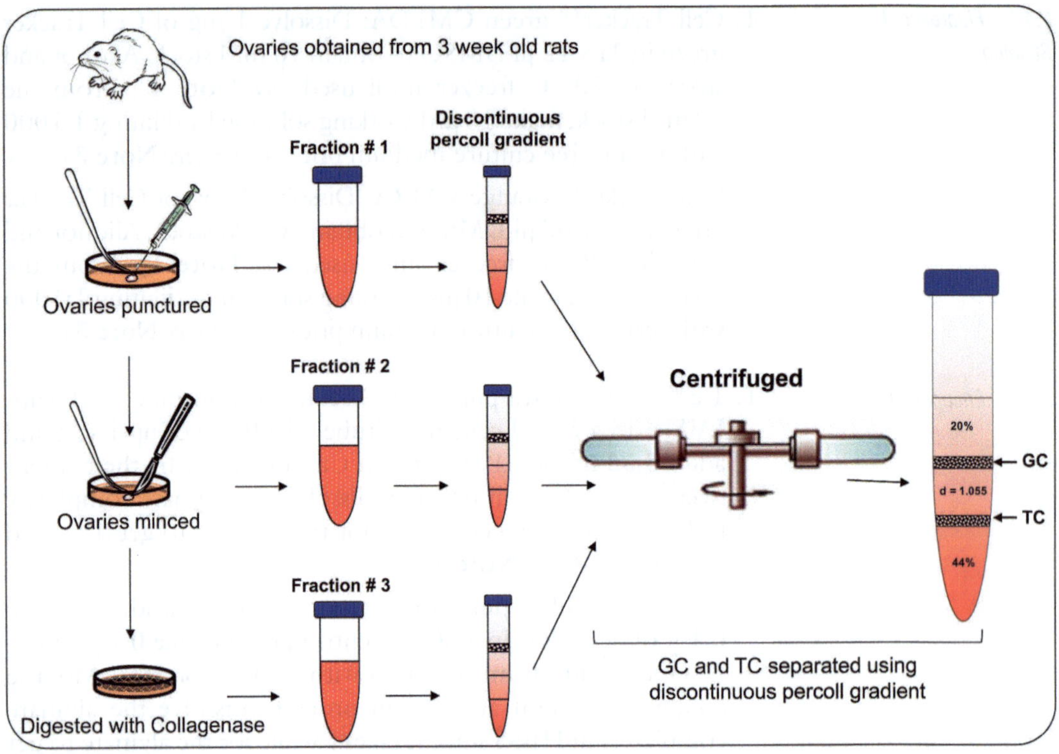

Fig. 1 Schematic of isolation of ovarian cells. Ovaries obtained from 3-week-old female rats were cleaned of extraneous tissue and punctured with syringe needle to liberate loosely packed cells (Fraction #1). Remaining tissues were chopped and the cells released during this process were collected (Fraction #2). The pieces of ovarian tissues were incubated with collagenase/DNase mixture followed by disruption by pipet to get all the remaining cells (Fraction #3). All the three fractions were loaded on to discontinuous percoll gradient and centrifuged to separate GCs from TCs

2. Remove the extraneous tissues, wash twice with ice-cold isolation medium, and puncture the ovaries gently with 27 G syringe needles while keeping the tissue in a culture dish in order to release the loosely packed cells (usually GCs) from the follicles; collect the released cells and keep them on ice as Fraction # 1.
3. Chop the remaining ovaries into fine pieces of ~0.25 mm^2 and collect the cells released during this process and keep them separately on ice as Fraction # 2.
4. Incubate the pieces of ovaries with freshly prepared collagenase (2 mg/ml)-DNase (10 μg/ml) mixture in isolation medium for 90 min at 37 °C with occasional gentle mixing.
5. Disperse the enzyme-digested pieces using a Pasteur pipette to obtain a single cell suspension and store them on ice as a separate fraction—Fraction # 3.
6. Load Fractions #1, # 2, and # 3 on top of three separate tubes with discontinuous percoll gradients and centrifuge at 400 × *g* for 20 min at 4 °C.

7. Recover the **GCs** which get separated at the first interphase (between 20% and $d = 1.055$ layers) and transfer in to a new centrifuge tube.
8. Similarly, recover the **TCs** which get separated at the second interphase (between $d = 1.055$ and 44% layers) and transfer in to another centrifuge tube.
9. Add extra isolation medium (~5×) to both the tubes with cells recovered from the discontinuous percoll gradients and centrifuge at 1500×rpm (500 × g) for 5 min in order to wash off the percoll.
10. As an optional step, the purity of each cell type could be assessed by flow cytometric analysis using cell-specific markers as previously reported by our group [6].
11. Use the cells isolated from ovaries to stain with Cell Tracker dyes for further steps.

Culture of these cells prior to fluorescent staining as described in Subheading 3.2 is optional and thus could be skipped. Since these cells have been shown to regain their physiological form and function once placed in 3D environment after culturing on 2D surfaces, they could be cultured on 2D dishes until confluency if these cells need recovery phase after isolation.

3.2 Culture of Granulosa and Theca Cells

Culture the purified GC and TC in McCoy's 5A medium [with respective hormone supplements as mentioned in materials section] until they reach confluency.

1. Use granulosa cell culture medium (supplemented with 50 ng/ml FSH and 100 nM E_2) for culturing GC and theca cell culture medium (supplemented with 50 ng/ml LH) for TC.
2. Incubate the cells at 37 °C under an atmosphere of 5% CO_2 in humidified air until they reach 80–90% confluency.
3. Upon reaching 80–90% confluent stage (~96 h), lift both GC and TC from the culture dish with a cell scrapper, pellet by centrifugation, and use for further steps (*see* **Note 6**).

3.3 Fluorescent Staining of GC and TC with Cell Tracker Dyes

Label the two cell types with vital dyes in order to differentiate them in the microcapsules and to visualize the compartmentalization of two cell types. Stain GC with Cell Tracker green and TC with Cell Tracker orange.

1. Wash the cells once with DPBS to remove any remnant of serum.
2. Resuspend GC in serum-free medium containing Cell Tracker green (10 μM).
3. Similarly, resuspend TC in serum-free medium containing Cell Tracker orange (10 μM).
4. Incubate both the cells at 37 °C for 15 min and protect them from light.

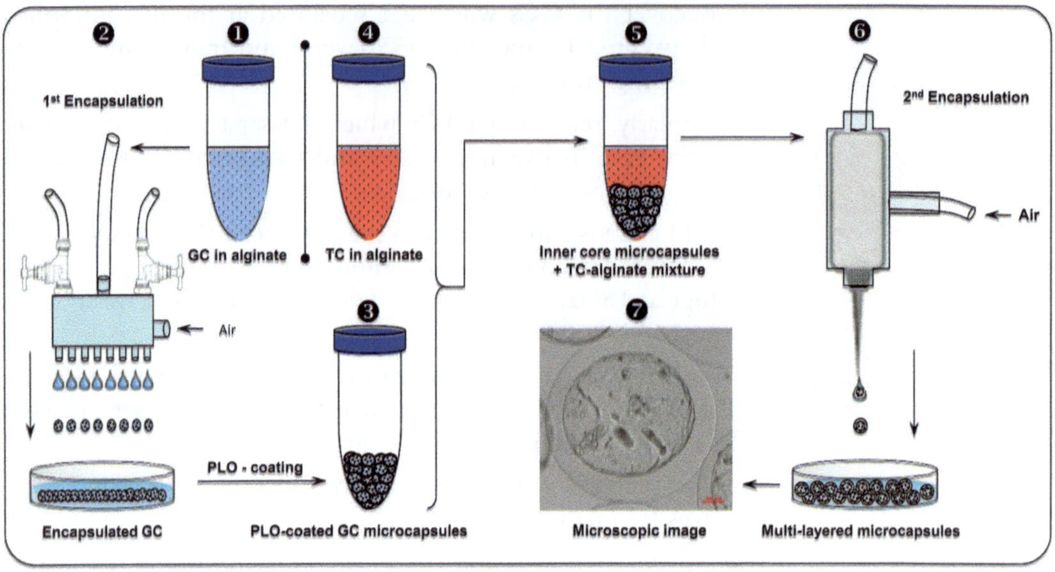

Fig. 2 Schematic of fabrication of multilayered microcapsules. (*1*) Cell Tracker green-stained granulosa cells (GC) mixed in 1.5 % up-LVM alginate; (*2*) GC-alginate mixture extruded through a multichannel microfluidic encapsulator device to generate the inner core with GC only. (*3*) Inner core microcapsules were coated with 0.1 % PLO for 20 min. (*4*) Cell Tracker orange-stained theca cells were mixed in 1.5 % up-LVM alginate. (*5*) Inner core microcapsules were mixed with TC-alginate mixture. (*6*) Inner core microcapsules-TC-alginate mixture was extruded through a single-channel microfluidic device generating the multilayered microcapsules. (*7*) Microscopic image of multilayered microcapsules (scale bar = 100 μm)

5. After 15 min, replace the cell tracker dyes with respective cell culture media and incubate again for another 30 min in order to allow the cells to metabolize the dye into fluorescent compound.
6. Pellet the cells by centrifuging at 1500 × rpm (500 × *g*) for 5 min at 4 °C.

3.4 Fabrication of Multilayered Alginate Microcapsules

The ovarian tissue constructs that mimic the follicular architecture are fabricated by using encapsulation techniques as described previously by our group [6] and depicted in Fig. 2.

1. Mix 5×10^6 GC with 1 ml of 1.5 % up-LVM alginate after removing any excess media from the cell pellet.
2. Extrude the GC-alginate mixture through the microfluidic encapsulator device into the cross-linking solution producing approximately 400–500 μm microcapsules (*see* **Note** 7).
3. After 15 min, sieve the microcapsules using 100 μm cell strainer and transfer into a 50 ml centrifuge tube.
4. Add 5–10 ml of 0.1 % PLO and incubate at RT by rocking for 20 min.
5. After 20 min of incubation on rocker, sieve the PLO-coated GC microcapsules using 100 μm cell strainer

6. Wash the PLO-coated GC microcapsules with sterile saline to remove any traces of Ca^{2+} and sieve them using 100 μm cell strainer.
7. Meanwhile, mix 5×10^6 TC with 1 ml of 1.5 % up-LVM alginate in a separate tube after removing excess media from the cell pellet.
8. Mix PLO-coated GC microcapsules with TC-alginate mixture and repeat **step 2** of Subheading 3.4 to obtain double alginate-layered microcapsules with a diameter of 700–800 μm.
9. Repeat **step 3** of Subheading 3.4 and resuspend in storage solution for imaging (or) for experiments.
10. In case of allogenic/xenogenic cells, repeat **step 4** of Subheading 3.4 to include additional layer of PLO.
11. Repeat **step 5** of Subheading 3.4 to wash off the PLO.
12. Mix the PLO-coated double alginate-layered microcapsules 0.5 % up-LVG (enough volume to cover the microcapsules) and incubate by rocking for 5 min at RT.
13. After 5 min, sieve the multilayered microcapsules using 100 μm cell strainer and resuspend in storage solution for imaging or for experiments.

These additional layers of PLO and alginate were included to provide immune-isolation for the encapsulated cells (Fig. 3c).

3.5 Imaging the Compartmentalized Cells by Confocal Microscopy

1. Use a confocal microscope to image the multilayered microcapsules and visualize the compartmentalized fluorescent-labeled GC and TC. As shown in Fig. 3d, e, green fluorescent-labeled GC could be located in the inner core of the multilayered microcapsules and orange fluorescent-labeled TC in the outer layer of the constructs.

4 Notes

1. Avoid freeze-thawing the aliquots, as they may lose their activity by the freeze-thaw cycle.
2. If the 44 and 20 % percoll solutions are made with phenol red-containing culture medium and $d = 1.055$ solution with phenol red-free isolation medium, it will help to differentiate clearly the layers in the discontinuous percoll gradient.
3. Cell Tracker dyes have to be prepared in serum-free medium, because serum might interfere with the staining process. Moreover, use of serum might affect the phenotype of the ovarian cells since the hormones in it might luteinize the ovarian cells. The working solutions of Cell Tracker dyes should be protected from light.

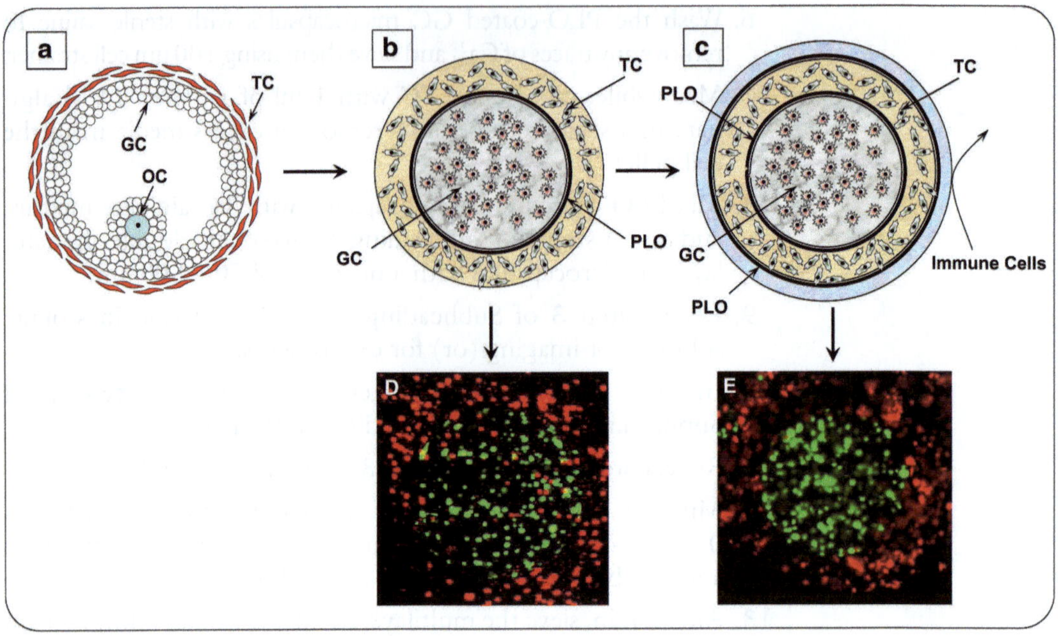

Fig. 3 Mimicking the 3D structure of ovarian follicle by microencapsulation. Schematic diagram of a native ovarian follicle, where GCs are found in the *center*, TCs in the periphery and separated by a basement membrane (**a**), compared to the compartmentalized microcapsules with GC in the central core and TC in the outer layer separated by perm-selective PLO layer (**b**). Inclusion of additional layer of PLO and alginate on the periphery modifies the construct for allogenic/xenogenic source of cells and provides immune-isolation from the host (**c**). Confocal images of 3D constructs (**d** and **e**) demonstrating the compartmentalization of different cell types within the constructs shown by Cell Tracker green-labeled GCs in the inner layer and Cell Tracker orange-labeled TCs in the outer layer. *GC* granulosa cells; *OC* oocyte; *TC* theca cells; and *PLO* poly-L-Ornithine

4. Manufacturers supply alginate as a sterile powder. Therefore, handle the alginate in a sterile cell culture hood while preparing the solution as it is impractical to sterile filter the alginate solution. Tare the weight of a sterile 50 ml conical tube on a chemical balance prior to addition of alginate to the tube. Take approximately one scoop of alginate using a sterile spatula in the sterile cell culture hood and close the tube tightly. Weigh the content and calculate the amount of HBSS needed according to the weight of the alginate. For every 1 mg of up-LVM, 66.66 μl of HBSS needs to be added to obtain 1.5 % (w/v) solution. Similarly for every 1 mg of up-LVG, 200 μl of HBSS needs to be added to obtain 0.5 % (w/v) solution.
5. It is always preferred to use freshly prepared cross-linking solution (or a solution not older than a week)
6. Avoid using trypsin to lift the cells from culture dishes since the use of serum for neutralizing trypsin might have hormones that might luteinize the ovarian cells. However, trypsinization could be used for other cell types when serum is not a problem.

7. The microfluidic encapsulator device is designed in our lab. Likewise any encapsulator device could be used. Size of the inner core GC microcapsules should preferably not be bigger than 500 μm since the overall size of multilayered microcapsule depends on it. Diffusion of nutrients, gases, and cellular secretory products and the viability will be affected by increasing the overall size of multilayered microcapsule.

References

1. Von der MK, Gauss V, von der MH, Muller P (1977) Relationship between cell shape and type of collagen synthesised as chondrocytes lose their cartilage phenotype in culture. Nature 267:531–532
2. Petersen OW, Ronnov-Jessen L, Howlett AR, Bissell MJ (1992) Interaction with basement membrane serves to rapidly distinguish growth and differentiation pattern of normal and malignant human breast epithelial cells. Proc Natl Acad Sci U S A 89:9064–9068
3. Benya PD, Shaffer JD (1982) Dedifferentiated chondrocytes reexpress the differentiated collagen phenotype when cultured in agarose gels. Cell 30:215–224
4. Emerman JT, Pitelka DR (1977) Maintenance and induction of morphological differentiation in dissociated mammary epithelium on floating collagen membranes. In Vitro 13:316–328
5. Lee EY, Parry G, Bissell MJ (1984) Modulation of secreted proteins of mouse mammary epithelial cells by the collagenous substrata. J Cell Biol 98:146–155
6. Sittadjody S, Saul JM, Joo S, Yoo JJ, Atala A, Opara EC (2013) Engineered multilayer ovarian tissue that secretes sex steroids and peptide hormones in response to gonadotropins. Biomaterials 34:2412–2420
7. Fang X, Sittadjody S, Gyabaah K, Opara EC, Balaji KC (2013) Novel 3D co-culture model for epithelial-stromal cells interaction in prostate cancer. PLoS One 8:e75187
8. Pertoft H (2000) Fractionation of cells and subcellular particles with Percoll. J Biochem Biophys Methods 44:1–30

Chapter 19

Primary Choroid Plexus Tissue for Use in Cellular Therapy

M.A. Sandrof, D.F. Emerich, and Chris G. Thanos

Abstract

The choroid plexus (CP) has been explored as a cellular therapeutic due to its broad-ranging secretome and demonstrated longevity in a variety of encapsulation modalities. While the CP organ is normally involved in disease repair processes in the brain, the range of indications that could potentially be ameliorated with exogenous CP therapy is widespread, including diseases of the central nervous system, hearing loss, chronic wounds, and others. The CP can be isolated from animal sources and digested into a highly purified epithelial culture that can withstand encapsulation and transplantation. Its epithelium can adapt to different microenvironments, and depending on culture conditions, can be manipulated into various three-dimensional configurations with distinct gene expression profiles. The cocktail of proteins secreted by the CP can be harvested in culture, and purified forms of these extracts have been evaluated in topical applications to treat poorly healing wounds. When encapsulated, the epithelial clusters can be maintained for extended durations in vitro with minimal impact on potency. A treatment for Parkinson's disease utilizing encapsulated porcine CP has been developed and is currently being evaluated in a Phase I clinical trial. The current chapter serves to summarize recent experience with CP factor delivery, and provides a description of the relevant materials and methods employed in these studies.

Key words Choroid plexus, Microcapsule, Alginate, Encapsulation, Cell transplantation

1 Introduction

The choroid plexus (CP) is a highly active epithelial structure that sits at the blood–brain barrier, supplied by a rich network of blood vessels at the basolateral interface (Fig. 1). A high density of tight junctions and aquaporin channels generates a convective filtration network capable of detoxification [1]. In humans, which have a CSF volume of approximately 80–150 mL with a turnover rate of 4–6 times per day, this filtration is driven by continuous production of new CSF via passive extravasation aided by an osmotic gradient [2]. The CP is responsible for monitoring for the presence of damaging compounds or invading pathogens in the CSF, and aids in subsequent recruitment of immune cells [3]. The CP monitors and responds to baseline levels of the extracellular milieu throughout the CNS. The CP of the developing embryo quickly assumes

Emmanuel C. Opara (ed.), *Cell Microencapsulation: Methods and Protocols*, Methods in Molecular Biology, vol. 1479,
DOI 10.1007/978-1-4939-6364-5_19, © Springer Science+Business Media New York 2017

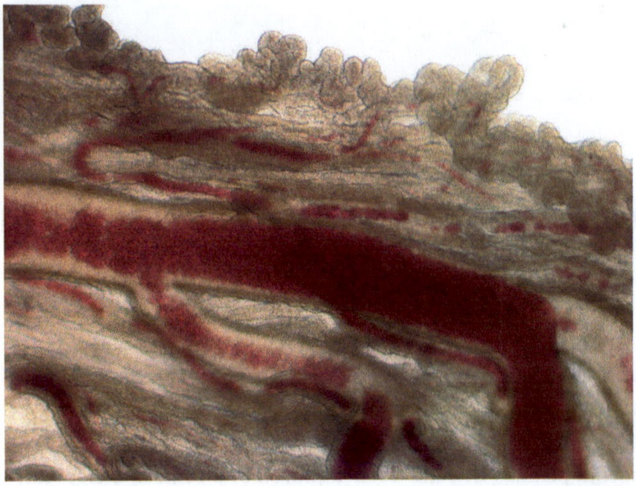

Fig. 1 Low-magnification macro photograph of neonatal porcine CP. Epithelial cells on the top portion overlie a dense highly vascularized stroma, with extension of capillaries into the epithelial fronds that extend from the surface

mediation of the developing extracellular environment [4] through secretion of morphogens, mitogens, and trophic factors that guide and pattern brain growth [5, 6].

As an encapsulated cell therapy, CP represents a unique candidate tissue that possesses both trophic and regenerative activity, achieved by simultaneous but relatively low-level secretion of many factors. In one sense, the wide array of biological activity that it possesses can act as a "shotgun-blast" approach for disease treatment, as it is possible that at least some combination of factors contained within its secretome will have biological relevance to a given indication. But perhaps more unique is the responsiveness of cultured CP to its microenvironment; it is possible to modulate the expression of distinct sets of growth factors through manipulation of culture conditions, or encapsulation parameters in the context of transplantation, thereby narrowing the "blast" to a region of potential interest [7]. To date, CP has been studied in models of neurodegenerative disease [8–11], wound healing [7, 12], aging [13–16], cochlear degeneration [17], and has recently entered clinical trials as a treatment for Parkinson's disease [8].

The CP is an abundant source of bioactive molecules that are neuroprotective in vitro and in vivo [6, 9–11, 18]. A number of bioactive proteins have been identified in the CP secretome [5, 19–22], and because of its wide range of biological activity, the therapeutic role of CP has been evaluated for a range of targets from neurodegeneration to hearing loss. A large number of inflammatory and growth factors are secreted from CP in culture [5, 12, 19–25]. Delivered in capsules to the brain, porcine CP epithelium protects against excitotoxin-induced degeneration in models of Huntington's disease [9, 11, 26], Parkinson's [27], and as a purified topical gel containing only secreted factors, CP proteins have been associated with improvements in the healing of cutaneous defects [12].

To date, the bulk of the work published on encapsulated CP has been conducted using tissue isolated from either the rat or the neonatal pig. In exploratory experiments, either tissue is a suitable cell source provided that if the transplant is discordant it is encapsulated or somehow immunoprotected. Transplanted xenogeneic cells must be protected from the host immune system by encapsulation within a semipermeable membrane that allows diffusion of nutrients, oxygen, and other desired molecules while denying passage to immune cells. Alginate, a common biomaterial candidate for microencapsulation, is a polysaccharide characterized by chains of guluronic (G) and mannuronic (M) acid linked by [1, 4]-alpha and [1, 4]-beta glycosidic bonds, the ratio of which defines physical macromolecular characteristics. Independent of the type and composition of the chosen alginate, further coating with a polycation is usually required to immunoprotect the capsule [28, 29]. Alginate encapsulation is a well-described method of encapsulating tissue in a biocompatible, immunoprotective environment [30]. It has been used to encapsulate an array of tissue types, including pancreatic [31], adipose [32], neuronal [22], and many others.

This cell encapsulation method, alginate-polycation-alginate microencapsulation (APA), has been used to deliver transplants of epithelial choroid plexus clusters in models of Huntington's and Parkinson's disease [6, 11, 33–39]. Alginate-encapsulated neonatal porcine CP transplants led to full improvement in motor responses and an 80% decrease in lesioned striatal area in a rat quinolinic acid HD model [9]. Alginate-encapsulated choroid plexus epithelial cells may also be protective against amyloid beta (Aβ)-induced toxicity in a rodent model of Alzheimer's disease. In this work, rather than maintaining the cluster geometry of CP epithelium, cells were cultured in attachment flasks, removed, and encapsulated within APA microcapsules as a suspension of individual cells. Aβ injection into rat brains, as in the QA model previously described, preceded APA capsule implantation by 1 week. In this model, encapsulated CP was associated with both behavioral and histological improvements [27].

Living Cell Technologies (LCT) has been developing encapsulated CP products for over a decade under the name NTCELL®. This product is derived from a source herd of pigs from Auckland Island, free from many of the viruses associated with domestic swine [40]. Following on positive data of neuroprotection in 6-OHDA lesioned rats [39] and MPTP-lesioned primates [38], a clinical trial of NTCELL® was initiated by LCT to treat patients with Parkinson's Disease [8]. Four patients aged 59–68 with Parkinson's disease received injections of NTCELL® into the affected brain areas. In addition to being well tolerated with no significant adverse events, after 26 weeks of implantation, all participants showed improvements in neurological ratings of movement abnormalities, physical symptoms, well-being, and ability to perform everyday tasks.

The current chapter describes methods to isolate, digest, and culture CP for encapsulation and subsequent intrastriatal implantation and assessment. While the disease paradigms described in this chapter focus on those that have already been associated with therapeutic efficacy with encapsulated CP, discussion of its therapeutic capability should extend beyond this scope [12].

Alginate-encapsulation techniques vary slightly but are well described in the literature and are detailed in protocol form elsewhere [30]. The methods employed in this work are based on a similar technique starting with a purified alginate solution containing a suspension of choroid plexus clusters.

2 Materials

2.1 Chemicals

1. Hank's Balanced Salt Solution (HBSS) (Corning, Tewksbury, MA).
2. Ciprofloxacin (Sigma, St. Louis, MO).
3. Human serum albumin.
4. Liberase HI (Roche, NJ).
5. RPMI-1640 (Sigma, St. Louis, MO).
6. Tribasic sodium citrate (Sigma, St. Louis, MO).
7. Neurobasal medium (Invitrogen, Carlsbad, CA).
8. Paraformaldehyde (Acros Organics, Geel, Belgium).
9. Neuron-specific enolase (Sigma, St. Louis, MO).
10. Quinolinic acid (Sigma, St. Louis, MO).
11. Zamboni's fixative (American Master Tech, Lodi, CA).
12. Mouse anti-neuronal nuclei monoclonal antibody (Chemicon, Temucula, CA).
13. Tris (Fischer Scientific).
14. Biotinylated horse anti-mouse IgG (Vector Labs, Burlingham, CA).
15. Avidin-biotin complex (Elite ABC Kits, Vector Labs).

2.2 Special Equipment

1. DTX-880 spectrophotometer (Beckman Coulter, Indianapolis, IN).
2. Concentric nozzle system (Kineteks).
3. Kopf stereotaxic apparatus (Kopf Instruments, Tujunga, CA).
4. Neurosurgical drill.

3 Methods

The methods described in this section highlight some of the most important processes, handling procedures, and analytical techniques that have been employed in the development of CP-based therapeutics. Isolation of choroid plexus from neonatal pigs is described, followed by culture and characterization. In both free and encapsulated formats, VEGF has been identified as a surrogate marker of potency that is consistent between transcriptomic and secretomic expression [22]. We also describe basic methods of evaluation in cell-based assays in vitro, and stereotaxic implantation. Finally, histological processing and histomorphometric analysis in lesioned brains is detailed. These methods require a basic understanding of tissue culture for CP, but when possible, relevant references have been cited for further detail.

3.1 Choroid Plexus Isolation

1. Neonatal Yorkshire pigs aged 7–10 days are anesthetized with ketamine (500 mg/kg) and xylazine (0.15 mg/kg), and maintained on 3 % isoflurane during controlled exsanguination.
2. Following surgical exposure, the brain is bisected to expose the CP in the lateral ventricles (Fig. 2a).
3. The choroid plexus is removed and placed in Hanks Balanced Salt Solution (HBSS, 0–4 °C) supplemented with ciprofloxacin (3 μg/mL) and 2 % human serum albumin (Fig. 2b).
4. Tissue is digested as previously described [9, 11]: transient mechanical disruption with scissors is followed by 20 min of light enzymatic digestion to release CP epithelial clusters (Liberase HI) (Fig. 3a).

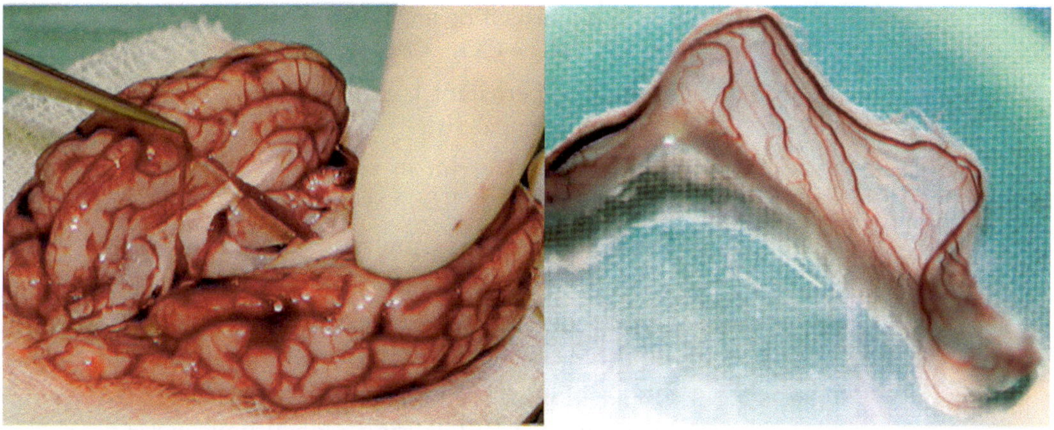

Fig. 2 (**a**) Bisected neonatal pig brain exposing the choroid plexus in the lateral ventricles. (**b**) Freshly harvested whole choroid plexus from neonatal piglets lifted to expose vascular architecture and epithelial surface area

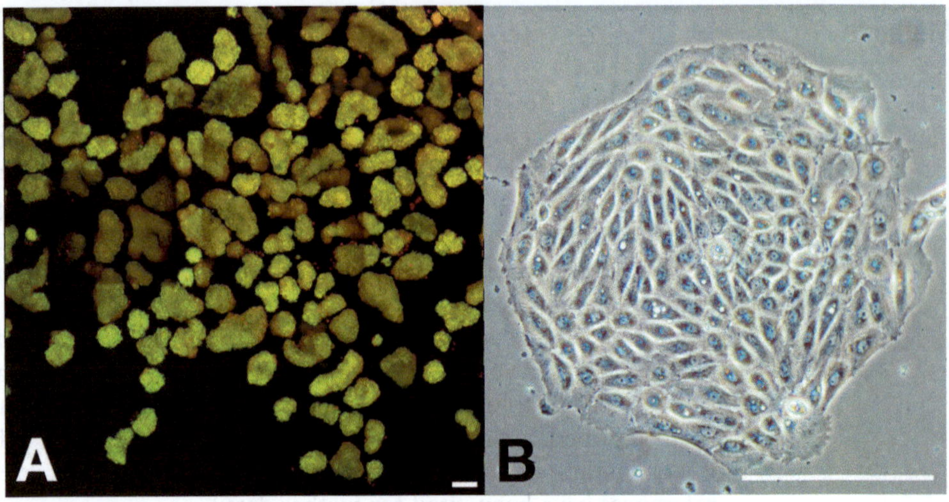

Fig. 3 (**a**) Fluorescent micrograph of cultured epithelial choroid plexus clusters stained with calcein AM (*green*) and ethidium homodimer (*red*). (**b**) Example of a cluster that attached and plated to the bottom of the culture flask during the initial stages of cell culture after isolation. Both scale bars are 100 μm

5. Epithelial clusters are washed and maintained in RPMI-1640 with 2% neonatal porcine serum and 3 μg/mL ciprofloxacin (RPMI-CPN) in low-attachment flasks at 37 °C with 5% CO_2.
6. Clusters that do not maintain suspension and plate to the bottom of the flask (Fig. 3b) are removed during media changes.
7. Cultures are fed after 24 and 48 h and every other day following until encapsulation

3.2 Evaluation of Encapsulated Choroid Plexus

1. Capsules are measured with image analysis (Scion Image, USA) and incubated in RPMI culture medium with ciproflaxocin.
   (a) Example capsules are 600 ± 20 μm in diameter (Fig. 4).
2. Cells are replenished by incubating with fresh RPMI-CPN for 24 h at 37 °C, 5% CO_2 (*see* **Note 1**).
3. For storage, supernatant is removed and capsules are frozen until subsequent analysis.
4. Thawed samples are run in triplicate using a human VEGF ELISA (Quantikine®, R&D Systems, USA) [41] (*see* **Note 2**).
   (a) Conditioned media samples are combined with assay diluent as indicated by ELISA protocol.
   (b) Optical density is measured at 450 nm on a Beckman Coulter DTX-880 spectrophotometer after development.
   (c) Obtained value is compared to a standard curve of recombinant human VEGF using a logarithmic curve fit.
   (d) Values are expressed as pg/mL/24 h/10^5 cells normalized to DNA.

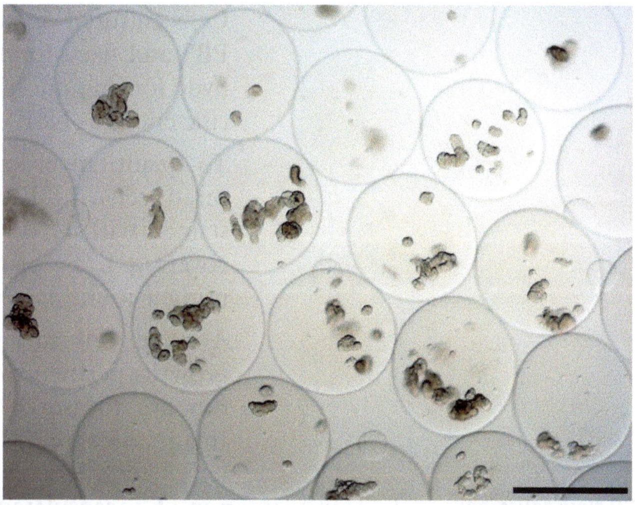

Fig. 4 Light micrograph of alginate-encapsulated choroid plexus clusters. Scale bar represents 500 μm

- Evaluation of potency as a function of cell density.
  - CP may also be seeded at densities of 1000–12,000 CP cell equivalents/mL media.
  - Supernatant is collected after 24 h incubation and analyzed for kinetic release of VEGF.

3.3 In Vitro Biological Activity of Choroid Plexus-Conditioned Media

1. Primary day 15 embryonic cortical neurons are added to media conditioned for 24 h with porcine CP cultures.
2. Effects of serum deprivation on neuronal survival, neurite numbers, and length are assessed as previously described [10]:
   (a) Homogeneous suspensions of neurons are cultured (QBM Cell Science, ON, Canada).
   (b) Neurons are resuspended in Neurobasal medium containing 10 % FBS and plated on PLL-coated plates.
   (c) On the 4th day of tissue culture, cells are seeded at 3 × 105 cells/cm2 on 13-mm cover slips in 24-well plates.
   (d) Cells are cultured for 2 days in neurobasal medium with 10 % FBS, without FBS, or without FBS but with 30 % CP-conditioned media (CP-CM supplement).
   (e) Neurite outgrowth and length are assessed on day 6 by microscopic analysis.
      - Neurites exceeding the cell diameter or their attached cell are quantified [42].
      - Cell survival is quantified on day 7.

- Cells cultured on coverslips are washed once with PBS and fixed for 10 min with 4 % paraformaldehyde for immunohistochemistry with neuron-specific enolase (NSE).
- To quantitatively assess the composition of cultured cells, NSE-positive neurons are counterstained with DAPI and counted in parallel strips covering 1/12 of the total area of the cover slip using a counting reticle.
- Average cell survival, number of neurite outgrowths per neuron, and neurite length per neuron are measured from cell images with NIH Image software.

3.4 CP Neurosurgical Implantation

1. Transplantation of encapsulated tissue and QA injections into nonhuman primates [43] (*see* **Note 3**).
   (a) Subjects are anesthetized with 10 mg/kg intramuscular ketamine and 4 % isoflurane.
   (b) Animals are placed in a Kopf Stereotaxic apparatus, and an incision is made to expose the skull directly over the right striatum.
   (c) Using a neurosurgical drill, a 2 × 3 cm craniotomy is made and the dura reflected to expose the cortex overlying the striatum.
   (d) Twenty cell-containing capsules loaded in a cannula are stereotaxically implanted into the head of the caudate nucleus as well as the putamen.
      - This technique is demonstrated in an in vitro gelatin model of capsule displacement where the tract is between 1 and 2 capsules in width (Fig. 5).
   (e) Following implantation, subcutaneous tissues, skull, and dura are stitched back in place with 5-0 vicryl suture.
   (f) All animals receive implants; the treatment group receives cell-containing capsules and controls receive empty capsules.
   (g) At 7 days post-implantation, animals are tranquilized and anesthetized and placed into the stereotaxic apparatus.
   (h) The skull cap is removed and the dura again reflected. Each animal is injected with 900 nmol QA in 5 μl approximately 2 mm posterior to the previous implant site [43].
   (i) The surgical site is closed as before.

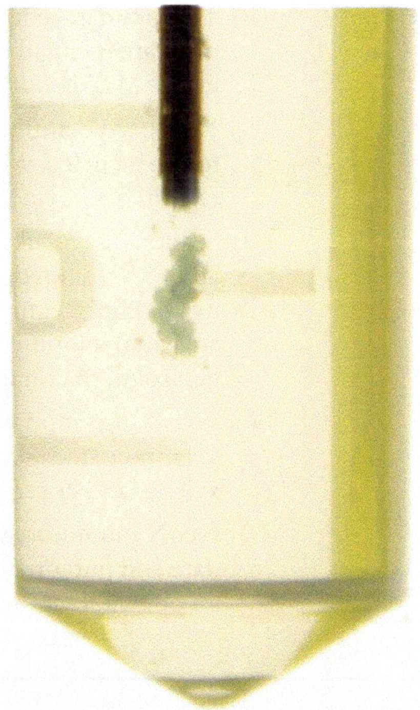

Fig. 5 Displacement of APA capsules into gelatin through a cannula loading system, modeling alginate capsule implantation in the brain. The capsule tract demonstrates compression of the 20-capsule stack into a shorter tract, with a width accommodating two capsules

3.5 Tissue Preparation of the Lesioned Striatum

1. Tissue isolation.
   (a) Four weeks following QA lesion, all animals are sacrificed.
   (b) Animals are anesthetized with pentobarbital (25 mg/kg, iv) before transcardial perfusion with 500 mL ice-cold heparinized saline (0.9%) followed by 1 L of 4% Zamboni's fixative.
   (c) The brain is removed and post-fixed in Zamboni's fixative for 48 h and cryoprotected in are-buffered saline solution (30% wt/vol sucrose/0.1 sodium).
   (d) Frozen sections are cut on a microtome at 40 μm for processing.
2. Immunohistochemistry of striatal neurons.
   (a) A mouse anti-neuronal nuclei (NeuN) monoclonal antibody is used, which was previously characterized against vertebrate neuron-specific nuclear protein [44].
      - NeuN identifies almost all CNS neuronal populations and distinguishes glia from atrophic neurons.

(b) To eliminate endogenous peroxidate activity, brain sections are incubated for 20 min in 0.1 M sodium peroxidate in 0.1 M Tris-buffered saline solution, pH 7.4.

(c) Nonspecific antibody binding is prevented by incubation for 1 h in 0.1 M PBS, pH 7.4, with 3% normal horse serum, 2% bovine serum albumin, and 0.1% Triton X-100.

(d) Sections are incubated for 48 h at 4 °C in the primary NeuN antibody (diluted 1:1,000), at room temperature in biotinylated horse anti-mouse IgG (1:200) for 1 h, and lastly in an avidin–biotin complex (1:500) for 75 min; the chromogen solution contains 0.05% DAB and 0.005% H2O2.

(e) Following immunostaining, sections are mounted on gelatin-coated slides, dehydrated in ethanol, dipped in xylene, and coverslipped.

(f) NeuN immunoreactive (NeuN-ir) neurons within the caudate and putamen nuclei are quantified stereologically with an optical fractionator unbiased sampling design [45–47].

4 Notes

1. To understand whether this phenomenon carried over to extended periods of storage in tissue culture and whether shelf-life impacts encapsulated CP potency, neonatal porcine CP was alginate-encapsulated and cultured in vitro for up to 7 months [35] prior to implantation into a rodent model of Huntington's Disease. In this experiment, CP microcapsules were held in culture medium for 1, 2, or 7 months and implanted into QA-lesioned striata of rats. More than 80% of encapsulated cells remained viable in each of the three groups prior to transplantation. Using behavioral testing for forelimb placement, control animals, which received QA injections without implantation, displayed impaired function of the contralateral forelimb along with significant weight-loss. However, CP-implanted animals maintained bodyweight levels and suffered no motor function impairment independent of the duration of in vitro cell culture. This result establishes the in vitro stability of encapsulated CP, and importantly, suggests that development into a therapeutic product may extend the shelf-life beyond a few weeks.
2. The CP is very sensitive to aging and displays many age-related phenotypes, including diminishing epithelial cell size and nuclei irregularity [19] and proliferation of Biondi Ring tangles and lipofuschin deposits [20]. CSF production is also compromised by up to 45% with a longer turnover [27]. To examine the effects of aging on CP potency, clusters harvested from rats at either 3-4 or 24 months of age were encapsulated, evaluated in

culture, and implanted into QA-lesioned rats [16]. In vitro, encapsulated CP from aged rats produced 33 % less VEGF than from young CP, and was more metabolically active.

3. In a study of nonhuman primates, isolated porcine CP cell clusters were confirmed to express transthyretin, zonula occludens 1 (ZO-1), and tubulin prior to encapsulation. In vitro, conditioned media from encapsulated CP protected cortical neurons from serum-deprivation-induced cell death and preserved the numbers and length of cortical neuronal neurites. In animals receiving QA injections, a lesion was created that extended throughout the caudate and putamen nuclei. Histologically, encapsulated CP transplants were associated with significant neuroprotective effects on QA-induced excitotoxicity in striatal neurons of cynomolgus primates, including preservation of intact striatal volume (672 vs. 447 mm^3) and stereological counts of NeuN-positive neurons (8 % vs. 43 % loss).

References

1. Iliff JJ, Wang M, Liao Y, Plogg BA, Peng W, Gundersen GA, Benveniste H, Vates GE, Deane R, Goldman SA, Nagelhus EA, Nedergaard M (2012) A paravascular pathway facilitates CSF flow through the brain parenchyma and the clearance of interstitial solutes, including amyloid beta. Sci Transl Med 4(147):147ra111. doi:10.1126/scitranslmed.3003748
2. Oshio K, Song Y, Verkman AS, Manley GT (2003) Aquaporin-1 deletion reduces osmotic water permeability and cerebrospinal fluid production. Acta Neurochir Suppl 86:525–528
3. Schwerk C, Tenenbaum T, Kim KS, Schroten H (2015) The choroid plexus – a multirole player during infectious diseases of the CNS. Front Cell Neurosci 9(80):1–11
4. Dziegielewska KM, Ek J, Habgood MD, Saunders NR (2001) Development of the choroid plexus. MicroscResTech52(1):5–20.doi:10.1002/1097-0029(20010101)52:1<5::AID-JEMT3>3.0.CO;2-J
5. Chodobski A, Szmydynger-Chodobska J (2001) Choroid plexus: target for polypeptides and site of their synthesis. Microsc Res Tech 52(1):65–82. doi:10.1002/1097-0029(20010101)52:1<65::AID-JEMT9>3.0.CO;2-4
6. Emerich DF, Skinner SJ, Borlongan CV, Vasconcellos AV, Thanos CG (2005) The choroid plexus in the rise, fall and repair of the brain. Bioessays 27(3):262–274
7. Thanos CG, Bintz BE, Goddard M, Boekelheide K, Hall S, Emerich DF (2011) Functional modulation of choroid plexus epithelial clusters in vitro for tissue repair applications. Cell Transplant 20(11-12):1659–1672. doi:10.3727/096368911X564985
8. ClinicalTrials.gov (2012) Open-label investigation of the safety and clinical effects of NTCELL in patients with Parkinson's disease. NIH. https://clinicaltrials.gov/ct2/show/NCT01734733. Accessed 15 Sept 2015
9. Borlongan CV, Skinner SJ, Geaney M, Vasconcellos AV, Elliott RB, Emerich DF (2004) Neuroprotection by encapsulated choroid plexus in a rodent model of Huntington's disease. Neuroreport 15(16):2521–2525
10. Borlongan CV, Skinner SJ, Geaney M, Vasconcellos AV, Elliott RB, Emerich DF (2004) Intracerebral transplantation of porcine choroid plexus provides structural and functional neuroprotection in a rodent model of stroke. Stroke 35(9):2206–2210
11. Emerich DF, Thanos CG, Goddard MB, Skinner SJ, Geaney M, Bell WJ, Bintz BE, Schneider PA, Chu Y, Babu RS, Borlongan CV, Boekelheide K, Hall S, Bryant B, Kordower JH (2006) Extensive neuroprotection by choroid plexus transplants in excitotoxin lesioned monkeys. Neurobiol Dis 23(2):471–480
12. Thanos CG, Emerich DF, Bintz BE, Goddard M, Mills J, Jensen R, Lombardi M, Hall S, Boekelheide K (2009) Secreted products from the porcine choroid plexus accelerate the healing of cutaneous wounds. Cell Transplant 18(12):1395–1409. doi:10.3727/096368909X12483162197402

13. Preston JE (2001) Ageing choroid plexus-cerebrospinal fluid system. Microsc Res Tech 52(1):31–37. doi:10.1002/1097-0029(20010101)52:1<31::AID-JEMT5>3.0.CO;2-T
14. Serot JM, Bene MC, Foliguet B, Faure GC (2000) Morphological alterations of the choroid plexus in late-onset Alzheimer's disease. Acta Neuropathol 99(2):105–108
15. Wen GY, Wisniewski HM, Kascsak RJ (1999) Biondi ring tangles in the choroid plexus of Alzheimer's disease and normal aging brains: a quantitative study. Brain Res 832(1-2):40–46
16. Emerich DF, Schneider P, Bintz B, Hudak J, Thanos CG (2007) Aging reduces the neuroprotective capacity, VEGF secretion, and metabolic activity of rat choroid plexus epithelial cells. Cell Transplant 16(7):697–705
17. Wise AK, Fallon JB, Neil AJ, Pettingill LN, Geaney MS, Skinner SJ, Shepherd RK (2011) Combining cell-based therapies and neural prostheses to promote neural survival. Neurotherapeutics 8(4):774–787. doi:10.1007/s13311-011-0070-0
18. Borlongan CV, Skinner SJ, Geaney M, Vasconcellos AV, Elliott RB, Emerich DF (2004) CNS grafts of rat choroid plexus protect against cerebral ischemia in adult rats. Neuroreport 15(10):1543–1547
19. Greenwood S, Swetloff A, Wade AM, Terasaki T, Ferretti P (2008) Fgf2 is expressed in human and murine embryonic choroid plexus and affects choroid plexus epithelial cell behaviour. Cerebrospinal Fluid Res 5(20)
20. Thouvenot E, Lafon-Cazal M, Demettre E, Jouin P, Bockaert J, Marin P (2006) The proteomic analysis of mouse choroid plexus secretome reveals a high protein secretion capacity of choroidal epithelial cells. Proteomics 6:5941–5952
21. Gard AL, Gavin E, Solodushko V, Pennica D (2004) Cardiotrophin-1 in choroid plexus and the cerebrospinal fluid circulatory system. Neuroscience 127(1):43–52
22. Thanos CG, Schneider PA, Bintz BE, Jensen R, Bryant B, Bell WJ, Hudak J, Emerich DF (2007) The in vitro expression and secretion of vascular endothelial growth factor from free and alginate-polyornithine encapsulated choroid plexus epithelium. Tissue Eng 13(4):747–756
23. Ericsson A, Liu C, Hart RP, Sawchenko PE (1995) Type 1 interleukin-1 receptor in the rat brain: distribution, regulation, and relationship to sites of IL-1-induced cellular activation. J Comp Neurol 361(4):681–698
24. Nadeau S, Rivest S (1999) Effects of circulating tumor necrosis factor on the neuronal activity and expression of the genes encoding the tumor necrosis factor receptors (p55 and p75) in the rat brain: a view from the blood-brain barrier. Neuroscience 93(4):1449–1464
25. Szmydynger-Chodobska J, Strazielle N, Zink BJ, Ghersi-Egea JF, Chodobski A (2009) The role of the choroid plexus in neutrophil invasion after traumatic brain injury. J Cereb Blood Flow Metab 29(9):1503–1516
26. Thanos CG, Bintz BE, Emerich DF (2009) Microencapsulated choroid plexus epithelial cell transplants for repair of the brain, vol 670, Advances in experimental medicine and biology: therapeutic applications of cell microencapsulation. Landes Bioscience, Austin, TX
27. Aliaghaei A, Digaleh H, Khodagholi F, Ahmadiani A (2015) Encapsulated choroid plexus epithelial cells actively protect against intrahippocampal abeta-induced long-term memory dysfunction; upregulation of effective neurogenesis with the abrogated apoptosis and neuroinflammation. J Mol Neurosci 56(3):708–721. doi:10.1007/s12031-015-0492-y
28. Lim F, Sun AM (1981) Microencapsulated islets in diabetic rats. Science 213(4512):1146. doi:10.1126/science.213.4512.1146-a
29. Omer A, Duvivier-Kali V, Fernandes J, Tchipashvili V, Colton CK, Weir GC (2005) Long-term normoglycemia in rats receiving transplants with encapsulated islets. Transplantation 79(1):52–58
30. Khanna O, Larson JC, Moya ML, Opara EC, Brey EM (2012) Generation of alginate microspheres for biomedical applications. J Vis Exp (66). doi: 10.3791/3388
31. Pareta R, McQuilling JP, Sittadjody S, Jenkins R, Bowden S, Orlando G, Farney AC, Brey EM, Opara EC (2014) Long-term function of islets encapsulated in a redesigned alginate microcapsule construct in omentum pouches of immune-competent diabetic rats. Pancreas 43(4):605–613. doi:10.1097/MPA.0000000000000107
32. Moya ML, Cheng MH, Huang JJ, Francis-Sedlak ME, Kao SW, Opara EC, Brey EM (2010) The effect of FGF-1 loaded alginate microbeads on neovascularization and adipogenesis in a vascular pedicle model of adipose tissue engineering. Biomaterials 31(10):2816–2826. doi:10.1016/j.biomaterials.2009.12.053
33. Fjord-Larsen L, Kusk P, Emerich DF, Thanos C, Torp M, Bintz B, Tornoe J, Johnsen AH, Wahlberg LU (2012) Increased encapsulated cell biodelivery of nerve growth factor in the brain by

transposon-mediated gene transfer. Gene Ther 19(10):1010–1017. doi:10.1038/gt.2011.178

34. Skinner SJ, Geaney MS, Rush R, Rogers ML, Emerich DF, Thanos CG, Vasconcellos AV, Tan PL, Elliott RB (2006) Choroid plexus transplants in the treatment of brain diseases. Xenotransplantation 13(4):284–288. doi:10.1111/j.1399-3089.2006.00310.x
35. Emerich DF, Thanos CG (2006) In vitro culture duration does not impact the ability of encapsulated choroid plexus transplants to prevent neurological deficits in an excitotoxin-lesioned rat model of Huntington's disease. Cell Transplant 15(7):595–602
36. Thanos CG, Bintz B, Emerich DF (2010) Microencapsulated choroid plexus epithelial cell transplants for repair of the brain. Adv Exp Med Biol 670:80–91
37. Borlongan CV, Thanos CG, Skinner SJ, Geaney M, Emerich DF (2008) Transplants of encapsulated rat choroid plexus cells exert neuroprotection in a rodent model of Huntington's disease. Cell Transplant 16(10):987–992
38. Luo XM, Lin H, Wang W, Geaney MS, Law L, Wynyard S, Shaikh SB, Waldvogel H, Faull RL, Elliott RB, Skinner SJ, Lee JE, Tan PL (2013) Recovery of neurological functions in non-human primate model of Parkinson's disease by transplantation of encapsulated neonatal porcine choroid plexus cells. J Parkinsons Dis 3(3):275–291. doi:10.3233/JPD-130214
39. Skinner SJ, Lin H, Geaney MS, Gorba T, Elliott RB, Tan PL (2011) Restoration of motor control and dopaminergic activity in rats with unilateral 6-hydroxy-dopamine lesions. Regen Med 6(3):319–326. doi:10.2217/rme.11.15
40. Thanos CG, Elliott RB (2009) Encapsulated porcine islet transplantation: an evolving therapy for the treatment of type I diabetes. Expert Opin Biol Ther 9(1):29–44
41. Sharma HS, Tang ZH, Gho BC, Verdouw PD (1995) Nucleotide sequence and expression of the porcine vascular endothelial growth factor. Biochim Biophys Acta 1260(2):235–238
42. Li SH, Cheng AL, Li H, Li XJ (1999) Cellular defects and altered gene expression in PC12 cells stably expressing mutant huntingtin. J Neurosci 19(13):5159–5172
43. Emerich DF, Winn SR, Hantraye PM, Peschanski M, Chen EY, Chu Y, McDermott P, Baetge EE, Kordower JH (1997) Protective effect of encapsulated cells producing neurotrophic factor CNTF in a monkey model of Huntington's disease. Nature 386(6623):395–399. doi:10.1038/386395a0
44. Wolf HK, Buslei R, Schmidt-Kastner R, Schmidt-Kastner PK, Pietsch T, Wiestler OD, Blumcke I (1996) NeuN: a useful neuronal marker for diagnostic histopathology. J Histochem Cytochem 44(10):1167–1171
45. West MJ, Slomianka L, Gundersen HJ (1991) Unbiased stereological estimation of the total number of neurons in the subdivisions of the rat hippocampus using the optical fractionator. Anat Rec 231(4):482–497. doi:10.1002/ar.1092310411
46. Kordower JH, Chu Y, Stebbins GT, DeKosky ST, Cochran EJ, Bennett D, Mufson EJ (2001) Loss and atrophy of layer II entorhinal cortex neurons in elderly people with mild cognitive impairment. Ann Neurol 49(2):202–213
47. Chu Y, Kompoliti K, Cochran EJ, Mufson EJ, Kordower JH (2002) Age-related decreases in Nurr1 immunoreactivity in the human substantia nigra. J Comp Neurol 450(3):203–214. doi:10.1002/cne.10261

Chapter 20

Microencapsulation of Stem Cells for Therapy

Shirae K. Leslie, Ramsey C. Kinney, Zvi Schwartz, and Barbara D. Boyan

Abstract

An increasing demand to regenerate tissues from patient-derived sources has led to the development of cell-based therapies using autologous stem cells, thereby decreasing immune rejection of scaffolds coupled with allogeneic stem cells or allografts. Adult stem cells are multipotent and are readily available in tissues such as fat and bone marrow. They possess the ability to repair and regenerate tissue through the production of therapeutic factors, particularly vasculogenic proteins. A major challenge in cell-based therapies is localizing the delivered stem cells to the target site. Microencapsulation of cells provides a porous polymeric matrix that can provide a protected environment, localize the cells to one area, and maintain their viability by enabling the exchange of nutrients and waste products between the encapsulated cells and the surrounding tissue. In this chapter, we describe a method to produce injectable microbeads containing a tunable number of stem cells using the biopolymer alginate. The microencapsulation process involves extrusion of the alginate suspension containing cells from a microencapsulator, a syringe pump to control its flow rate, an electrostatic potential to overcome capillary forces and a reduced Ca^{++} cross-linking solution containing a nutrient osmolyte, to form microbeads. This method allows the encapsulated cells to remain viable up to three weeks in culture and up to three months in vivo and secrete growth factors capable of supporting tissue regeneration.

Key words Stem cells, Microencapsulation, Hydrogel, Alginate, Cell-based therapy, Microbeads, Degradation

1 Introduction

The therapeutic nature of stem cells derives from their ability to self-renew as well as maintain the ability to differentiate along various lineages. In addition to differentiating into tissue specific cells and contributing directly to tissue repair, stem cells can create a microenvironment to support the regeneration of tissues or organs by secreting factors that elicit a regenerative response from surrounding cells. Both of these roles require stem cells to be delivered to the desired site and to remain at the site long enough to achieve the therapeutic goal.

The most common method of delivery is to infuse a stem cell suspension via the vasculature [1]. Even when cell suspensions are injected directly to the intended site, most of the cells are gone

Emmanuel C. Opara (ed.), *Cell Microencapsulation: Methods and Protocols*, Methods in Molecular Biology, vol. 1479,
DOI 10.1007/978-1-4939-6364-5_20, © Springer Science+Business Media New York 2017

within 24 h. There are a number of reasons for this, including cell migration and necrosis resulting from the harsh environment. Additionally, the host's immune system may reject the newly introduced stem cells. A delivery system that shields and creates a temporary home for the cells can improve their viability and increase their residence time at the delivery site.

Porous polymer hydrogels like alginate and gelatin have been used as cell-delivery vehicles for some time [2]. Cells can be incorporated into the gels by simple dripping, vibration-assisted dripping, and emulsion-based encapsulation [3]. These methods result in microbeads that are generally about 500 μm in diameter. Microbeads of this size can limit the exchange of nutrients and waste products and may not be injectable. To overcome these limitations, we developed an electrostatic encapsulation method that results in microbeads that are 200 μm in diameter, are easily injectable, and do not have reduced diffusivity of nutrients and waste products [4].

Alginate is an accessible, block co-polymer that consists of pure mannuronic acid (M) blocks or pure guluonric acid (G) blocks, or alternating MG blocks [5]. Due to its ability to instantaneously cross-link after exposure to divalent cations, alginate is used as a hydrogel. We have successfully produced microbeads using a variety of alginate formulations, including high mannuronate content alginate with low, medium, and high viscosities [4] (Fig. 1).

Here we report the microencapsulation of cells using low viscosity, high mannuronate content alginate (Fig. 2). The cross-linking solution contains calcium chloride and glucose and is buffered using 4-(2-hydroxyethyl)-1-piperazineethanesulfonic acid (HEPES). Calcium chloride provides divalent cations to cross-link alginate between mannuronate residues. Glucose is added as nutrient osmolyte in order to maintain physiological osmolality of the cross-linking solution for the cells, thus enabling reduced Ca^{++} content and enhancing cell viability [4]. HEPES prevents in vivo

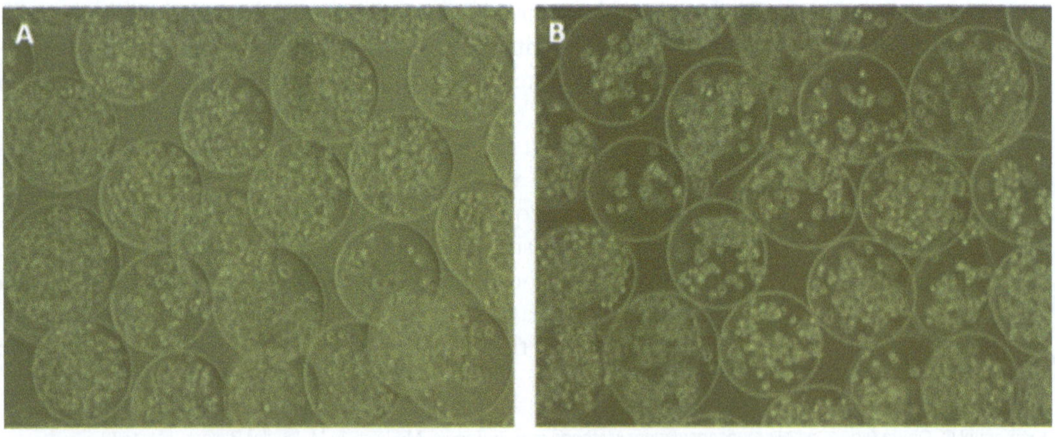

Fig. 1 Microencapsulated ASCs derived from (**a**) rat and (**b**) human

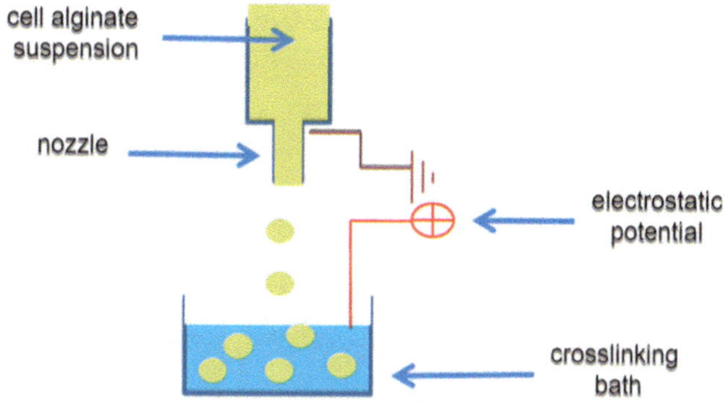

Fig. 2 Schematic of electrostatic potential-based microencapsulation

calcification by maintaining pH at or below pH 7.3 [6]. This method produces microbeads with a high surface area to volume ratio, which is beneficial for molecular transport. Using this method adipose stem cells (ASCs) retain high viability during microencapsulation and remain viable for at least 3 weeks in culture [7].

Depending on the desired therapeutic use of the microencapsulated cells, the residence time for cells within the microbeads may vary. In order for this to be possible the material used to encapsulate the cells must be able to undergo controlled degradation in vivo. For example, a nondegradable polymer can be used if the therapeutic use of microencapsulated stem cells is insulin production in the case of diabetic patients [8]. A degradable polymer can be used if it is required for the stem cells to be released to support tissue regeneration. To control the rate of cell release, we have developed a method by which the degradation of alginate can be modulated enzymatically (Fig. 3). Alginate-lyase cleaves the glycosidic bonds in alginate specifically between the mannuronate and guluronate residues through a β-elimination reaction [9].

2 Materials

2.1 Alginate Microbeads

1. 2% (w/v) Alginate: (PRONOVA™ UP, ultrapure sodium alginates, FMC BioPolymer) made up in filter sterilized saline (0.9% w/v) (Ricca Chemical, Arlington, TX).
2. Calcium cross-linking solution:
   (a) Formulation 1: 50 mM $CaCl_2$, 150 mM glucose with 15 mM 4-(2-hydroxyethyl)-1-piperazineethanesulfonic acid (HEPES) (Sigma Aldrich) dissolved in ultrapure water, pH with 1 M NaOH to 7.3 and sterile filtered.
   (b) Formulation 2: 75 mM $CaCl_2$, 90 mM glucose with 10 mM HEPES dissolved in ultrapure water, pH with 1 M NaOH to 7.3 and sterile filtered.

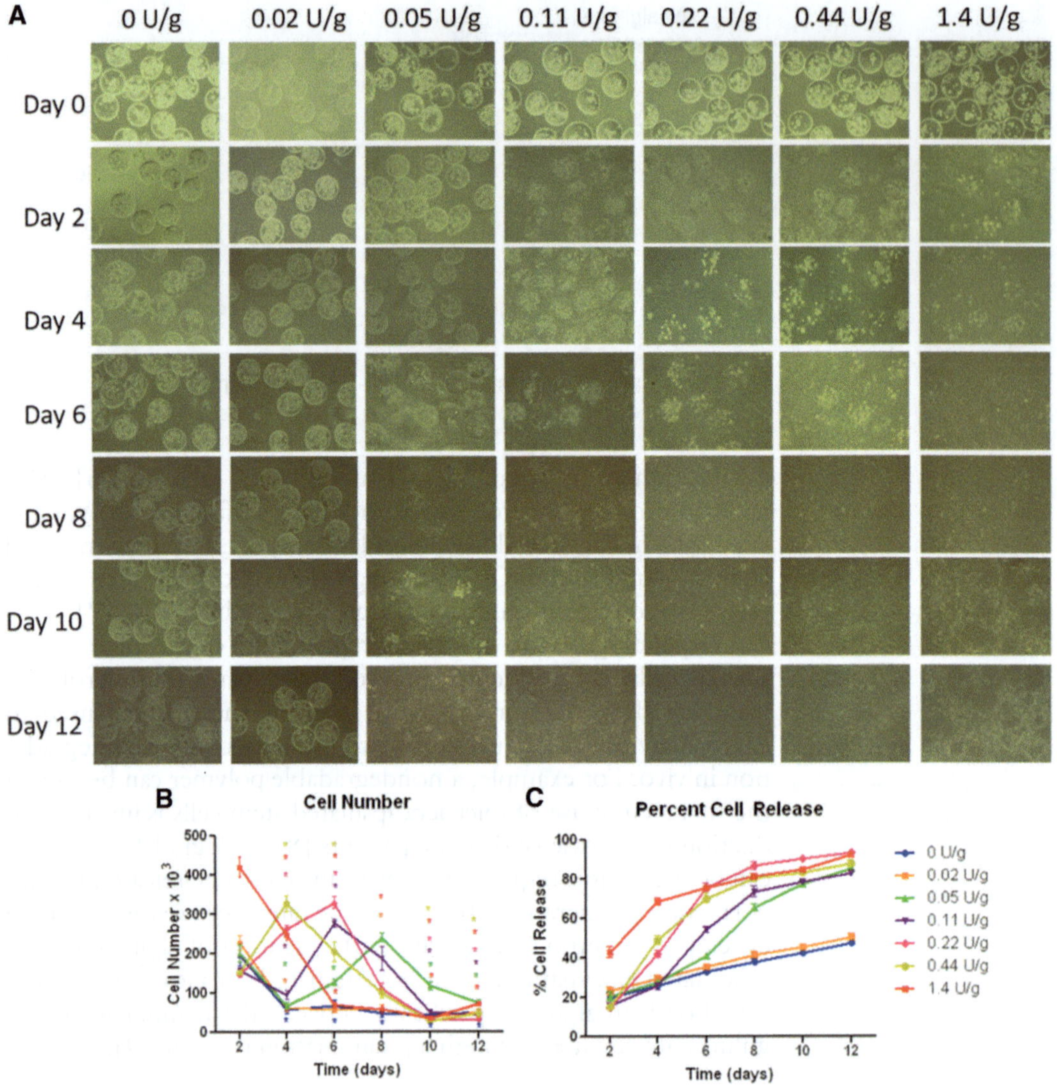

Fig. 3 Release of adipose stem cells (ASCs) from alginate (medical grade) microbeads with different formulations of alginate-lyase (U):alginate (g) over a 12-day period. For this study a 0.12 mm inner diameter nozzle was used, resulting in microbeads that were 200 μm in diameter. Each microbead contained 37 cells. (**a**) Microbeads under an inverted light microscope (mag. = 1×x). (**b**) The number of released ASCs per unit time. (**c**) Percent cell release. Data were analyzed using one-way ANOVA followed by Bonferroni's modification of Student's *t*-test. * $p < 0.05$ vs. day 2 per group. The data shown are similar to previously published data using microbeads generated using a 0.175 mm inner diameter nozzle [10]

3. Calcium supplemented saline: 1.8 mM Ca^{++} using $CaCl_2$ (Sigma Aldrich) dissolved in saline (0.9% w/v) (Ricca Chemical) and filter sterilized.
4. Dulbecco's modified Eagle medium (DMEM) (Invitrogen, Carlsbad, CA).

5. Nisco Encapsulator VAR V1 LIN-0043 (Nisco Engineering AG, Zurich, Switzerland) with a nozzle:
   (a) Formulation 1: 0.175 mm (inner diameter).
   (b) Formulation 2: 0.12 mm (inner diameter).
6. KD scientific syringe pump (Holliston, MA).

2.2 Controlled Release Microbeads

1. Alginate-lyase (28000 U/g, Sigma Aldrich).
2. Alginate-lyase stock solution (10 U/ml) in ultrapure water followed by filter sterilization. Dilutions are made in ultrapure water to achieve a concentration needed for a particular degradation profile.
3. 4 % (w/v) Alginate made up in filter sterilized saline (0.9 % w/v).
4. Calcium cross-linking solution: 75 mM $CaCl_2$, 90 mM glucose and 10 mM HEPES in ultrapure water; pH with 1 M NaOH to 7.3 and sterile filtered.
5. Nisco Encapsulator VAR V1 LIN-0043 with a 0.12 mm nozzle.

3 Methods

The methods used in the Boyan-Schwartz laboratory to microencapsulate ASCs are provided below.

All detachable portions of the microencapsulator (nozzle, arm supporting the nozzle, plastic tube connecting the syringe and nozzle, pyrex dish used for the cross-linking solution) are cleaned and autoclaved prior to microencapsulation. The pump and microencapsulator are sterilized with 70% ethanol and UV-sterilized overnight in the cell culture hood. On the day of encapsulation all components are assembled in the cell culture hood.

3.1 Alginate Microbeads

1. Cells are encapsulated in UV-sterilized research-grade alginate at a concentration of 10×10^6 cells/ml and cross-linked in a calcium solution containing glucose.
2. Medical-grade alginate is dissolved in 0.9% (w/v) saline at a concentration of 40 mg/ml.
3. Cells are suspended in saline.
4. After combining equal volumes of the cell suspension and 4% alginate, the alginate–cell suspension is transferred to a sterile syringe pump, which is attached to the Nisco Encapsulator. Microbeads incorporating cells are created using a 4 ml/h flow rate and an electrostatic potential of 6 kV. The microbeads are then cross-linked in a solution containing 50 mM $CaCl_2$ and 150 mM glucose or another nutrient osmolyte so that the ionic strength of the suspension is maintained at 300–330 mOsm [4, 6].

5. The encapsulator is equipped with a nozzle with an inner diameter of 0.175 mm, which together with the flow rate and the electrostatic potential, controls the microbead diameter.
6. The number of cells per bead is controlled by the cell density of the starting suspension (*see* **Note 1**).
7. Alginate microbeads are allowed to cross-link for an additional 10 min.
8. After washing with 0.9% (w/v) saline to remove the excess Ca^{++}, the microbeads are suspended in DMEM, or other suitable medium.

Parameters can be modified to achieve the desired microbead size when properties of the alginate differ in terms of chain length, chemical composition, or viscosity, or when components of the encapsulator differ. To generate microbeads using medical-grade low viscosity, high mannuronate (LVM) alginate, we used a cross-linking solution containing 75 mM $CaCl_2$, 90 mM glucose, and 10 mM HEPES. The flow rate was 5 ml/h and the nozzle had an inner diameter of 0.12 mm [10] (*see* **Note 2**). Alginate microbeads were allowed to cross-link for an additional 10 min. After washing with 0.9% (w/v) saline to remove the excess Ca^{++} the microbeads were suspended in saline containing 1.8 mM Ca^{2+} (*see* **Notes 3** and **4**).

3.2 Controlled Release Microbeads

Humans do not possess enzymes that can metabolize alginate. Some degradation of the hydrogel does occur as Ca^{++} is lost through diffusion. To control the rate of degradation, the enzyme alginate-lyase is incorporated into microbeads at various ratios of alginate-lyase (units [U] of enzyme activity) to alginate [g] (*see* **Note 5**). Equal volumes of 4% LVM alginate, and alginate-lyase 0.9% (w/v) saline are mixed at 132 × *g* for 1 min to form a final 2% (w/v) alginate solution. Prior to the combination, both components are kept at 4°C (*see* **Note 6**). For example, to form microbeads with a 1.75 U/g formulation, 1.5 ml of 4% (w/v) alginate is combined with 1.5 ml of 5 U/ml alginate-lyase, and then mixed for 1 min at 132 × *g*. Cells are suspended in the mixture at 10×10^6 cells/ml and microbeads are produced using a Nisco Encapsulator VAR V1 LIN-0043 at a 5 ml/h flow rate, nozzle with an inner diameter of 0.12 mm, an electrostatic potential of 6 kV, followed by cross-linking in a solution containing 75 mM $CaCl_2$, 90 mM glucose and 10 mM HEPES [10]. Degradable alginate microbeads are allowed to cross-link for an additional 10 min. After washing with 0.9% (w/v) saline to remove the excess Ca^{++} the microbeads are suspended in saline containing1.8 mM Ca^{2+}.

3.3 Advantages

The small diameter of these microbeads allows them to be delivered by injection using an 18 G needle (*see* **Note 7**). The encapsulated stem cells are viable for at least 3 weeks in vitro [10]. The microencapsulated stem cells can secrete growth factors that are important in osteogenesis, chondrogenesis, and angiogenesis [11]. The number of cells per bead can be controlled by the number of cells in the starting solution. Cells are distributed homogenously throughout the bead. The polymer matrix of the microbeads allows growth factors to diffuse from the alginate microbeads to the surrounding area while retaining a portion of the growth factors within the microbeads. The production of these growth factors can be controlled by the culture media used, for example: mesenchymal stem cell growth medium (MSCGM), osteogenic differentiation medium (OM), or chondrogenic differentiation medium (CM) (*see* **Note 8**). The microbeads localize cells at the delivered site. Controlled release of cells from the microbeads can be achieved using alginate-lyase.

3.4 Methods for Analysis

3.4.1 Quantifying Growth Factors Secreted by Microencapsulated Stem Cells

Following incubation, conditioned media are removed and the microbeads are collected in a conical test tube. The microbeads containing cells are uncross-linked with 5 ml of 82.5 mM sodium citrate, frozen at −80 °C, and lyophilized. Lyophilized samples are reconstituted in 0.75 ml of 1 U/ml alginate-lyase solution and incubated for 1 h at 37 °C. The digest is sonicated for 0.5 min and assayed for growth factors of interest using enzyme-linked immunosorbent assays (ELISAs). The growth factors released into the conditioned media are also quantified using ELISAs.

3.4.2 Quantifying mRNA Levels of Microencapsulated Cells

Cells are collected by centrifugation after uncross-linking the microbeads. TRizol® [Invitrogen] is added to the resulting cell pellet and homogenized using a Qiashredder (QIAGEN, Valencia, CA). RNA is extracted using chloroform (Sigma Aldrich) and isopropanol (Sigma Aldrich), and quantified with the Nanodrop Spectrophotometer (Thermo Scientific, Waltham, MA). Samples are converted to cDNA by reverse transcription of 1 μg RNA with random primers (Applied Biosystems, Warrington, UK) and Multiscribe Reverse Transcriptase (Applied Biosystems). The cDNA is analyzed for mRNA levels by real-time PCR using the StepOne Plus PCR System (Applied Biosystems).

3.4.3 Cell Release

Microbeads are suspended in culture medium in cell strainers so that released cells are able to enter the surrounding media. Cells are collected by centrifugation at 2000 × *g* for 10 min, followed by two consecutive washes with saline.

3.4.4 Live/Dead Analysis

Microbeads (200 μl aliquots) are transferred to a chamber slide and incubated for 30 min at 37 °C in 0.3 ml 0.9 % saline containing 4 μM ethidium homodimer-1 and 2 μM calcein-AM. Images are obtained using a Zeiss LSM 700–405 confocal microscope (Carl Zeiss MicroImaging, Inc., Thornwood, NY). At least three images are obtained per sample. Red (dead) and green (live) cells are counted using Image-Pro Plus software (version 4.5.1.22, Media Cybernetics, Rockville, MD) to determine the percent viability.

3.4.5 In Vivo Imaging

Prior to microencapsulation, ASCs are labeled with XenoLight DiR (1,1′-dioctadecyltetramethyl indotricarbocyanine iodide) fluorescent dye (PerkinElmer, Waltham, MA), after first determining the optimal dye concentration for labeling the cells while maintaining viability. We have determined labeling conditions for rat ASCs over a 2-week period with dye concentrations ranging from 4.2 to 135 μg/ml. Briefly, a million cells were incubated in the respective dye concentration for 30 min at 37 °C. After labeling was complete, cells were washed with phosphate buffered saline twice. Cells were microencapsulated using LVM alginate and a 75 mM $CaCl_2$, 90 mM glucose, and 10 mM HEPES cross-linking solution. Following injection of the microencapsulated cells into the gastrocnemius or deltoid of athymic nude mice, Odyssey Infrared Imaging System (Li-Cor, Lincoln, NE) was used to identify their presence over time.

4 Notes

1. The number of cells per bead is controlled by the number of cells per ml of suspension.
2. One of the obstacles in this microencapsulation method stems from the presence of cell clumps clogging the nozzle. However, this can be reduced by ensuring the cell suspension is homogenously mixed by pipetting up and down to separate any cell clumps. In the case that the nozzle becomes clogged during encapsulation, stop the pump, pull the alginate cell suspension into the syringe, remove the nozzle, and flush the nozzle by attaching the nozzle to a syringe already filled with sterile saline. Alternatively, the nozzle can be flushed with sodium citrate after which saline should be used to flush the system at least three times to ensure that all the sodium citrate is removed.
3. Once microbeads are made it is very important that they are stored in a medium with at least 1.8 mM Ca^{++} to prevent alginate microbeads uncross-linking. In addition, beads are not placed in media with any form a chelator (EDTA, sodium citrate, phosphate), which would result in the microbeads being uncross-linked.
4. Collecting microbeads immediately after production or while changing the media should never involve centrifugation. Instead, allow the microbeads to settle in a conical tube and then remove liquid after which the microbeads can be resuspended in an appropriate media.
5. Incorporating alginate-lyase into alginate microbeads to achieve degradation is specific to the type of alginate used and alginate-lyase. Therefore, an optimization step should be done with a range of ratios (units of alginate-lyase, U/ alginate g) to find the best ratio needed for the desired release profile.

Different alginates have varying compositions of G and M units. In addition, alginate-lyases are specific to the glycosidc bonds whether it is between MG units or MM or GG units.

6. In the production of degradable alginate microbeads it is very important all components—alginate, alginate-lyase solution, and cross-linking solution are kept on ice. This will temporarily decrease the alginate-lyase's activity, thus allowing the formation of the microbeads and preventing premature degradation. Once the beads are formed they should be transferred to a 4 °C refrigerator in between the washing steps.
7. To achieve the maximum number of microbeads for an injection, be sure to transfer microbeads to a syringe using at least an 18 G needle, invert the syringe and remove any of the extra liquid once the microbeads have settled.
8. There is a 20% expansion in microbead diameter after moving the beads out of the cross-linking solution into culture media.

References

1. Kurtz A (2008) Mesenchymal stem cell delivery routes and fate. Int J Stem Cells 1(1):1–7
2. Wilson JL, McDevitt TC (2013) Stem cell microencapsulation for phenotypic control, bioprocessing, and transplantation. Biotechnol Bioeng 110(3):667–682. doi:10.1002/bit.24802
3. Steele JA, Halle JP, Poncelet D, Neufeld RJ (2014) Therapeutic cell encapsulation techniques and applications in diabetes. Adv Drug Deliv Rev 67–68:74–83. doi:10.1016/j.addr.2013.09.015
4. Boyan BD, Kinney RC, Schwartz Z (2012) Microencapsulation of cells in hydrogels using electrostatic potentials. US Patent 8,202,701 B2, 19 June 2012
5. Rowley JA, Madlambayan G, Mooney DJ (1999) Alginate hydrogels as synthetic extracellular matrix materials. Biomaterials 20(1):45–53
6. Lee CS, Moyer HR, Gittens RA, Williams JK, Boskey AL, Boyan BD, Schwartz Z (2010) Regulating in vivo calcification of alginate microbeads. Biomaterials 31(18):4926–4934. doi:10.1016/j.biomaterials.2010.03.001
7. Moyer HR, Kinney RC, Singh KA, Williams JK, Schwartz Z, Boyan BD (2010) Alginate microencapsulation technology for the percutaneous delivery of adipose-derived stem cells. Ann Plast Surg 65(5):497–503
8. de Vos P, Faas MM, Strand B, Calafiore R (2006) Alginate-based microcapsules for immunoisolation of pancreaticislets. Biomaterials 27(32):5603–5617. doi:10.1016/j.biomaterials.2006.07.010
9. Yamasaki M, Moriwaki S, Miyake O, Hashimoto W, Murata K, Mikami B (2004) Structure and function of a hypothetical Pseudomonas aeruginosa protein PA1167 classified into family PL-7: a novel alginate lyase with a beta-sandwich fold. J Biol Chem 279(30):31863–31872. doi:10.1074/jbc.M402466200
10. Leslie SK, Cohen DJ, Sedlaczek J, Pinsker EJ, Boyan BD, Schwartz Z (2013) Controlled release of rat adipose-derived stem cells from alginate microbeads. Biomaterials 34(33):8172–8184. doi:10.1016/j.biomaterials.2013.07.017
11. Lee CS, Burnsed OA, Raghuram V, Kalisvaart J, Boyan BD, Schwartz Z (2012) Adipose stem cells can secrete angiogenic factors that inhibit hyaline cartilage regeneration. Stem Cell Res Ther 3(4):35. doi:10.1186/scrt126

Chapter 21

Microencapsulated Cells for Cancer Therapy

L. Saenz del Burgo, J. Ciriza, R.M. Hernández, G. Orive, and J.L. Pedraz

Absract

The microencapsulation of different types of cells that are able to produce therapeutic factors is being investigated for the treatment of several human diseases. Most efforts are focused on chronic and degenerative diseases as this strategy could become an alternative to some commonly used parenteral treatments that need to be repeatedly administered. But, this approach has also been investigated in the field of oncology with the aim of providing immunomodulatory antibodies that are able to enhance the patient's inherent immune response against the tumor. These kind of treatments would provide the patient with the therapeutic drug produced in situ, de novo, and in a sustained way, making the therapy more comfortable.

Although different devices are nowadays available to produce cell-enclosing alginate-microcapsules, here, we describe the most important steps and advices in order to fabricate alginate-poly-L-lysine-alginate microcapsules containing hybridoma cells for cancer management using an electrostatic bead generator, and how to evaluate the viability of those cells over the time.

Key words Cell microencapsulation, Electrostatic bead generator, Alginate, Hybridoma cells, Controlled drug delivery system, Microtechnology, Cancer

1 Introduction

Cell microencapsulation is an advanced technology by which small beads of semisolid material (between ~ 100 μm to few millimeters), containing the desired type of cell, are surrounded by a continuous layer of polymeric material in order to obtain an effective long-term drug delivery system. The enclosed cells such as primary cells, transfected cells, mesenchymal stem cells, etc., will produce in situ, de novo, and in a sustained manner the therapeutic agent/s of interest, making this strategy really interesting for the treatment of chronic diseases avoiding the need of repeated administrations or even immunosupression [1]. The inner component of the microcapsules is known as core or matrix and is where the encapsulated cells will live and proliferate in a controlled manner. Although, over the past couple of decades different materials have been tested for cell encapsulation, sodium alginate

L. Saenz del Burgo and J. Ciriza contributed equally with all other contributors.

Emmanuel C. Opara (ed.), *Cell Microencapsulation: Methods and Protocols*, Methods in Molecular Biology, vol. 1479, DOI 10.1007/978-1-4939-6364-5_21, © Springer Science+Business Media New York 2017

has been by far one of the most used one [2–5]. This naturally occurring soluble salt of the alginic acid is a biocompatible polysaccharide that is safe for human application [6]. When microcapsules are fabricated as a continuous drug delivery system for internal use, it is necessary to provide them with an external wall which is also known as coating or membrane. This membrane is essential in order to protect the cells when they are implanted as it will avoid the entry of most of the elements of the host immune system. Also, it can be modified in order to control the pore size and therefore, the molecules that can enter and exit the capsule, or to give higher mechanical stability to the capsule. In addition, it needs to assure that there is a correct flow of nutrients as well as O_2 to the inside core in order to nourish the cells, while the waste products and therapeutic molecules that they produce are easily released [7]. Importantly, microcapsules can be configured so that they can be removed and replaced when necessary [1, 8].

Hearteningly, numerous clinical trials have been conducted and also are ongoing nowadays in order to determine the efficacy of this advanced therapy on several human diseases. Probably, the management of the type 1 diabetes mellitus is the field where cell microencapsulation has developed more widely. The pancreatic islets are protected from the host patient thanks to the microenvironment that the capsule provides [9, 10]. Nevertheless, this technology is showing its potential in many other different fields. Even though the implantation of these microcapsules in places such as the brain in order to provide the central nervous system with, for example, trophic factors needs to be made by invasive means, the capacity of the encapsulated cells to fabricate the therapeutic factors in a long term, in a stable manner and at sufficient quantities could be an enormous advance for chronic diseases such as Alzheimer's disease [11], Huntington's disease [12], amyotrophic lateral sclerosis [13] or Parkinson's disease [14].

Another clinic discipline where cell microencapsulation is being tested lately is oncology. In fact, alginate microcapsules can be designed to work as immunotherapeutic organoids where capsules can act as cell factories that will secrete in a sustained manner tumor targeting antibodies [15]. These antibodies would be able to redirect the patient's immune system effector cells towards the tumor mass increasing the effectiveness of the inherent immune response against the tumor [16, 17]. Also, a more complete strategy consist on the design of bispecific antibodies in order to both convert tumor-associated antigens into recruitment points for the immune system, and, at the same time, promote the activation of effector T cells [18, 19]. These cells producing these bispecific antibodies can be microencapsulated in alginate-poly-L-lysine-alginate (APA) microcapsules to work as continuous drug delivery systems [15].

A different approach that also has been studied in cancer management is the microencapsulation of hybridoma cells that are able to produce immunostimulatory monoclonal antibodies [20]. These antibodies are capable of upregulating the immune response of the patient against the tumor by different means and some of them have already entered clinical trials in cancer patients [17, 21–24]. Interestingly, a combination of different kind of antibodies seems to induce synergic antitumor responses in cancer animal models [25]. Therefore, the microencapsulation of immunostimulatory antibodies producing hybridoma cells seems an alternative strategy in order to increase the efficacy of this approach as the production of the antibodies would be sustained over the time avoiding the need of repeated intravenous administrations. Our group showed that the microcapsules acted as cells factories for the production of anti-CD137 and anti-OX40 monoclonal antibodies which gave rise to immune-mediated tumor rejections when implanted under the skin of mice with established CT26 colon carcinomas [20]. Therefore, these APA capsules could provide an alternative to the exogenous and repeated infusion of naked immunoglobulins for enhancing the tumor-specific cellular immunity.

These two tumor-directed therapies based on the use of antibodies would, therefore, increase the specificity of the immune response against the tumor and could be used in combination with other approaches, such as chemotherapy. In fact, the combined therapy (traditional chemotherapy and antibody producing microcapsules) could suppose a reduction on the administered chemotherapeutic drug doses. In consequence, many of the problems of toxicity arising from the use of conventionally administered anti-tumor drugs, which often limits the dose to be administered, could be avoided.

In this chapter, we describe the protocol used in our laboratory in order to elaborate therapeutic molecules producing APA microcapsules. Nowadays, there are a few different options or devices to elaborate these cell containing capsules. Here we describe the use of the electrostatic bead generator [26]. In this kind of devices it has been studied that the microbead size is a function of several factors such as the needle diameter, electrode geometry, distance from the collecting solution, the electrostatic potential and also, the cell suspension flow rate, density, and viscosity [27, 28]. The basic concept behind the droplet generator involves the use of an electrostatic force, which acts to disrupt a liquid surface, to form a charged stream of fine droplets from the needle tip to the gelling bath. An advantage of this system is that once it is correctly adjusted the resulting drops, and therefore microcapsules, are reproducible with a narrow size distribution. Also, the production rate can be quite high and it is possible to work under complete sterile conditions, which makes this strategy feasible for scaling-up for clinical needs. Importantly, although a high electric field is used, cells are not damaged and therefore, there is not a reduction in cell viability once the process is completed. Here, we give the reader the details of a ready to use protocol that can be applied for the encapsulation of different cell types in alginate matrixes using an electrostatic bead generator.

Afterwards, these capsules are covered by a double layer membrane consisting on a first poly-L-lysine layer and a second alginate layer at a lower concentration. In addition, we explain how to evaluate cell viability and metabolic activity over the time.

2 Materials

2.1 Hybridoma Cell Lines

Culture medium: Dulbecco's modified Eagles's medium (RPMI 1640 + Glutamax) supplemented with 10% heat-inactivated fetal bovine serum, 1% penicillin/streptomycin solution, 50 μmol/l 2-mercaptoethanol and 1% BM Condimed (*see* **Note 1**).

2.2 Cell Encapsulation

All the solutions used for the elaboration of microcapsules enclosing cells are prepared in 1% (w/v) mannitol and, afterwards, they need to be sterilized under a laminar flow cabinet by means of filtration through a 0.2 μm filter in order to avoid future contaminations (Fig. 1). Then, they are kept at 4 °C until the day the microencapsulation (*see* **Note 2**).

The amounts of solutions displayed in the present protocol are sufficient for obtaining one batch of microcapsules (approximately 4 ml) (*see* **Note 3**).

1. 1% (w/v) mannitol solution: dissolve 2.5 g of mannitol into 250 ml of distillated water or scale up to the required volume. Stir until the solution is clear indicating that the mannitol has already dissolved.
2. Gelling solution, 55 mM calcium chloride solution: for 100 ml solution, dissolve 0.61 g of calcium chloride (≥96% purity) into 100 ml of 1% mannitol solution. Scale up the amounts of

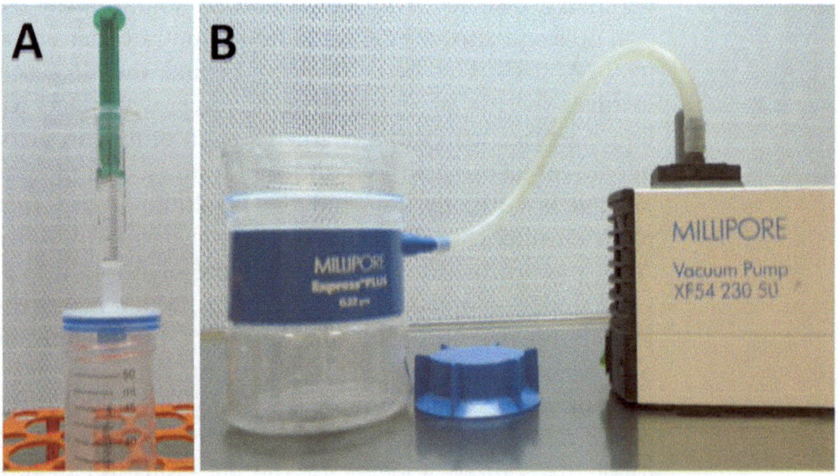

Fig. 1 Filtration procedure depending on the solution: (**a**) Syringe filter for small volumes; (**b**) Stericup® vacuum system for higher volumes

calcium chloride to higher volumes as needed. Stir until calcium chloride is dissolved (*see* **Note 4**).

3. 2 % (w/v) sodium alginate LVG (NovaMatrix) solution for the microcapsule matrix: we use ultra pure low-viscosity and high glucuronic acid alginate (LVG). For 10 ml, dissolve 0.2 g of alginate into 10 ml of 1 % mannitol solution and stir (*see* **Note 5**). For higher volumes of alginate solution, scale up as required. Lower or higher alginate concentrations will require to modify the weight of alginate dissolved into 1 % mannitol. When alginate solution is transparent, filter through a 0.22 μm syringe filter (Fig. 1a) (*see* **Note 6**).

2.3 Microcapsule Coatings

Similarly, the solutions that are used for coating the microcapsules are prepared in 1 % (w/v) mannitol and, afterwards, need to be sterilized in order to avoid future contaminations under a laminar flow cabinet by means of filtration through a 0.2 μm filter (Fig. 1) (*see* **Note 7**). Then, they are kept at 4 °C until needed.

1. 0.05 % (w/v) poly-L-lysine solution (PLL solution): for 10 ml, dissolve 5 mg of poly-L-lysine (molecular weight: 15,000–30,000 Da, Sigma) into 10 ml of 1 % mannitol solution or scale up to the required volume.
2. 0.1 % (w/v) sodium alginate LVG (NovaMatrix) solution: for 10 ml, dissolve 10 mg alginate LVG into 10 ml of 1 % mannitol. For higher volumes of alginate solution, scale up as required.

2.4 Assessment of the Viability of Encapsulated Cells

The solutions that will be used for the assessment of the viability or the metabolic activity of the cells are prepared at the moment of the assay and, therefore, they do not need to be sterilized as the experiments will last for just a few hours (*see* **Note 8**).

1. Quantification of viable cells: alginate lyase solution: prepare 1 mg/ml alginate lyase (Sigma) solution in complete medium.
2. Calcein/ethidium staining solution for microscopy (0.5 μM/0.5 μM): dilute 0.25 μl of ethidium and 0.125 μl of calcein provided in the LIVE/DEAD® Viability/Cytotoxicity Kit for mammalian cells (Life-Technologies) into 1 ml of DPBS with calcium and magnesium. For higher volumes, scale up as required (*see* **Note 9**).
3. Determination of the metabolic activity of the cells: prepare a 5 mg/ml solution of 3-[4,5-dimethylthiazol-2yl]-2,5-diphenyltetrazolium bromide (MTT) (Sigma) in PBS.

3 Methods

All the following steps need to be carried out under aseptic conditions in a flow laminar cabinet and all the material needs to be sterilized in autoclave previously.

3.1 Hybridoma Cells Culture

1. Cell lines are stored in 2 ml cryovial tubes in a liquid nitrogen tank and they should be thawed quickly at 37 °C in a water bath. Afterwards, cells are added to a 50 ml conical centrifuge tube with 10–20 ml of complete culture medium in order to dilute the DMSO that was used as cryoprotectant, as it is toxic for the cells.
2. Spin the cell suspension and resuspend the cell pellet with 20 ml of culture medium. Then, place the suspension into a T-175 flask and incubate at 37 °C in a humidified 5 % CO_2 / 95 % air atmosphere.
3. Cells are cultivated in suspension in as many flasks as needed depending on the cell density that will be used for encapsulation and the final volume of the batch that will be fabricated. Cells should be passaged every 2–3 days and plated with a 1:3 dilution.

3.2 Preparation of the Hybridoma Cell Suspension in Alginate for Microencapsulation

1. Remove cell suspension from the T-175 flasks and collect it into a 50 ml conical tube.
2. Filter the cell suspension through a 40 μm filter in order to remove cell aggregations and count the cells with an automatic cell counter (TC20™ cell counter, Bio-Rad, or similar).
3. Collect the volume of cell suspension in order to get the number of cells required for cell encapsulation.
4. Spin the cell suspension collected and carefully discard supernatant.
5. These hybridoma cells are encapsulated at a 3×10^6 cells per ml of alginate (for a 4 ml batch, 12×10^6 cells are needed). Add the volume of 2 % alginate LVG required to get the desired cell/alginate concentration, in this case 4 ml. Carefully resuspend hybridoma cells in the alginate solution with the help of a sterile spatula (*see* **Note 10**).

3.3 Hybridoma Cells Microencapsulation

1. Transfer the alginate cell suspension into a 10 ml sterile syringe avoiding the formation of bubbles. Place the syringe into a peristaltic pump (Fig. 2a).
2. Connect the syringe to the stainless steel needle tip (inner diameter 0.17 mm) (*see* **Note 11**) through a silicone tube (Fig. 2b) (*see* **Note 12**).
3. Fill an 80 mm diameter sterile beaker containing a small magnetic stirrer with 80 ml of the sterile calcium chloride gelling solution and place it under the needle. Set up agitation to 200 rpm. This would keep the beads separated during the gelling procedure.
4. The needle holder should be place at 7 cm of height (Fig. 2b). As it is shown on the picture, the distance between the needle tip and the calcium chloride solution will be 2 cm.

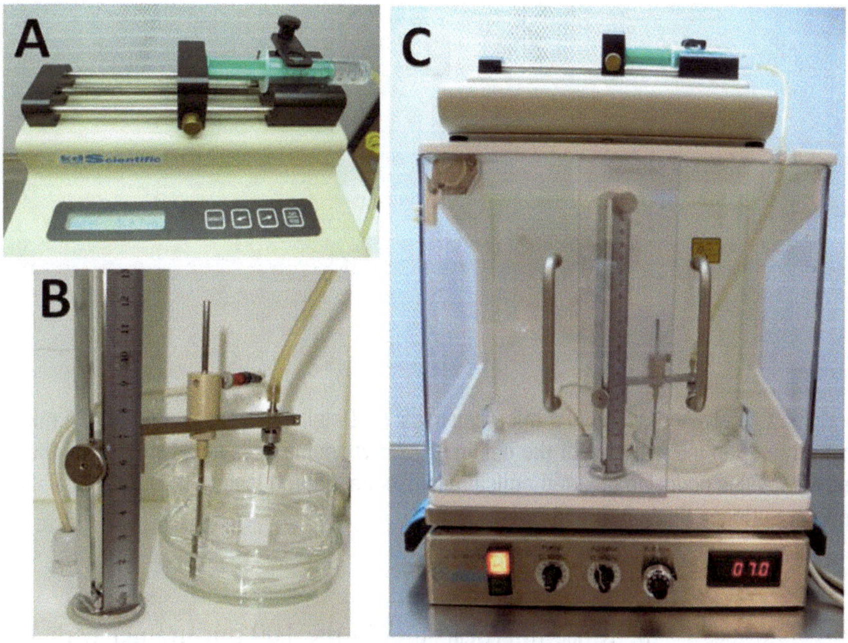

Fig. 2 Cell microencapsulation in detail: (**a**) Peristaltic pump set at 5.9 ml/h; (**b**) Detail of the set up of the needle and its holder; (**c**) Complete set up of the Nisco® bead generator

5. Place the stainless steel stick inside the gelling bath and connect the system as in Fig. 2c.
6. Set up the conditions of voltage (7 kV) and flow rate (5.9 ml/h) for 480 ± 10 μm diameter alginate microcapsules (*see* **Note 13**).
7. Close the electrical safe cage (otherwise the system will not work).
8. Push the syringe containing alginate until the alginate suspension reaches the tip of the nozzle and start the peristaltic pump. Then, let the alginate solution to be pushed out into the beaker through the nozzle until the syringe is empty.
9. Finally, stop the flow of the pump and let the alginate microcapsules stir in the calcium chloride solution for at least 7–10 min for complete gellation.

3.4 Alginate-Poly-L-Lysine-Alginate (APA) Microcapsule Formation: Coating Application

1. Collect microcapsules from calcium chloride solution by filtering the solution through a 100 μm strainer (Fig. 3a).
2. Invert the 100 μm strainer and collect alginate microcapsules into a conical tube with 10 ml of the sterile 0.05 % PLL solution.
3. Place the conical tube on a roller for 5 min at room temperature.
4. Collect microcapsules from the 0.05 % PLL solution by filtering them through a 100 μm strainer.

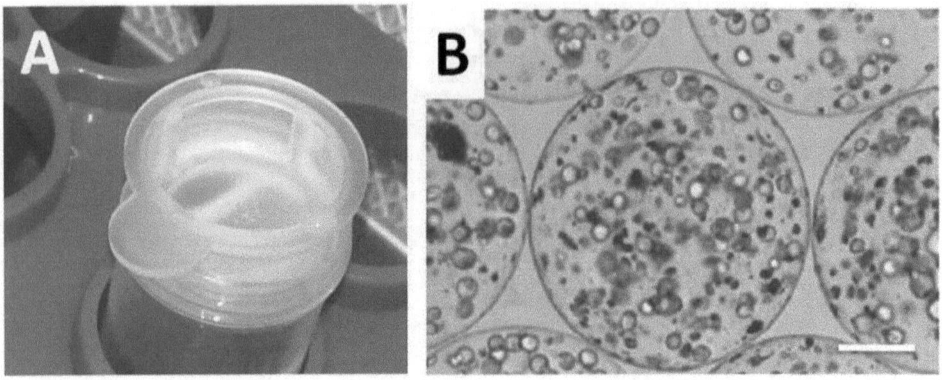

Fig. 3 Hybridoma 2A cells containing APA microcapsules: (**a**) Detail of the recovering of APA microcapsules in a 40 μm cell strainer on the last step of the coatings; (**b**) Hybridoma cells encapsulated in 480 μm diameter size APA microcapsules under an inverted optical microscope. Reproduced from Dubrot et al. 2010 [20] with permission from Springer. Scale bar: 125 μm

5. Rinse microcapsules in the 100 μm strainer with 10 ml of 1 % mannitol solution.
6. Invert the 100 μm strainer and collect alginate-poly-L-lysine microcapsules into a conical tube with 10 ml of the sterile 0.1 % alginate LVG solution.
7. Place the conical tube on a roller for 5 min at room temperature.
8. Collect microcapsules from the 0.1 % alginate LVG solution by filtering the solution through a 40 μm strainer.
9. Rinse the APA microcapsules in the 40 μm strainer with 10 ml of 1 % mannitol solution and then, with another 10 ml of complete medium.
10. Invert the 40 μm strainer and collect APA microcapsules into a conical tube with 10 ml of complete culture solution, then, place the APA microcapsules in a T-75 flask with a pipette and add another 10 ml of complete medium.
11. Observe the microcapsules under an inverted microscope and keep the flask in the incubator for further studies (Fig. 3b).

3.5 Viability Assessment of Encapsulated Cells Under Fluorescent Microscope

1. Collect the suspension of microcapsules from the T-75 flask and place it into a 1.5 ml Eppendorf tube. Let it sediment by gravity and remove the medium. Repeat this process until a volume of approximately 20 μl of microcapsules is reached.
2. Rinse the microcapsules three times with 1 ml of DPBS containing Ca^{2+} and Mg^{2+}.
3. Dilute rinsed microcapsules with 300 μl of the calcein/ethidium staining solution for microcopy.

4. Dispense this staining solution containing the microcapsules into 3 wells of a 96 well-plate (approximately 100 μl solution/well).
5. Incubate at room temperature for 45 min protected from light.
6. Assess under a fluorescent microscope the viability of encapsulated cells. Calcein (in green, for viable cells) and ethidium (in red, for dead cells) can be viewed simultaneously with a conventional fluorescein longpass filter. The fluorescence from these dyes may also be observed separately; calcein can be viewed with a standard fluorescein bandpass filter and ethidium can be viewed with filters for propidium iodide or Texas Red® dye (*see* **Note 14**).

3.6 Viability Assessment of Encapsulated Cells in an Automated Cell Counter

1. Collect the suspension of microcapsules from the T-75 flask and place it into a 1.5 ml Eppendorf tube. Let it sediment by gravity and remove the medium. Repeat this process until a volume of approximately 50 μl of microcapsules is reached.
2. Wash the microcapsules twice with 1 ml of PBS.
3. Soak the microcapsules in 1 ml of the alginate lyase solution.
4. Incubate for 30 min at 37 °C (*see* **Note 15**).
5. Pipette up and down the mixture in order to obtained the complete disruption of the microcapsules. When capsules are degraded completely, filter the cell suspension through a 40 μm strainer in order to remove the rests of alginate.
6. Centrifuge the released cells at 450 × *g* for 10 min.
7. Aspirate the alginate lyase solution and resuspend the cell pellet in PBS.
8. Mix 15 μl of cells with 15 μl of 0.4 % trypan blue and place the mixture in a counting slide.
9. Read the slide in order to get the cell viability in the TC™ cell counter (Bio-Rad) (*see* **Notes 16** and **17**).

3.7 Determination of the Metabolic Activity of the Enclosed Hybridoma Cells

1. Collect the microcapsule suspension from the T-75flask and place it into a 1.5 ml Eppendorf tube. Let it sediment by gravity and remove medium. Repeat this process until a volume of 25 μl of encapsulated cells is reached.
2. Resuspend the encapsulated cells (25 μl) into 500 μl of complete culture solution.
3. Dispense this suspension containing the microcapsules into 5 wells of a 96 well-plate (100 μl solution/well) (*see* **Note 18**).
4. Add 10 μl of the MTT solution into each well including controls.
5. Incubate the 96 well-plate at 37 °C for 4 h.
6. Remove the MTT solution by vacuum or pipette aspiration.

7. Add 100 μl of dimethylsulfoxide into each well including controls. Swirl gently.
8. After 5 min, measure the absorbance in each well including controls, of the formed purple solution at 560 nm with reference wavelength at 690 nm on a microplate reader.
9. Determine the average values from the five readings of each condition and substract the average value for the blank.

4 Notes

1. BM Condimed (Roche) contains a complex mixture of growth factors and cytokines that have a marked stimulatory effect on the growth of hybridoma cells.
2. It is recommended that all the solutions used for cell microencapsulation are prepared the day previous to the procedure.
3. For making a batch of microcapsules, 4 ml of sterile alginate solution is enough. However, because it gets stuck on the beaker, spatula, etc., and also because during the subsequent filtration some of it is lost, it is highly recommended to prepare a higher volume of the alginate solution. For example, 10 ml as mentioned on the protocol.
4. The 1% mannitol and gelling solutions are filtered using a Stericup® vacuum system which has a 0.22 μm filter as usually higher volumes are needed and it is easier and faster (Fig. 1).
5. Alginate used for the matrix, because of its high concentration and viscosity, needs several hours stirring for getting dissolved. Also, bear in mind that sometimes alginate gets stuck on the bottom or the walls of the beaker and requires to be taken off using a spatula. Other times alginate gets agglomerated on the top of the solutions and again the agglomerate requires to be broken using a spatula.
6. We recommend filtering the alginate solution using a 2 ml syringe connected to a 0.20 μm syringe filter (Minisart Plus, Sigma), and in a few steps since higher volumes require a high strength due to its high viscosity (Fig. 1).
7. These solutions are filtered using a 10 ml syringe connected to a 0.20 μm syringe filter when prepared in low volumes.
8. All these experiments are recommended to be performed at least every 2–3 days during the first couple of weeks. Afterwards, once a week could be enough for in vitro assays.
9. Calcein/ethidium staining solution for microcopy must be freshly prepared the moment it is needed and light protected previous to its use.

10. This is a critical step for the success of the encapsulation. Cell suspension in alginate should be completely homogeneous without the presence of bubbles. Try to remove any bubble with a micropipette. The presence of bubbles and the presence of cell aggregation might imply the failure of the cell encapsulation.
11. For 480 μm diameter alginate microcapsules we use the 0.17 mm needle but for lower of higher diameter microcapsules different needles with different inner diameters should be used (Fig. 2b).
12. We usually clean the needle and the silicone tube pushing through 1% mannitol by means of a syringe. Then, we place both in a beaker with water and soap and sonicate them for 20 min at 20 °C. Afterwards, we clean them with water and sterilize in autoclave before the next use.
13. With the described conditions we obtain 480 ± 10 μm diameter alginate microcapsules that will slightly increase in diameter after coating. The bead diameter will decrease with smaller needle diameter and increasing voltage.
14. We usually observe samples under a Nikon TMS confocal microscope with the excitation/emission settings for calcein AM (excitation 495 nm, emission 515 nm) and ethidium homodimer (excitation 495 nm, emission 635 nm). Next, we analyze images with Eclipse Net software, version 1.20.0.
15. In order to get a complete disruption of the microcapsules, every 10 min during the 30 min incubation, microcapsules are pipetted up and down with a micropipette.
16. In case of not having an automated cell counter, a hemocytometer can be used.
17. Viable cells will appear round and clear. However, nonviable cells will have an asymmetrical appearance and will absorb the dye, therefore appearing blue.
18. Also include three control wells of medium alone to provide the blanks for absorbance readings.

References

1. Orive G, Santos E, Pedraz JL, Hernandez RM (2014) Application of cell encapsulation for controlled delivery of biological therapeutics. Adv Drug Deliv Rev 67–68:3–14
2. de Vos P, Lazarjani HA, Poncelet D, Faas MM (2014) Polymers in cell encapsulation from an enveloped cell perspective. Adv Drug Deliv Rev 67–68:15–34
3. de Vos P, Faas MM, Strand B, Calafiore R (2006) Alginate-based microcapsules for immunoisolation of pancreatic islets. Biomaterials 27(32):5603–5617
4. Lee KY, Mooney DJ (2012) Alginate: properties and biomedical applications. Prog Polym Sci 37(1):106–126
5. Gasperini L, Mano JF, Reis RL (2014) Natural polymers for the microencapsulation of cells. J R Soc Interface 11(100):20140817
6. Orive G, Ponce S, Hernandez RM, Gascon AR, Igartua M, Pedraz JL (2002) Biocompatibility of microcapsules for cell immobilization elaborated with different type of alginates. Biomaterials 23(18):3825–3831

7. De Castro M, Orive G, Hernandez RM, Gascon AR, Pedraz JL (2005) Comparative study of microcapsules elaborated with three polycations (PLL, PDL, PLO) for cell immobilization. J Microencapsul 22(3):303–315
8. Acarregui A, Murua A, Pedraz JL, Orive G, Hernandez RM (2012) A perspective on bioactive cell microencapsulation. BioDrugs 26(5):283–301
9. Calafiore R, Basta G (2014) Clinical application of microencapsulated islets: actual prospectives on progress and challenges. Adv Drug Deliv Rev 67–68:84–92
10. Basta G, Montanucci P, Luca G, Boselli C, Noya G, Barbaro B et al (2011) Long-term metabolic and immunological follow-up of nonimmunosuppressed patients with type 1 diabetes treated with microencapsulated islet allografts: four cases. Diabetes Care 34(11):2406–2409
11. Ferreira D, Westman E, Eyjolfsdottir H, Almqvist P, Lind G, Linderoth B et al (2015) Brain changes in Alzheimer's disease patients with implanted encapsulated cells releasing nerve growth factor. J Alzheimers Dis 43(3):1059–1072
12. Bachoud-Levi AC, Deglon N, Nguyen JP, Bloch J, Bourdet C, Winkel L et al (2000) Neuroprotective gene therapy for Huntington's disease using a polymer encapsulated BHK cell line engineered to secrete human CNTF. Hum Gene Ther 11(12):1723–1729
13. Zurn AD, Henry H, Schluep M, Aubert V, Winkel L, Eilers B et al (2000) Evaluation of an intrathecal immune response in amyotrophic lateral sclerosis patients implanted with encapsulated genetically engineered xenogeneic cells. Cell Transplant 9(4):471–484
14. Emerich DF, Orive G, Thanos C, Tornoe J, Wahlberg LU (2014) Encapsulated cell therapy for neurodegenerative diseases: from promise to product. Adv Drug Deliv Rev 67–68:131–141
15. Saenz Del Burgo L, Compte M, Aceves M, Hernandez RM, Sanz L, Alvarez-Vallina L et al (2015) Microencapsulation of therapeutic bispecific antibodies producing cells: immunotherapeutic organoids for cancer management. J Drug Target 23(2):170–179
16. Beck A, Wurch T, Bailly C, Corvaia N (2010) Strategies and challenges for the next generation of therapeutic antibodies. Nat Rev Immunol 10(5):345–352
17. Melero I, Hervas-Stubbs S, Glennie M, Pardoll DM, Chen L (2007) Immunostimulatory monoclonal antibodies for cancer therapy. Nat Rev Cancer 7(2):95–106
18. Compte M, Blanco B, Serrano F, Cuesta AM, Sanz L, Bernad A et al (2007) Inhibition of tumor growth in vivo by in situ secretion of bispecific anti-CEA x anti-CD3 diabodies from lentivirally transduced human lymphocytes. Cancer Gene Ther 14(4):380–388
19. Kontermann R (2012) Dual targeting strategies with bispecific antibodies. MAbs 4(2):182–197
20. Dubrot J, Portero A, Orive G, Hernandez RM, Palazon A, Rouzaut A et al (2010) Delivery of immunostimulatory monoclonal antibodies by encapsulated hybridoma cells. Cancer Immunol Immunother 59(11):1621–1631
21. Murillo O, Arina A, Tirapu I, Alfaro C, Mazzolini G, Palencia B et al (2003) Potentiation of therapeutic immune responses against malignancies with monoclonal antibodies. Clin Cancer Res 9(15):5454–5464
22. Korman AJ, Peggs KS, Allison JP (2006) Checkpoint blockade in cancer immunotherapy. Adv Immunol 90:297–339
23. Vonderheide RH, Flaherty KT, Khalil M, Stumacher MS, Bajor DL, Hutnick NA et al (2007) Clinical activity and immune modulation in cancer patients treated with CP-870,893, a novel CD40 agonist monoclonal antibody. J Clin Oncol 25(7):876–883
24. Suntharalingam G, Perry MR, Ward S, Brett SJ, Castello-Cortes A, Brunner MD et al (2006) Cytokine storm in a phase 1 trial of the anti-CD28 monoclonal antibody TGN1412. N Engl J Med 355(10):1018–1028
25. Uno T, Takeda K, Kojima Y, Yoshizawa H, Akiba H, Mittler RS et al (2006) Eradication of established tumors in mice by a combination antibody-based therapy. Nat Med 12(6):693–698
26. Manojlovic V, Djonlagic J, Obradovic B, Nedovic V, Bugarski B (2006) Immobilization of cells by electrostatic droplet generation: a model system for potential application in medicine. Int J Nanomedicine 1(2):163–171
27. Bugarski B, Li QL, Goosen MFA, Poncelet D, Neufeld RJ et al (1994) Electrostatic droplet generation – mechanism of polymer droplet formation. Aiche J 40(6):1026–1031
28. Poncelet D, Babak VG, Neufeld RJ, Goosen MFA, Burgarski B (1999) Theory of electrostatic dispersion of polymer solutions in the production of microgel beads containing biocatalyst. Adv Colloid Interf Sci 79(2-3):213–228

Chapter 22

Microencapsulation of Bacterial Cells by Emulsion Technique for Probiotic Application

Surajit Mandal and Subrota Hati

Abstract

Probiotics are dietary concepts to improve the dynamics of intestinal microbial balance favorably. Careful screening of probiotic strains for their technological suitability can also allow selection of strains with the best manufacturing and food technology characteristics. However, even the most robust probiotic bacteria are currently in the range of food applications to which they can be applied. Additionally, bacteria with exceptional functional heath properties are ruled out due to technological limitations. New process and formulation technologies will enable both expansion of the range of products in to which probiotics can be applied and the use of efficacious stains that currently cannot be manufactured or stored with existing technologies. Viability of probiotics has been both a marketing and technological concern for many industrial produces. Probiotics are difficult to work with, the bacteria often die during processing, and shelf life is unpredictable. Probiotics are extremely susceptible environmental conditions such as oxygen, processing and preservation treatments, acidity, and salt concentration, which collectively affect the overall viability of probiotics. Manufacturers have long been fortifying products with probiotics; they have faced significant processing challenges regarding the stability and survivability of probiotics during processing and preservation treatments, storage as well during their passage through GIT. Application of microencapsulation significantly improves the stability of probiotics during food processing and gastrointestinal transit.

Key words Microencapsulation, Alginate, Lactobacillus, Probiotics, Emulsion technique

1 Introduction

Probiotic foods, the important discipline of functional foods, are defined as "foods containing live microorganisms, which actively enhance the health of consumers by improving the balance of microflora in the gut when ingested live in sufficient numbers" [1]. Probiotic foods come under functional foods which, in addition to their basic nutritive value and natural being, will contain the proper balance of ingredients, which will help as functional, therapeutic, and nutraceutical agents enabling the human to prevent the illness and diseases. Problems associated with probiotics incorporation into foods are survival and stability during processing, preservation, storage and GI transient and probiotic foods should include

Emmanuel C. Opara (ed.), *Cell Microencapsulation: Methods and Protocols*, Methods in Molecular Biology, vol. 1479,
DOI 10.1007/978-1-4939-6364-5_22,

probiotic strains at suitable level until the time of consumption. General approach to improve probiotics survivability is the selection of suitable probiotic strains for food application, i.e., a "probabilistic approach," as it is very difficult to select a probiotic strain having all properties by applying classical probabilistic approach. Stress adaptation method is another approach; however, stress responses are strain-specific. A number of techniques have been developed to protect the probiotics form environmental stresses in food matrices, processing, and storage and GIT passage. Among these, microencapsulation has been found the most suitable and accessible technology to protect the tiny living organisms.

Microencapsulation is a process where droplets of liquids, solids, or gases (core) are coated by thin films (coatings), which protect the core until it is needed [2]. During microencapsulation, specially designed equipment coats probiotic bacteria in a matrix. This will increase formulation possibilities, broadening the range of ingredients with which probiotics can be blended. In order to make this technology successful to entrap probiotics, the protective wall materials should be such that it would afford protection to the probiotics against the processing and storage conditions, GIT transient but release them in lower part of small instestine and large instestine. It has been investigated that some lipid coatings are not only an efficient protective barrier against chemical entities, such as moisture, oxygen, and acids but also a good protector against short exposure to high temperature and pressure [3].

Encapsulation in hydrocolloid beads entraps or immobilizes the cells within the bead matrix, which in turn provides protection in such an environment. The cells are retained within an encapsulating membrane to reduce cell injury or cell loss. The encapsulation techniques applied to probiotics for the use in fermented milk products or biomass production can be classified in to two groups: extrusion (droplet method) and emulsion or two-phase system. Both extrusion and emulsion techniques increase the survival of probiotics by up to 80–95 % [4–10].

Microencapsulation of probiotics in hydrocolloid beads has been tested for improving their viability in food products and during GIT transit [11, 12]. The most widely used matrix for microencapsulation is alginate [13]. Alginate beads have been found to increase the survival of probiotics [8, 12, 14, 15].

2 Materials

2.1 Chemicals and Cultures

1. *Lactobacillus* cultures (freeze-dried ampoule).
2. MRS broth.
3. Sodium alginate.
4. Calcium chloride.
5. Soybean oil.

6. Tween 80.
7. Whatman filter paper.

2.2 Equipment

1. Magnetic stirrer with temperature controlling facility.
2. Refrigerated centrifuge.
3. Phase-contrast microscopy.
4. Incubator (37 °C).
5. Laminar air flow system for culture transfer.

3 Methods

3.1 Preparation of Bacterial Cell Suspension

1. Freeze-dried *Lactobacillus casei* NCDC 298 culture is activated in chalk litmus milk at 37 °C/ 24 h and subcultured twice in MRS broth (37 °C/15–18 h) (*see* Fig. 1; **Note 1**).
2. The activated and refrigerated stored *L. casei* NCDC 298 culture in chalk litmus milk is reactivated with 2–3 transfers in MRS broth (Himedia, Mumbai, India).
3. 100 ml of MRS broth is inoculated with active Lactobacillus culture (at 1 %) and incubated at 37 °C for 15 h to get stationary phase culture.
4. Cell biomass from the culture is separated by centrifugation at 10,000 rpm for 10 min at 4 °C (Kubota High Speed Centrifuge, Osaka, Japan) and pellet is washed twice before suspended in normal saline.
5. The cell concentration is adjusted to ~1.0×10^{11} cfu/ml for microencapsulation.

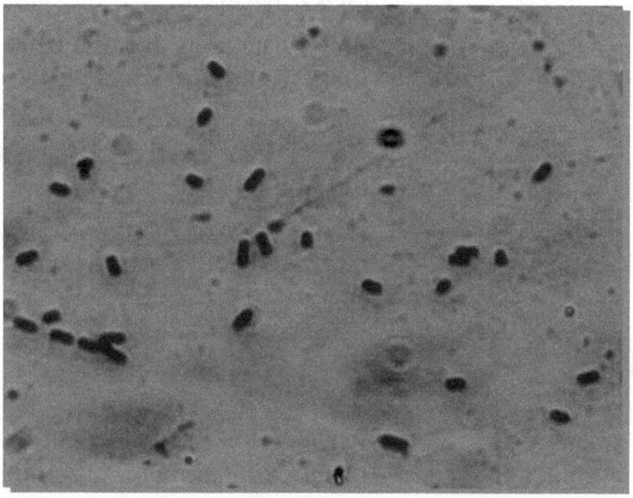

Fig. 1 Light microscopic view of Gram's stained *Lactobacillus casei* NCDC 298

3.2 Preparation of Alginate Solution and Soybean Oil

1. Sodium alginate solution (4%) is prepared by mixing 4 g of sodium alginate in 100 ml distilled water by warming at 50–60 °C. The alginate solution is sterilized by autoclaving (120 ° for 15 min) and cooled to 38–40 °C (*see* **Note 2**).
2. Soybean oil is added with 0.2% Tween 80 (emulsifier) and sterilized by autoclaving (121 °C for 20 min) followed by drying of sterilized oil in hot air oven (160 °C for 3–4 h).

3.3 Calcium Chloride Solution

1. 0.1 M calcium chloride solution is prepared in distilled water and sterilized by autoclaving (120 ° for 15 min) and cooled to 4 °C.

3.4 Microencapsulation

1. The *L. casei* NCDC 298 cells are encapsulated in sodium alginate as described by Sheu et al. [8] and modified by our group [12].
2. Twenty milliliters of sterile sodium alginate solution and 4 ml of cell suspension are transferred into a centrifuge tube (40 ml).
3. The content is mixed for uniform distribution of cells in alginate matrix.
4. One hundred milliliter sterilized soybean oil containing 0.2% Tween 80 (emulsifier) is taken in a sterilized beaker (500 ml).
5. The alginate–cell mixture is then added dropwise to the continuous oil phase that is magnetically stirred (600 rpm) after 5 min and a uniformly turbid emulsion is formed (*see* Fig. 2, Steps I and II; **Notes 3** and **4**).
6. The emulsion is broken by quickly addition of cooled 0.1 M calcium chloride (100 ml) and the capsules are formed by hardening of alginate droplets by calcium ions (*see* Fig. 2, Step III, **Note 5**).
7. The capsules are harvested by gentle centrifugation at 1000 rpm, 10 min at 4 °C (Kubota High Speed Centrifuge, Osaka, Japan).
8. The capsules are washed twice with 0.1 M calcium chloride solution and finally with distilled water (*see* **Note 6**).
9. The beads were separated by filtration (Whatman filter paper No. 1), transferred to sterile petri dish and stored in refrigerator until use.
10. The morphology microcapsules are studied by phase-contrast microscopy (40× objective) (Fig. 3).

4 Notes

1. *Lactobacillus* cells must be harvested at early stationary phase culture for getting maximum survival and performance.

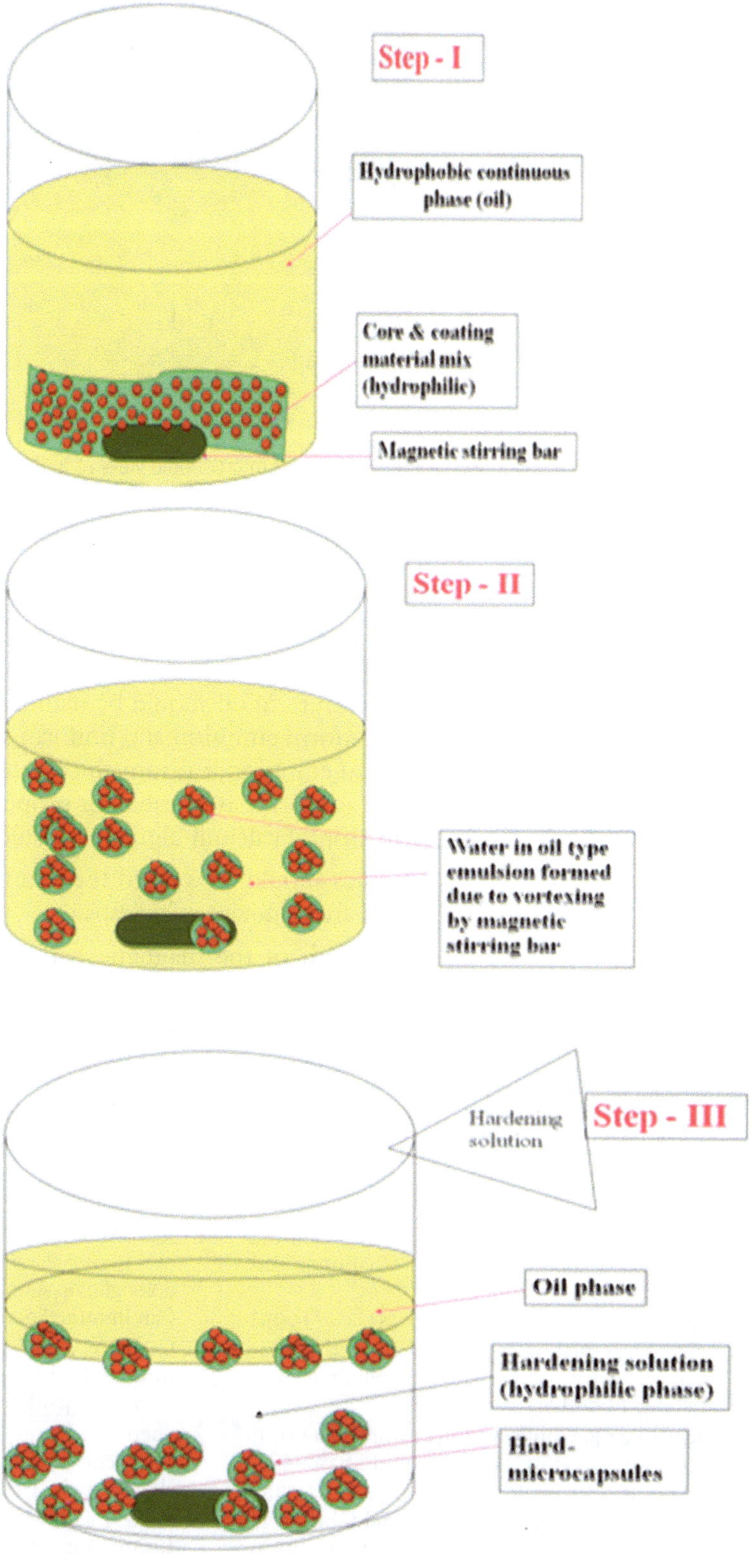

Fig. 2 Microencapsulation by emulsion technique using alginate

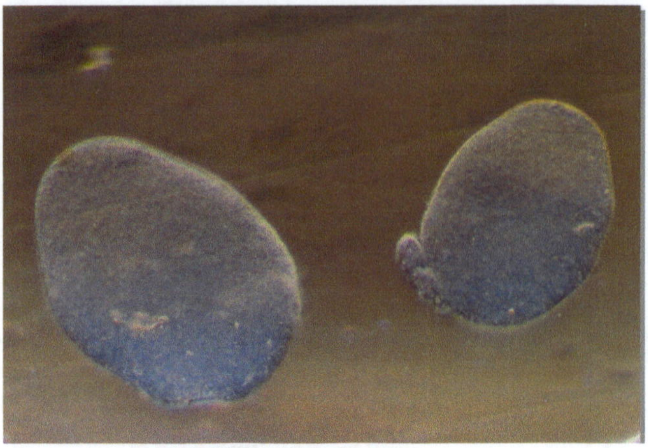

Fig. 3 Phase-contrast microscopic view of alginate capsules

2. Sodium alginate should be properly dissolved in water before sterilization; there should be no particulates/lumps in the solution. Otherwise, nonuniform distribution of microcapsules will occur.
3. Temperature of oil should be maintained at 35–37 °C for getting uniform emulsion and uniform size particle. After getting uniform emulsion, calcium chloride chilled solution should be added as quickly as possible for proper and complete and quick formation of calcium-alginate capsules.
4. Capsules should be washed thoroughly as much as possible to get oil-free alginate microcapsules.
5. The length of the magnetic bar used for stirring should be more than 1/4th of the diameter of beaker.
6. The beaker of long length and short diameter and round bottom should be used for getting good performance of the emulsion encapsulation technique.

References

1. Fuller R (1992). Probiotics: The scientific basis. Ed. Fuller R. London: Chapman and Hall, London, pp 209–224.
2. Sheu TY, Rosenberg M (1995) Microencapsulation by spray drying ethyl caprylate in whey protein and carbohydrate wall systems. J Food Sci 60:98–103
3. Suita-Cruz P, Goulet J (2001) Improving probiotic survival rates. Food Technol 55:36–42
4. Audet P, Céline P, Christophe L (1988) Immobilized growing lactic acid bacteria with k-carrageenan—locust bean gum gel. Appl Microbiol Biotechnol 29(1):11–18
5. Rao AV, Shiwnarain N, Maharaj I (1989) Survival of microencapsulated Bifidobacterium pseudolongum in simulated gastric and intestinal juices. Can Institute Food Sci Technol J 22:345–349
6. Sheu TY, Marshall RT (1991) Improving culture viability in frozen dairy desserts by microencapsulation. J Dairy Sci 74:107–111
7. Sheu TY, Marshall RT (1993) Microencapsulation of lactobacilli in calcium alginate gels. J Food Sci 58:557–561
8. Sheu TY, Marshall RT, Heymann H (1993) Improving survival of culture bacteria in frozen desserts by micro-entrapment. J Dairy Sci 76:1902–1907

9. Jankowski T, Zielinska M, Wysakowska A (1997) Encapsulation of lactic acid bacteria with alginate/starch capsules. Biotechnol Tech 11:31–34
10. Kebary, K.M.K., Hussein, S.A. and Badausi, R.M. 1998. Improving viability of bifidobacteria and their effect on frozen ice milk. Egypt. J. Dairy Sci. 26:319–337
11. Krasaekoopt W, Bhandari B, Deeth H (2003) Evaluation of encapsulation techniques of probiotics for yoghurt. Int Dairy J 13:3–13
12. Mandal S, Puniya AK, Singh K (2006) Effect of alginate concentrations on survival of microencapsulated Lactobacillus casei NCDC-298. Int Dairy J 16:1190–1195
13. Kailasapathy K (2002) Microencapsulation of probiotic bacteria: technology and potential applications. Curr Issues Intest Microbiol 3:39–48
14. Annan NT, Borza AD, Hansen LT (2008) Encapsulation in alginate-coated gelatin microspheres improves survival of probiotic Bifidobacterium adolescentis 15703T during exposure to simulated gastro-intestinal conditions. Food Res Int 41(2):184–193
15. Mandal S, Hati S, Basu S, Puniya AK, Singh R, Singh K (2011) Effect of k-carrageenan concentrations on survival of microencapsulated Lactobacillus casei NCDC 298. Indian J Dairy Sci 63:470–473

Part V

Cell Microencapsulation in Clinical Applications

Chapter 23

Microencapsulation of Islets for the Treatment of Type 1 Diabetes Mellitus (T1D)

Riccardo Calafiore, Giuseppe Basta, and Pia Montanucci

Abstract

Microencapsulation technology, based on use of alginic acid biopolymers, has been devised many years ago. However, when intended for enveloping human islets for transplantation purposes, the method needs to be up-scaled and implemented with care being taken to comply with simple but important measures. It is almost indispensable to rely on an ultrapurified alginic polymers: in fact, any, even minimal, alginate contamination with endotoxins, pyrogens, and proteins could provoke the host's inflammatory reaction upon graft, with heavy adverse consequences on the capsules immunoprotective properties, hence on graft survival per se. Care should be taken in ensuring fabrication of reproducible microspheres, in terms not only of shape and size, but also consistency of the peripheral layers around the central alginate gel core, where the islets are immobilized. Once the product is well defined and stable, care should also be taken in accurately selecting patients with T1D that are candidate for encapsulated islet cell transplantation with no general immunosuppression. A series of pre- and post-intraperitoneal transplant metabolic, chemical, and immunological parameters are to be monitored, in conjunction with image analysis of the abdomen, in order to assess efficacy of the intervention according to well defined grading scale.

Key words Human islets, Alginate, Microencapsulation, Hyperglycemia, Insulin, Transplantation

1 Introduction

1.1 Mapping the Field of Interest

This chapter focuses on cell therapy of T1D as a potential strategy to one day cure this chronic, severe metabolic disorder and its attending multiorgan secondary complications. This approach, based on graft of pancreatic islets of Langerhans, retrieved from cadaveric human donor organs, and born almost three decades ago, has so far missed the final goal of radically and substantially reversing the hyperglycemic syndrome typical of the disease. Nevertheless, considerable progress has been made in ascertaining T1D pathogenesis and disease process, with special regard to autoimmune pathways leading to selective pancreatic islet β-cell killing. As far as this issue is concerned, increasing evidence can now be marshaled upon the fact that the recipient's general immunosuppression, to be instituted in order to counteract the

Emmanuel C. Opara (ed.), *Cell Microencapsulation: Methods and Protocols*, Methods in Molecular Biology, vol. 1479,
DOI 10.1007/978-1-4939-6364-5_23, © Springer Science+Business Media New York 2017

islet graft-directed immune destruction, adversely affects the islet graft survival and functional life span, other than inducing severe systemic side effects. Hence new technologies to immunoprotect the islet grafts, with no recipient's general pharmacologic immunosuppression, are required.

1.2 Diabetes Mellitus: A Dangerous Metabolic Disease and a Global Epidemics

Diabetes mellitus, a chronic metabolic disorder hallmarked by exceedingly high blood glucose (BG) levels, is actually being defined a "global epidemics," with steady, steep rise in prevalence and incidence worldwide. In fact, it is estimated that over 150 million people cumulatively suffer for T1D and type 2 diabetes (T2D), but the number is destined to expand to over 300 million people by 2025 [1]. Chronic hyperglycemia leads to severe multiple organ damage, including often life-threatening secondary complications, that accounts for severe impact of the disease on the patient's wellness. Diabetic complications essentially comprise macroangiopathy/microangiopathy, with special regard to retinopathy, leading to premature blindness; nephropathy leading to end-stage renal disease that can only be treated with hemodialysis or kidney transplantation; disabling neuropathy, and cardiovascular disease (diabetes is today the leading cause of myocardial infarction and stroke) [2]. The main forms of diabetes mellitus are T1D and T2D: the latter is the most prevalent (90–95 %), and it mainly strikes on overweight subjects, at middle age, typically on a background of peripheral insulin resistance, followed by progressive pancreatic β-cell failure, through the possible abrogation of insulin secretory function [1]. Hence T2D, after an initial phase where insulin resistance prevails, and the disease is treated with oral hypoglycemic agents, may in the end require exogenous insulin supplementation to correct hyperglycemia. T1D, while much less common (5–10 %) than T2D, and with prevalent onset in the young age, requires intensive exogenous insulin supplementation since diagnosis. Insulin dependency, a constant feature of T1D, is associated with the inevitable attending choir of strictly monitoring BG levels by finger-prick capillary blood sampling, several times a day in order to assess how many insulin injections are necessary to keep a fair daily mean BG control. However, even the most accurate insulin therapy regimen does not grant for prevention of such complications, thereby reflecting the fact that it is almost impossible to reproduce or mimic the stimulus-coupled minute-by-minute physiological insulin secretory patterns of the normal islet β-cells [3].

1.3 Insulin Therapy Regimens for the Therapy of Diabetes

According to guidelines promulgated by the American Diabetes Association [4] the injection of insulin is essential for management of patients with T1D, but it may also be needed by patients with T2D for intermittent or continuous glycemic control. Today new insulin molecules are available, that are associated with very flexible pharmacokinetics and pharmacodynamics, with special regard to both short-acting (i.e., insulin *lispro*, insulin *aspart*, insulin *glulysine*) and long-acting (insulin *glargine*, insulin *detemir*, insulin *degludec*) analogic preparations. These molecules have resulted in considerable improvement of metabolic control, primarily in patients with T1D but also in those with insulin-dependent T2D. The Diabetes Control Clinical Trial (DCCT) data group has unequivocally demonstrated that the most effective approach to try to prevent/attenuate though not completely eliminate the risk for contracting secondary complications of T1D is to apply intensive insulin therapy regimens, based on at least 4 daily administration of both short- and long-acting preferably analogic, insulin preparations. This is today the best way to maintain a balanced BG control which results in improved values of glycated hemoglobin (HbA1c), a known mid/long-term marker of metabolic control (the higher the HbA1c values the worse the degree of metabolic control), although at expense of strict blood glucose monitoring to decide, every day, the insulin regimen schedules. Intensive insulin therapy has certainly helped achievement of better daily mean blood glucose levels in patients with T1D, but on the other hand, they are associated with two major drawbacks: (1) treated patients are exposed to higher risk of hypoglycaemia, which positively correlates with decrease of HbA1c levels; (2) the improved metabolic control may reduce but not eliminate the risk for developing secondary complications of the disease. Last but not least, compliance of those T1D patients undertaking intensive insulin regimens with daily both, strict blood glucose monitoring and multiple insulin injections is scarce, especially in the younger segments of the diabetic population.

1.4 New Strategies for the Cure of T1D

New strategies to possibly cure this metabolic disease have been actively sought for many years and still are in progress. The initial idea was to replace the diseased/destroyed islet β-cells, upon the autoimmune attack, with fresh and viable endocrine tissue, retrieved from cadaveric human donors. The first attempt, on this purpose, was to graft the entire pancreas retrieved from a donor into T1D patients usually also requiring kidney graft, because of terminal kidney failure as a consequence of diabetes [5]. The recipients were under strict and heavy general immunosuppression, to avoid the transplanted organ rejection, with all the many attending risks. Besides, this was a major surgery, burdened with heavy mortality and morbidity of the treated patients. That is why most of the pancreatic grafts, reported in the literature, were performed in combination with kidney grafts, in patients with T1D and end-stage renal disease, in order to minimize the impact of general immunosuppression, virtually evitable for T1D subjects (since they could always continue on insulin regimen therapy) but ineludible for renal transplantation. Almost in parallel, and immediately after the entire organ, islets of Langerhans retrieved from donor pancreases were considered for transplantation in T1D recipients.

The immediate advantage was that the islets evidently account for incomparably lower mass as compared to the whole pancreas, hence they can be implanted (usually in the liver) by minimally invasive procedures, and furthermore, only the islet β-cells are required for cell substitution therapy in T1D. Nevertheless, the isolated islets, upon disconnection from their native pancreatic extra-cellular matrix, are extremely vulnerable to noxious, immune or not immune, environmental agents, and are more difficult to engraft, especially in a stringent organ tissue like the liver. Moreover, the isolated islets, like the whole pancreatic grafts, are the target of not only immune rejection but also autoimmune recurrence of the disease.

Looking at clinical data on islet transplantation (TX), grafted intra-hepatically, the outcome of the islet grafts has steadily improved throughout the past few decades, with special regard to 2007–2010, when, according to the data of the Clinical Islet Transplant (CIT) Consortium, 65 % of the treated patients enjoyed insulin-independency at 1 year of post-TX. Less consistent was the maintenance of the accomplished insulin independence at 5 years post TX, that ranged on as low as 5–6 %, with few exceptions [6]. The reasons for such as globally poor outcome of the islet graft clinical trials are multiple, and only partially identified, at this juncture. One reason certainly lies on the progressive decline of the functionally active β-cell mass, by the time of retrieval from donor pancreases, to the grafting procedures and the immediate post-transplant phase [7]. Moreover, the islet engraftment process is adversely affected by several immune and not immune specific factors, mainly based on local activation of the coagulation system (i.e., IBMIR: instant blood mediated inflammatory reaction) as well as the unfavorable liver cell tissue microenvironment [8]. Finally, general immunosuppression results in both local and systemic toxicity which, ultimately and invariably, impairs the grafted islet cell survival and functional competence.

1.5 The Concept of Immunoisolation of Islet Cell Grafts

Because of the many attending risks and disadvantages, associated with general immunosuppression of the recipients, in cell grafting procedures, a potential solution could be represented by enveloping the cell grafts within selective permeable and artificial polymeric membranes so as to prevent any physical contact between the entrapped cells and the host's immune system. In detail, if the grafted islets were enveloped within artificial membranes that selectively regulate cross-permeability of noxious soluble factors, while preventing access to immunoactive cells and molecules, general immunosuppressive treatment of the recipients could theoretically be obviated. The physical shield surrounding the islets would anyway attenuate the impact of acute and chronic immune rejection pathways on the grafted tissue, while also offering the opportunity to employ less toxic and possibly locally delivered agents that would render the procedure more acceptable. Moreover, microencapsulation might allow for use of nonhuman tissue as a resource for donor islets, an increasingly appealing option, in light of the evident shortage of cadaveric donor pancreases, with respect to the real need for transplantable tissue.

The immunoprotection approach by physical barriers may be pursued by use of macrodevices or microcapsules, both based on use of highly selective and nontoxic artificial membranes, variably configured, and comprising highly purified constituent biopolymers. We have selected, from the beginning of our research activities on islet immunoprotection, microcapsules basically made of alginic acid derivatives, complexed with aminoacidic polycations [2].

1.6 Microencapsulation: General Principles

Microencapsulation consists of entrapping cells/tissue within polymeric and non-cytotoxic membranes that constitute immunoprotective barriers. Based on this concept, a wide spectrum of cells, other than pancreatic islets, have been immobilized within the microcapsules which may expand applicability of this strategy beyond diabetes. However, it has recently been postulated that success of this approach may require detailed review of multiple issues concerning biocompatibility and bio-performance of the raw microcapsules [9–11]. Microcapsules offer a three-dimensional architecture matrix, and upon formulation for instance with highly purified and almost endotoxin and protein-free alginates, they have been shown to promote better growth, differentiation, and maturation of different cell types, from pancreatic islets to mesenchymal stem cells, to embryonic stem cell (ESCs), to neural stem cells and hepatocytes [12–14].

The vast majority of alginate-based microcapsules historically measured 600–800 μm in diameter, although generation of smaller-size alginate microbeads has actively been pursued over the past few years. Since most of successful in vivo graft studies, including pilot clinical trials, have been conducted only with alginate-based microcapsules, these deserve better attention. Undoubtedly, the only polymer approved for fabricating microcapsules for human use remains alginate in its sodium salt [15]. Hence, despite the fact that other biopolymers have been used to formulate microcapsules for graft purposes (i.e., agarose, polyethylene glycol, collagen, dextran, hyaluronic acid, chitosan etc.) this chapter covers only alginates.

1.7 Alginates for Microencapsulation

Alginic acid, a polysaccharide originally extracted from brown seaweeds, and its salts have historically represented the most common material to fabricate microcapsules used to envelop, mainly, although not solely, pancreatic islet cells (Fig. 1; Table 1).

Alginates are linear copolymers composed of two building units, β -D-mannuronic (M) and α -L-guluronic (G) acids, mainly patterned through the entire molecule in the form of MM or GG or MG dimeric blocks. The proportions and sequential arrangement of the uronic acids depend on the species of algae and the

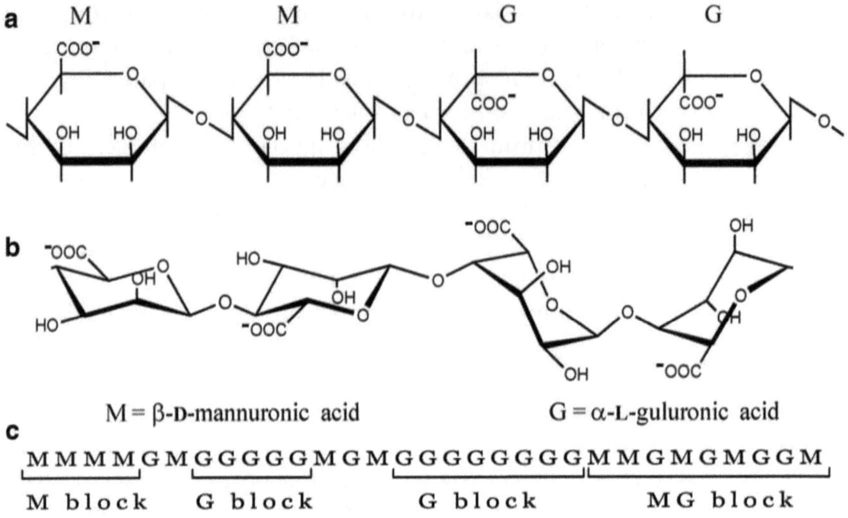

Fig. 1 Molecular patterns of alginate (AG): (**a**) AG monomers; (**b**) chemical chains configuration; (**c**) blocks distribution

Table 1
Typical alginate composition from various seaweeds

Algae species	F_G (%)	F_M (%)	Pgas	Pmas	M/G ratio
Macrocystis pyrifera	39	61	17.7	40.6	1.56
Laminaria digitalia	41	59	12.7	60.5	1.45
Laminaria hyperborea	69	31	18.5	61.2	0.45

specific algal tissue (stem or leaf) from which the alginate is extracted. The most relevant characteristic of the alginates is the selective binding to multivalent cations, a property that allows for formation of alginate gel beads [16]. The gel formation always implies a process of specific ion exchange. The starting point is a water soluble alginate salt made of monovalent cations like sodium or potassium as counter-ions that have to be exchanged with divalent cations in order to activate the gelling process by a monocation displacement process. The affinity of alginates for divalent cations depends on their composition [17–20]. Guluronic acid-based alginate is more prone to ion binding compared to the mannuronic acid-based product, while the affinity for the alkaline earth metals changes by the following order: $Mg \ll Ca < Sr < Ba$ [16–20]. The divalent metals diffuse into an alginate solution and the cation-binding crosslinks the alginate in a highly cooperative manner, thereby forming a gel, with the crosslinking density being based on the original ion concentration. It has been described that Ca^{2+}

cations bind both to G sequences and to alternating GM dimeric blocks but not to M-blocks only; Sr^{2+} ions bind well to polyG and not at all to polyM, while a very limited binding is detected for polyMG; Ba^{2+} ions bind to separate M and G blocks but not to hybrid MG sequences [21].

1.8 The University of Perugia Microencapsulation Technology for Human Application

Upon completion of long-term preclinical trials of microencapsulated allogeneic/xenogeneic islet transplantation in either rodents or higher mammal animal models of either spontaneous (i.e., NOD mice; insulin-dependent diabetic dogs) or artificially induced (i.e., mice upon i.p. injection of streptozotocin) diabetes, and having assessed full safety and significant functional performance of our system [22, 23], we moved into humans. We were confident to pursue this goal, mainly due to the availability of our in-house method for alginate ultrapurification ("clinical grade" see pending patent below), and well tested reproducible microencapsulation technology [24]. After a long series of audits, we were granted official permission by the Italian Ministry of Health to initiate a closed, pilot phase-I clinical trial of microencapsulated human islet allo-TX into nonimmunosuppressed patients with long-standing T1D, on intensive insulin therapy regimens (4 insulin injections daily). Upon enrolment of ten patients within the established and approved study inclusion criteria, thus far we have completed four patients that have been monitored and followed-up throughout 5 years post-TX.

The primary end-point of the study was to determine long-term safety of the encapsulated human islet (HI) TX. The following parameters were examined:

1. TX-related adverse reactions;
2. TX directed immune reactivity;
3. the host's sensitization to grafted encapsulated foreign islet cell antigens (since donors and recipients were fully mismatched).

Only excellent human islet cell preparations that appeared to be morphologically intact, upon imaging examination, and functionally viable, upon static incubation with glucose, were considered for donor islets, and underwent microencapsulation prior to TX.

We examined the following TX recipient patient-related metabolic parameters:

1. Changes in daily exogenous insulin consumption;
2. Levels of post-TX serum C-peptide levels;
3. Changes in severe nocturnal hypoglycemia, as defined by blood glucose (BG) <40 mg/dL;
4. Changes in HbA1c plasma levels.

All the four patients were grafted intraperitoneally, under local anesthesia and ultrasound echography guidance, except one case receiving three subsequent grafts, the last of which was delivered under general anesthesia, by laparoscopic surgery. The goal, in this specific instance, was to deliver the encapsulated islets into sites (i.e., mesentery) that would be receiving higher blood supply than the great peritoneal cavity. In all cases, we observed no adverse reactions with respect to the grafting procedure, nor evidence of immune sensitization. In terms of metabolic outcome, all patients showed decline of exogenous daily insulin consumption that resulted nearly halved, except one case that temporarily suspended the insulin injections. C-peptide levels were detected in all cases and lasted for at least 100 through 480 days, reaching peaks of 1.8 ng/ml in basal and after stimulation. As a unique finding, no anti-MHC class I–II or anti-GAD65 antibodies or islet cell antibodies were detected in any of the transplanted patients throughout the 5 years of post-TX follow-up [2, 23]. Hence, microcapsules prevented any detection of foreign antigens by the host, despite the lack of immunosuppressive treatment. In our opinion, this was the most important result, clearly dependent upon use of endotoxin-/and protein-free basic constituent alginate polymers.

2 Materials

2.1 Examination of Isolated Islet Morphology

- *Staining solution (diphenyl-thiocarbazone-DTZ-) to assess cell morphology:* Dissolve 50 mg dithizone in 10 ml DMSO, dilute to 50 ml with D-PBS, filter the solution using a 0.45 mm nylon filter. Expiration date of the solution: 24 h after preparation (to be stored at 6 °C after wrapping the containing tube with aluminum foil, to prevent light exposure).
- *Staining solution to assess cell viability:* Prepare immediately before use a solution comprising 2.5 ml ethidium bromide (0.2 mg/ml in PBS) and 12.5 μl diacetate fluorescein (5 mg/ml in acetone). Use 500 μl of the mixture for 50 μl of islets suspension. Leave the islets for 5 min in the dark with the mixture and then pick them up for observation under a fluorescence microscope.
- *Krebs buffer (1×):* 118 mM NaCl, 4.7 mM KCl, 1.2 mM KH_2PO_4, 1.2 mM $MgSO_4$, 25 mM $NaHCO_3$, 2.5 mM $CaCl_2{\cdot}2H_2O$; buffer the solution to pH 7.4, filter with 0.22 μm nylon filters.
- *Krebs buffer/1 % BSA (1×)*: previous + 1 % bovine serum albumin (BSA).
- *Acid/alcohol solution:* 75 % ethanol, 15 % HCl, 10 % H_2O
- *Powdered alginate* purchased from Monsanto-Kelco featuring the following properties: molecular weight = 120,000–190,000 kDa; mannuronic acid (M) and guluronic acid (G) = M fraction (*F* M) 61 %; G fraction (*F* G) 39 % is a "high-M" alginate.

All solutions are to be filtered in order to grant for sterility. All machinery used for microencapsulation should be treated with *'E-Toxa-Clean Concentrate'* (Sigma-Aldrich), for 16 consecutive hours and then autoclaved for 20 min at 120 °C and 2 atm.

3 Methods

3.1 Alginic Polymers Qualification for Production of Microcapsules for Human Use

The raw alginate powder available in the market ("pharmaceutical grade") must undergo an ultrapurification process, with elimination of several contaminants in order to obtain a chemically stable, endotoxin/pyrogen free, and low protein-content final product that can be stored for long-term, with no loss of the basic desirable physical–chemical properties. This process is uniquely active in our Center and also reported by few other laboratories worldwide [25, 26].

As mentioned above, we have obtained a "clinical-grade" basic sodium alginate solution for microcapsules fabrication, using a special process based on multiple, sequential filtration process through positively charged filters that allow ultrapurification of the initial raw powder, upon homogeneous solution, and sequential adjustments in pH and osmolality (patent number WO 2009/093184 A1). Endotoxins are almost completely removed from the original preparation, by the process, while physical, chemical, and molecular properties of the original align product remain unchanged. Whatever the method of purification of sodium alginate (*see* **Note 1**), the following evaluations are essential:

1. *Inspection*—The first step of the product assessment, perhaps trivial to describe, is the QC inspection, to determine if the solution appears homogeneous, with no precipitates of any kind, transparent in absence of floating micro-particles. Quantitative evaluations will ensue.
2. *Protein content*—Total protein content in the alginate samples is measured, according to guidelines from FDA by NanoOrange kit (Molecular Probes, Invitrogen or similar fluorescent dyes (*see* **Note 2**). In particular, the 1.8 % (w/v) alginate solution is diluted to final concentration of 1 % and then tested, according to the manufacturer's instructions, using a linear standard curve generated with bovine serum albumin.
3. *Endotoxin content*—The test of the lysate of *Limulus polyphemus amebocyte* (LAL), is indicated for the quantitative determination of endotoxin in the final product in "injectable drugs" for human use and this is the only test recommended by FDA for the alginates. The *Endosafe PTS* (Charles River) was used to quantify the total endotoxin content of the alginates. In particular, 1.8 % (w/v) alginate samples were diluted tenfold in (LAL) Reagent water (Charles River), thereafter fivefold in Endosafe-specific buffer (Charles River). The test assay was then performed following the manufacturer's instructions.

Table 2
Alginate molecular specifications

1. Viscosity: 2%—100–300 cps (Brookfield 25 °C, speed 3, 60 rpm)					
2. pH: 6.4–8.0					
3. Protein content: <0.5% EU/ml					
Chemical analysis:					
Ca:	<100 ppm	Mg:	<40 ppm	Mn:	<10 ppm
Cu:	<40 ppm	Zn:	<40 ppm	Sr:	<40 ppm
Fe:	<60 ppm	Pb:	<50 ppm	As:	<100 ppb
Hg:	<40 ppb	Si:	<10 ppm		
8. Endotoxin level—measured by LAL test (at University of Perugia): <0.5 EU/ml					
9. Molecular weight: 120,000–190,000 kDa					
10. Mannuronic acid (M) content: M fraction (F_M) 61%					
11. Guluronic acid (G) content: G fraction (F_G) 39%					

Storage is in endotoxin-free polystyrene bottles at 4 °C and no light exposure

4. *pH, Osmolarity, Viscosity*—For measuring solution pH, viscosity and osmolarity standard laboratory devices may be used with no particular recommendation. Care should only be taken to ensure that the alginate solution complies with required parameters as for those listed in Table 2.
5. *Sterility*—It must be tested by culture of samples of solution to exclude the presence of microbial agents (gram-positive and gram-negative bacteria, Mycoplasma, yeasts and fungi).
6. *Heavy metals content*—We recommend delegate to a certified laboratory evaluation of the content of heavy metals. Heavy metal content should range below the recommended cutoff, listed in the Table 2.
   Other finer tests can be performed, for instance testing the product's MRI spectra: they may serve to precisely assess evaluate both, the M/G ratio and the possible degree of acidic hydrolysis (an unwanted phenomenon that may break down the original algin molecular configuration and patterns)
7. *1H nuclear magnetic resonance spectra*—20 mg of lyophilized purified sodium alginate are dissolved in 1 ml of D2O and analyzed in a Bruker NMR Avance 400 MHz instrument (Fig. 2). All the spectra are recorded without the suppression of the water signal using the native (control) or purified alginate without any preliminary treatment at 338 K. We found that at 338 K the low field spectral region, assigned to the ano-

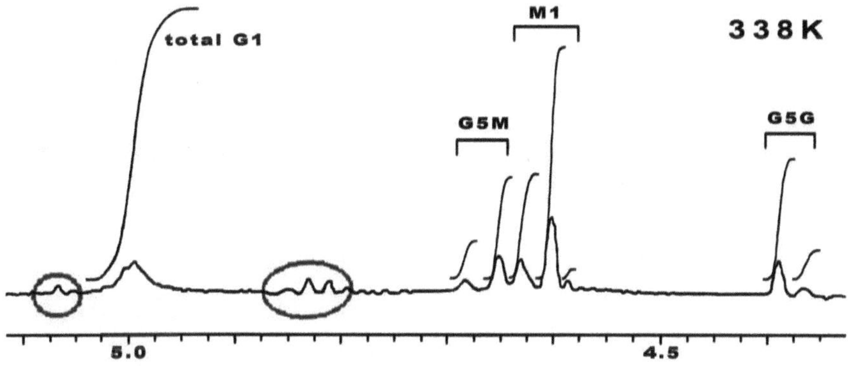

Fig. 2 A representative ^{1}H-spectrum of purified sodium alginate

Table 3
Chemical composition of native versus purified alginate

	M/G	Monads	Diads
Native alginate	1.50	FG 0.4 FM 0.6	FMM 0.39 FMG 0.21 FGM 0.21 FGG 0.19
Purified alginate	1.50	FG 0.4 FM 0.6	FMM 0.39 FMG 0.21 FGM 0.21 FGG 0.19

meric protons, can be recorded with a good resolution, giving important information about the composition of the polymer, and allow to extrapolate information on the M/G ratio as well as the molar fractions of monads, diads and some triads of nearest neighbors along the interacting copolymer chain (Table 3) [27]. The less intense resonances (circled in Fig. 2) were assigned to the anomeric protons of the reducing end-groups and give valuable information on the rate and grade of hydrolysis.

3.2 Islet Cells Quality Control

An even more stringent issue is to exclude and prevent any contamination of the islet tissue to be grafted. Hence QC procedures are to be undertaken on islet cell culture specimens after isolation and before encapsulation.

In our laboratory, human islets (Fig. 3) are isolated from single donor pancreases, according to the Edmonton protocol with minor changes [28]. The Edmonton protocol does not need to be described here, with focus being required only on the procedures to be adopted at the end of the islet isolation. The islets are

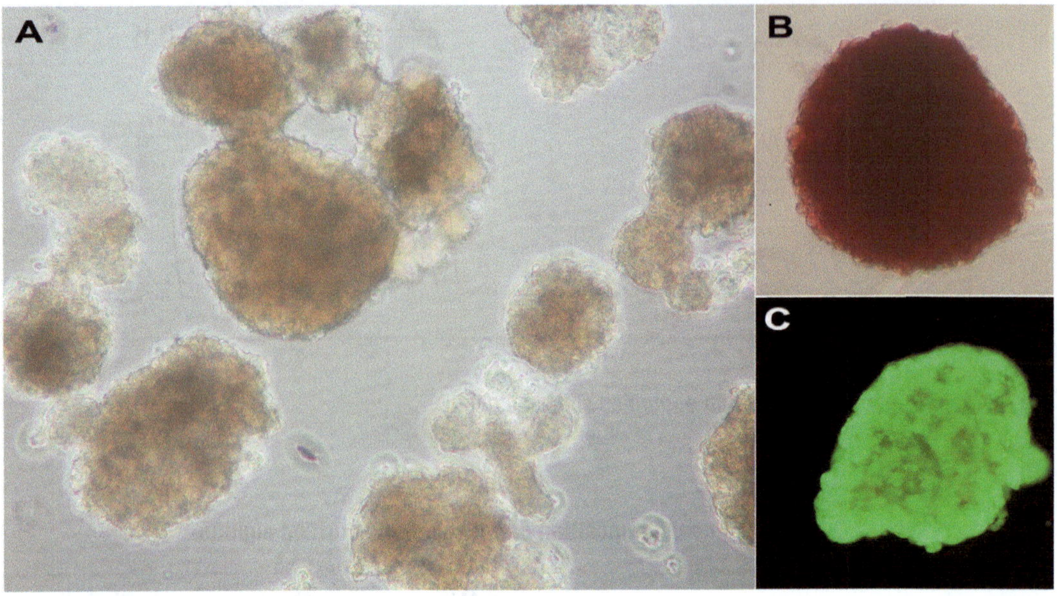

Fig. 3 (**a**) Human islets as they appear after the isolation procedure. (**b**) DTZ staining of an isolated islet. (**c**) Islet viability assay (EB/FD)

cultured for 24 h in HAM F12 (Celbio,), supplemented with antibiotics and 1.25 % human albumin (Kedrion) at 37 °C in 95 % air/CO_2.

Three days after isolation, the cell cultures are tested for microbiological contamination by:

- *General microbial screening*: for gram-positive and gram-negative bacteria, Mycoplasma spp., and fungi (*laboratory testing:* University of Perugia Hospital, Division of Microbiology).
- *Bacterial endotoxin testing*: by LAL test [any level below 5 EU/ml in this test is considered endotoxin-free].
- *Islet purity*: it is determined by DTZ (final concentration 100 μg/ml) staining of the cells, as mentioned above. DTZ is a zinc-chelating agent and a supravital stain that selectively binds to zinc of the insulin molecule, hence evidencing in red the insulin containing islet β-cells: such a distinctive red color will identify the islet cells over unstained pancreatic exocrine acinar cells. Practically, 100 μl from 100 ml of islet suspension are drawn. The picked islets are placed on 3 cm^2 dish where 30 μl (3 drops) of the DTZ solution are added; 1–2 min should be allowed to complete the staining for 1–2 min, with the addition of an extra D-PBS (1 ml). The islets are finally examined under light microscope.
- *Viability* of the islet cells is assessed by using ethidium bromide (EB) and fluorescein diacetate (FD) (Sigma-Aldrich) (Life/Dead Solution) under fluorescence microscopy. FD is a fluo-

rescent dye that readily crosses the cell membranes to stain the cytoplasm and nucleus. Bright green fluorescence in both the nucleus and cytoplasm, upon exposure to ultraviolet (UV) light, denotes intact live cells. Conversely, propidium iodide (PI) is a fluorescent dye that cannot cross intact membranes. It emits a bright red fluorescence when exposed to UV light, and the presence of PI in a cell nucleus indicates severe cell damage or death. Using these fluorescent dyes in combination, the islet viability can be assessed by the FD/EB ratio. Isolated islets from healthy donors generally are associated with 90–95 % viability, or green FD-staining of 90–95 % and red EB staining detectable in only 5–10 % of the islet constituent cells.

3.3 In Vitro Glucose-Stimulated Insulin Release (GSIR) Assay

To measure GSIR, islets are cultured in a "low" glucose concentration, typically 2.8 mM (50 mg/ml), to measure the amount of insulin secreted into the media under "basal" or "unstimulated" conditions. Stimulated insulin release is measured by exposing the islets to higher glucose concentration such as 8.3 mM (150 mg/ml) (half-maximal) and/or 16.7 mM (300 mg/ml) (maximal). In response to the glucose stimulation, the time course of the islet response is biphasic, consisting of a rapid spike in insulin release (first phase) followed by a decline to a prolonged second phase plateau of insulin that remains stable throughout the duration of the stimulus. Determination of the in vitro insulin secretory capacity in response to these stimuli is undertaken on both free islets (after 3 days in culture) and upon subsequent encapsulation (*see* Subheading 3.4, below). For better accuracy of this test, 200 μl of microencapsules islets (Cps) or precise amount of free islets (i.e., 50) are put into 24-multiwell following this procedure:

- Wash Cps or free islet with Krebs solution.
- Pre-incubate for 30 with Krebs solution (to be then discarded).
- Wash with Krebs solution + 1 % BSA.

Stimulus 1–I 50: 60 with 300 μl Krebs solution containing glucose 50 mg/ml (recovered and store at −80 °C, pending assay).

Wash with Krebs solution + 1 % BSA.

Stimulus 2–150: 60 with 300 μl Krebs solution containing glucose 150 mg/ml (recovered and store at −80 °C pending assay).

Wash with Krebs solution + 1 % BSA.

Stimulus 3–300: 60 with 300 μl Krebs solution containing glucose 300 mg/ml (recovered and store at −80 °C pending assay).

Wash with Krebs solution + 1 % BSA.

Stimulus–4 50: 60 with 300 μl Krebs solution containing glucose 50 mg/ml (recovered and store at −80 °C pending assay).

Wash with Krebs solution + 1 % BSA.

Acid/ethanol extraction: Incubate o/n + 4 °C with 300 μl solution Acid/ethanol (recover and store at −80 °C pending assay).

Recovery of the cells through lysis solution for DNA isolation kit (commercial) or by rupture of the capsules, centrifugation and subsequent lysis procedure for the isolation of DNA (*see* **Note 3**).

The concentration of insulin in the collected fractions is measured by RIA assay and expressed as μ Insulin Units (IU)/ml. To obtain normalization of the values, in order to possibly compare repeated experiments or different experiments, it is necessary to divide this value by the DNA content expressed as μg/ml. In this way our GSIR is expressed as μUI/μg DNA. It is possible to convert this value into ng of insulin/μg DNA taking into account the known equation: 28 μUI = 1 ng of I.

It is also possible to express GSIR values as "glucose stimulation index" (SI): the ratio of stimulated-to-basal insulin secretion (Fig. 4). Healthy islets have an SI of 2–20 depending on several factors including strain, age, and body weight.

3.4 Alginate Encapsulation Procedure

The islets are washed with 0.9 % NaCl and thoroughly mixed with 1.8 % endotoxin- and pyrogen-free sodium alginate so as to make a homogeneous alginate/islet cell suspension (the ratio is flexible, depending upon purity of the islet preparation which mandates careful use of alginate in the dilution suspension process). Initially, a "dry run" with saline should always to be scheduled so as to

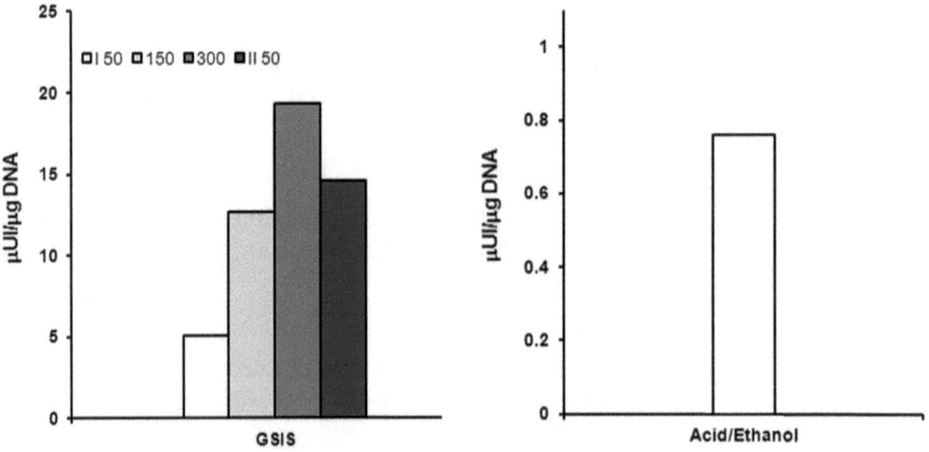

Fig. 4 Static incubation with glucose where the obtained values are normalized for μg/DNA

make sure that the system is operative (*see* **Note 4**). The suspension is then extruded through a microdroplet generator. The AG/islet mixture is continuously aspirated by a peristaltic pump at a flow rate of approximately 0.5 ml/min, with care been taken to adjust distance between needle tip and surface of the collection bath. This comprises 200 ml of 1.2% $CaCl_2$ in 250 ml glass beaker. Extrusion of the mixture through the jet-head device is driven, as above mentioned, by both mechanical pressure (from the peristaltic pump) and air shearing forces (air flow rate: 3 nl/min) applied to the external adaptor of the jet needle. Upon contact on the Calcium solution the AG/islet microdroplets will immediately turn into islet containing gel microspheres. Upon 10 min, the beads pellet is carefully washed in saline. The beads, containing 1–2 islets, and measuring an average 500 μm in diameter, are sequentially double-coated with 0.12 and 0.06% poly-L-ornithine (Sigma) in 0.9% NaCl. The microspheres are resuspended in a polyaminoacidic solution volume that is double as compared to their own, and then placed on a rocking plate for 10 min (coating 1) and 6 min (coating 2). The microsphere algin gel central core is then liquefied by the addition of 55 mM sodium citrate for 3′. Preparation is ready to be washed again with saline and finally incubated with a solution of 0.04% AG, in order to make a biocompatible outer shell around the spheres. Final washes with saline and culture media will make the preparation ready for culture, at 37 °C in 95% air/CO_2, overnight prior to use for transplant or other purposes (Fig. 5).

QC as far as sterility, viability and function of the islets are to be repeated after microencapsulation (Fig. 6).

Fig. 5 Empty capsules measuring an equatorial diameter of approximately 500 μm

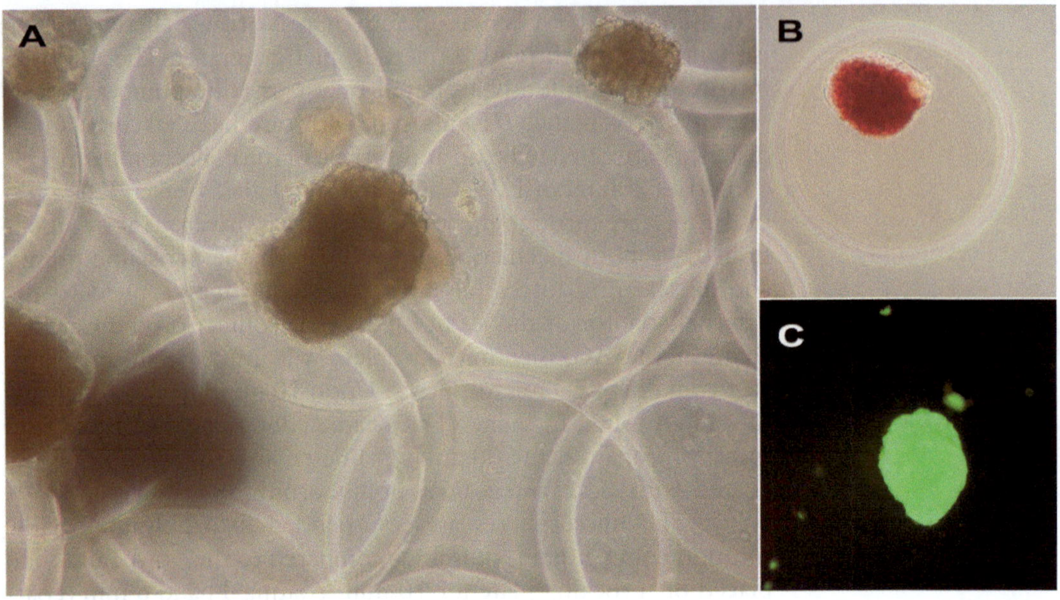

Fig. 6 (**a**) Microencapsulated human islets as they appear after the encapsulation procedure. (**b**) DTZ staining of an islet. (**c**) Islet viability assay (EB/FD)

3.5 Transplantation Procedures

Transplantation of immunoprotected (alginate-encapsulated) human islet cells is undertaken by intraperitoneal deposition. As above mentioned, the procedure may take place by direct puncture of the abdomen, under local anesthesia, under ecography guidance or by laparoscopy, under general anesthesia. In the first instance, the radiologist, upon visualization of the area, will inject saline to constrast the graft site. Delivery time by a 60 ml syringe tipped with 12G needle, will go over 10 min, with care being taken to avoid capsules clumping at the syringe luer level. In the second instance, the procedure is conducted by a transplant surgeon upon laparoscopic visualization of the selected abdominal sites. Should any signs of peritoneal inflammation be detected the transplant will not take place.

3.6 Outcome Measures for Assessing Efficacy

Only patients with no detectable residual islet β-cell function should be enrolled into the clinical trial, to clearly assess islet graft function with no interferences. This is assessed by serum RIA ultrasensitive C-peptide assay. Insulin is normally produced from proinsulin in the endocrine pancreas and for every mole of insulin produced, an equimolar amount of C-peptide is also generated. In patients with no residual islet cell function, no C-peptide is detectable and insulin dependency is full. In fact, withdrawal of exogenous insulin supplementation is followed by severe diabetic symptoms within 24 h, which may result in severe ketoacidosis, still a life-threatening acute complication of T1D. Finally, there are no reported instances of spontaneous remission of T1D, except as a transient phenomenon shortly after initiation of pharmacologic treatment (the so called "honey moon phenomenon").

Prior to TX, patients will undergo a 6-week run-in period, when conventional insulin treatment will be optimized. This may possibly involve changes in the frequency of insulin injections and self-monitoring of blood glucose, together with more frequent physician contacts and optimization of exercise and dietary regimens. During this time, the following parameters will be recorded:

(a) Daily blood glucose levels (at least four tests daily, measured by a glucose meter, with the time and concentration recorded);

(b) HbA_{IC} levels at monthly intervals;

(c) Recording occurrence and frequency and severity of hypoglycemic events: a detailed diary of hypoglycemic episodes in the 6-week run-in period preceding the transplant will be recorded and used as the baseline. In addition, a history of severe hypoglycemic episodes (WHO classification) and episodes of hypoglycemia unawareness (not included in the WHO criteria) over the preceding year will also be obtained;

(d) The dose (number of units) and type of insulin administered.

(e) Abdominal ultrasound scan/MRI.

The final 2 weeks of the pretransplant period will be regarded as the baseline.

Following TX, the above reported parameters will be monitored for at least 6 months. If after this time no significant improvements have occurred, restoration of the previous conventional treatment will be applied in full. On the other hand, if there has been a statistically significant decrease in the daily insulin dose as compared to baseline, the follow-up will continue as long as the daily insulin is reduced.

The main measures that will indicate (or not) success of the human islet cell graft in treated patients include:

1. Daily blood glucose levels (at least four determinations daily, measured by a glucose meter, with the time and concentration recorded) (range: 80–150 mg/dl);
2. The average daily insulin consumption to achieve identical or better glucose control versus baseline;
3. The appearance of serum C-peptide;
4. C-peptide levels will be measured prior to transplantation, and at regular intervals post-transplantation, under the same conditions for either continuation or termination of the study, as above described for HbA_{IC} levels. sCPR samples, by RIA (sCPR sensitivity: 0.06 ng/ml [intra-assay CV 3.7–4.5%, interassay CV 4.4–5%]);
5. HbA_{IC} levels at monthly intervals;
6. Reduction in frequency/severity of hypoglycemic episodes:

   Recording and grading all hypoglycemic episodes after transplantation in compliance with WHO criteria, including eventual episodes of hypoglycemia unawareness;

7. Eventual changes in diabetic complications:

   Although patients with moderate to severe diabetes complications will be excluded from the study, baseline observations in all patients will be made at the initial pre-transplant visit (at –6 weeks). Microalbuminuria will be tested 3 months after transplantation, and an overall assessment of diabetes complications will be repeated at the last scheduled visit (at 6 months). Because of the reported background/mild (pre-proliferative) retinopathy worsening upon rapid improvements of blood glucose levels, such possibility will be taken into careful account. Evidence of macrovascular complications eventually detected after transplantation will be recorded. In men, the occurrence of erectile dysfunction will be recorded both before and after transplantation.

8. A 75-g oral glucose tolerance test (OGTT) is scheduled for treated patient at 60 days post-TX.
9. Abdominal ultrasound scan is to be performed at 30 days post-TX
10. Quality of life changes (as assessed by the ADDQoL questionnaire; see further below) (Table 4: parameters to be recorded for grafted patients).

Table 4
Outcome measures for assessment of efficacy

Duration of T1D (years)	Pre-TX HbA1c (%)
	Post-TX HbA1c (%) 16 weeks
Pre-TX severe hypoglycemia (events/week)	Post-TX HbA1c (%) 24 weeks
Post-TX severe hypoglycemia (events/week)	Post-TX HbA1c (%) 48 weeks
Pre-TX mean blood glucose (mg/dl)	Anti-GAD65 ab pre- and post-TX
Post-TX mean blood glucose (mg/dl) 6 months	Anti-Islet cell ab pre- and post-TX
Post-TX mean blood glucose (mg/dl) 12 months	Class I HLA pre- and post-TX
	Class II HLA pre- and post-TX
Pre-TX serum C-peptide (ng/ml)	
Post-TX serum C-peptide (ng/ml) 6 months	Post-TX sCPR (ng/ml) 90 days/12 months/36 months
Post-TX serum C-peptide (ng/ml) 12 months	0 min
Pre-TX daily exogenous insulin (IU)	30 min
Post-TX daily exogenous insulin (IU) 4 weeks	60 min
Post-TX daily exogenous insulin (IU) 24 weeks	90 min
Post-TX daily exogenous insulin (IU) 48 weeks	120 min
	180 min
	240 min

3.6.1 Quantitative Measures for Assessment of Efficacy

Day-to-day variations of up to 10% of the usual total daily dose of insulin are regularly encountered in any given diabetic patients, and are related to variations in food intake, exercise and intercurrent illness. Long-term changes of 10% in dosage may relate to continued changes in lifestyle, but changes of >20% long term are generally unlikely. Therefore, a reduction in the average total daily dose of insulin of >20% will be maintained as the ***minimum presumptive evidence*** of transplant function. This must be accompanied by:

1. No objective evidence of deterioration of diabetes control (daily blood glucose and monthly HbA_{1C} levels)
2. No evidence of major weight loss or ketoacidosis; and
3. Commensurate quantitative appearance of C-peptide.

The quantitative extent of function can be estimated from the same measurements—up to complete absence of insulin requirement. The duration of such evidence of transplant function can then be recorded. The provisional degree of ***useful*** function is a 50% or greater reduction in insulin requirements to achieve identical or better glucose control versus baseline.

Thus, the clinical response after transplantation at each dose level can be rated as follows:

1. *Excellent response:* complete absence of insulin requirements over the period of observation.
2. *Very good response:* ≥75 to <100% reduction in the average total daily dose of insulin over the period of observation.
3. *Good response:* ≥50 to <75% reduction in the average total daily dose of insulin over the period of observation.
4. *Fair response:* ≥20 to <50% reduction in the average total daily dose of insulin over the period of observation.
5. *No response:* <20% reduction in the average total daily dose of insulin over the entire period of observation.

3.6.2 Outcome Measures for Assessment of Safety

Safety endpoints include the following:

1. Clinical and laboratory evidence of ***infection.***
2. Clinical and laboratory evidence of hypoglycaemic or perioperative reactions.
3. Clinical and laboratory evidence of other adverse events, including adverse psychological impact.

4 Notes

1. Various methods have been reported to purify alginates, by knocking down endotoxins levels so as to comply with FDA regulations. Alternative to ours, a method is based on precipitation of the product in ethanol and chloroform, fol-

lowed by lyophilization [29]. Another procedure consists of dialysis of the alginate solution follwed by lyophilization. We do believe that these alternative methods are burdened with low yield and excessive manipulation of the final purified product.

2. Method sensitivity ranges on 10 and 10 μg/ml of protein concentration. This level is considerably better than that achieved with the BCA method, Bradford assay, Lowry assay, or absorption at 280 nm; furthermore, the last methods are difficult to apply to alginate solutions. For instance, Bradford solution contains phosphoric acid (85 %), hence it is clearly acidic and may induce precipitation of alginate together with eventual proteins (to be assessed
3. Microcapsules formulated with PLO are quite strong which makes it difficult their breakage to retrieve their content. However, and limited to in vitro studies, the alginate core may be coated with a single 0.05 % PLO layer. The capsules will be "less resistant" and suitable for mechanical disruption.
4. Our long-standing experience suggests that empty capsules should be made before using them for envelopment of human islets, in order to avoid unwanted technical failure.
5. Limitations, critique, and outlook.

   It is more and more evident that graft of pancreatic human islets, separated and purified from cadaveric donor organs, and eventually immunoprotected within highly biocompatible and selectively permeable, alginate-based microcapsules may not represent the future of cell and molecular therapy for T1D. As above briefly touched, the organ donation rate, also in consideration of the fact that the donor–recipient ratio usually exceeds 1:1 (meaning that 2 or more pancreases are required to harvest sufficient islet mass to graft a single recipient) largely is insufficient. Consequently, the already very high costs of the procedure will go further up, to unbearable levels, should the procedure represent the final cure available to all patients with T1D. If use of microencapsulation may mitigate problems associated with the recipients' general pharmacologic immunosuppression, still the organ availability issue, plus costs, would continue to represent a major bottleneck for the future of this approach. Under these circumstances, the majority of T1D patients would be better off continuing their daily exogenous insulin injections. Microcapsules could theoretically permit access to nonhuman islets, i.e., porcine, derived from special Specific Pathogen Free (SPF) pig herds (not contaminated by adventitious microbial agents), that besides clear analogies between human and pork insulin, would offer the opportunity of virtually unlimited procurement at much lower costs. However, bans on operative employment of these islets are still effective in the majority of world Countries which makes porcine islets not immediately usable at this time. The next generation approaches for cell and molecular therapy of T1D, aiming at cure of this disease, could envision use of new technologies employing stem cells for insulin producing cells in the place of traditional pancreatic islets. While a number of restrains still applies to human embryonic stem cells, due to their intrinsic potential unsafety, in spite of their clear supposed effectiveness, and besides still active ethical bans, in most of the western world Countries, adult stem cells could fulfill this gap. While less plastic than embryonic, these cells are undoubtedly safer and virtually suitable for cell substitution therapy in many chronic disorders, T1D included. Among others, adult human mesenchymal stem cells (MSC) that are retrievable in bulk from many sources likely represent a good compromise. In fact, these cells do not express class II MHC antigens, which makes them poorly prone to rejection induction, and have been proven able to differentiate into many specific cell lineages. Differentiation studies of MSC into β-like insulin producing cells are underway. Finally, these cells are fully suitable for microencapsulation. In parallel, development of the next generation of microcapsules, based on the same biocompatible above, described polymers but associated with new designs to better integrate with the host tissues hold promises that new biohybrid constructs, with new cells and materials, may really help providing for the final cure of T1D.

Acknowledgments

Altucell Inc., Astor Court 3, Dix Hills, New York, NY (USA) is kindly acknowledged.

References

1. Basta G, Montanucci P, Calafiore R (2015) Islet transplantation versus stem cells for the cell therapy of type 1 diabetes mellitus. Minerva Endocrinol 40(4):267–282
2. Basta G, Montanucci P, Luca G, Boselli C, Noya G, Barbaro B, Qi M, Kinzer KP, Oberholzer J, Calafiore R (2011) Long-term metabolic and immunological follow-up of nonimmunosuppressed patients with type 1 diabetes treated with microencapsulated islet allografts: four cases. Diabetes Care 34(11): 2406–2409
3. Israni AK, Skeans MA, Gustafson SK, Schnitzler MA, Wainright JL, Carrico RJ et al (2014) OPTN/SRTR 2012 annual data report: pancreas. Am J Transplant 14:45–68
4. Cefalu WT (2013) State of the union. Diabetes Care 36:1–3
5. Morath C, Zeier M, Süsal C, Döhler B, Opelz G (2015) Response to a reassessment of the survival advantage of simultaneous kidney-pancreas versus kidney-alone. Transplantation 99:e175
6. CITR Research Group (2009) 2007 Update on allogeneic islet transplantation from the Collaborative Islet Transplant Registry (CITR). Cell Transplant 18:753–767
7. Balamurugan AN, Naziruddin B, Lockridge A, Tiwari M, Loganathan G, Takita M et al (2014) Islet product characteristics and factors related to successful human islet transplantation from the Collaborative Islet Transplant

Registry (CITR) 1999–2010. Am J Transplant 14:2595–2606

8. Nilsson B, Ekdahl KN, Korsgren O (2011) Control of instant blood-mediated inflammatory reaction to improve islets of Langerhans engraftment. Curr Opin Organ Transplant 16:620–626
9. Murua A, Portero A, Orive G, Hernández RM, de Castro M, Pedraz JL (2008) Cell microencapsulation technology: towards clinical application. J Control Release 132:76e83
10. de Vos P, Andersson A, Tam SK, Faas MM, Hallé JP (2006) Advances and barriers in mammalian cell encapsulation for treatment of diabetes. Immunol Endocr Metabol Agents Med Chem 6:139e53
11. Thanos CG, Calafiore R, Basta G, Bintz BE, Bell WJ, Hudak J et al (2007) Formulating the alginate-polyornithine biocapsule for prolonged stability: evaluation of composition and manufacturing technique. J Biomed Mater Res A 83:216e24
12. Hermann A, Liebau S, Gastl R, Fickert S, Habisch HJ, Fiedler J et al (2006) Comparative analysis of neuroectodermal differentiation capacity of human bone marrow stromal cells using various conversion protocols. J Neurosci Res 83:1502e1418–1502e1421
13. Dean SK, Yulyana Y, Williams G, Sidhu KS, Tuch BE (2006) Differentiation of encapsulated embryonic stem cells after transplantation. Transplantation 82:1175e84
14. Montanucci P, Pennoni I, Pescara T, Basta G, Calafiore R (2013) Treatment of diabetes mellitus with microencapsulated fetal human liver (FH-B-TPN) engineered cells. Biomaterials 34:4002–4012
15. de Vos P, Faas MM, Strand B, Calafiore R (2006) Alginate-based microcapsules for immunoisolation of pancreatic islets. Biomaterials 27:5603–5617
16. Draget KI, Smidsrød O, Skjak-Bræk G (2005) Alginates from algae. In: Steinbuchel A, Rhee KS (eds) Polysaccharides and polyamides in the food industry. properties, production and patents. Willey, Weinheim, Germany, pp 1–30
17. Smidsrød O, Haug A (1968) Dependence upon uronic acid composition of some ion-exchange properties of alginates. Acta Chem Scand 22:1987–1997
18. Smidsrød O (1970) Solution properties of alginate. Carbohydr Res 13:359–372
19. Smidsrød O (1974) Molecular basis for some physical properties of alginates in the gel state. J Chem Soc Faraday Trans 57:263–274
20. Haug A, Smidsrød O (1970) Selectively of some anionic polymers for divalent metal ions. Acta Chem Scand 24:843–854
21. Montanucci P, Terenzi S, Santi C, Pennoni I, Bini V, Pescara T, Basta G, Calafiore R (2015) Insights in behavior of variably formulated alginate-based microcapsules for cell transplantation. Biomed Res Int 2015:965804
22. Petruzzo P, Pibiri L, De Giudici MA, Basta G, Calafiore R, Falorni A, Brunetti P, Brotzu G (1991) Xenotransplantation of microencapsulated pancreatic islets contained in a vascular prosthesis: preliminary results. Transpl Int 4:200–204
23. Brunetti P, Basta G, Faloerni A, Calcinaro F, Pietropaolo M, Calafiore R (1991) Immunoprotection of pancreatic islet grafts within artificial microcapsules. Int J Artif Organs 14: 789–791
24. Calafiore R, Basta G, Luca G, Lemmi A, Racanicchi L, Mancuso F, Montanucci MP, Brunetti P (2006) Standard technical procedures for microencapsulation of human islets for graft into nonimmunosuppressed patients with type 1 diabetes mellitus. Transplant Proc 38:1156–1157
25. Calafiore R, Basta G (2007) Artificial pancreas to treat type 1 diabetes mellitus. Methods Mol Med 140:197–236
26. Duvivier-Kali V, Fernandes J, Tchipashvili V, Colton CK, Weir GC (2005) Long-term normoglycemia in rats receiving transplants with encapsulated islets. Omer A Transplant 79:52–58
27. Santi C, Coppetta D, Santoro S, Basta G, Montanucci P, Racanicchi L, Calafiore R (2008) NMR analysis of non hydrolyzed samples of sodium alginate 12th international electronic conference on synthetic organic chemistry ECSOC-12
28. Shapiro AM, Lakey JR, Ryan EA et al (2000) Islet transplantation in seven patients with type 1 diabetes mellitus using a glucocorticoidfree immunosuppressive regimen. N Engl J Med 343:230–238
29. De Vos P, De Haan BJ, Wolters GHJ, Strubbe JH, Van Schilfgaarde R (1997) Improved biocompatibility but limited graft survival after purification of alginate for microencapsulation of pancreatic islets. Diabetologia 40:262–270

Chapter 24

Immunological Challenges Facing Translation of Alginate Encapsulated Porcine Islet Xenotransplantation to Human Clinical Trials

Rahul Krishnan, David Ko, Clarence E. Foster III, Wendy Liu, A.M. Smink, Bart de Haan, Paul De Vos, and Jonathan R.T. Lakey

Abstract

Transplantation of alginate-encapsulated islets has the potential to treat patients suffering from type I diabetes, a condition characterized by an autoimmune attack against insulin-secreting beta cells. However, there are multiple immunological challenges associated with this procedure, all of which must be adequately addressed prior to translation from trials in small animal and nonhuman primate models to human clinical trials. Principal threats to graft viability include immune-mediated destruction triggered by immunogenic alginate impurities, unfavorable polymer composition and surface characteristics, and release of membrane-permeable antigens, as well as damage associated molecular patterns (DAMPs) by the encapsulated islets themselves. The lack of standardization of significant parameters of bioencapsulation device design and manufacture (i.e., purification protocols, surface-modification grafting techniques, alginate composition modifications) between labs is yet another obstacle that must be overcome before a clinically effective and applicable protocol for encapsulating islets can be implemented. Nonetheless, substantial progress is being made, as is evident from prolonged graft survival times and improved protection from immune-mediated graft destruction reported by various research groups, but also with regard to discoveries of specific pathways involved in explaining observed outcomes. Progress in the latter is essential for a comprehensive understanding of the mechanisms responsible for the varying levels of immunogenicity of certain alginate devices. Successful translation of encapsulated islet transplantation from in vitro and animal model testing to human clinical trials hinges on application of this knowledge of the pathways and interactions which comprise immune-mediated rejection. Thus, this review not only focuses on the different factors contributing to provocation of the immune reaction by encapsulated islets, but also on the defining characteristics of the response itself.

Key words Alginate, Islet, Diabetes, Biocompatibility, Antigenicity, Rejection, Transplant, PAMPs, DAMPs, MSCs, Encapsulation, Xenotransplantation

1 Introduction

Over the last decade, human islet transplantation has emerged as a viable alternative to conventional management strategies to treat type 1 diabetes. With appropriate immunosuppression regimens,

Emmanuel C. Opara (ed.), *Cell Microencapsulation: Methods and Protocols*, Methods in Molecular Biology, vol. 1479,
DOI 10.1007/978-1-4939-6364-5_24, © Springer Science+Business Media New York 2017

patients that are able to remain insulin independent for up to 8 years post-transplant [1]. Unfortunately, a scarcity of donor organs, low islet yields after isolation, graft destruction due to the instant blood-mediated inflammatory reaction [IBMIR] and graft dysfunction due to concomitant immunosuppressive therapy impede further progress. Alternative tissue sources like islets from designated pathogen free [DPF] pigs have demonstrated great promise in greatly expanding the donor tissue pool. However, porcine islet xenografts are prone to hyperacute graft rejection within minutes of transplantation. Encapsulation of islets within biocompatible hydrogels is a solution proposed by biomedical engineers to address this uniquely frustrating issue. Cell encapsulation involves enveloping cells in a selectively permeable biocompatible matrix that allows for the diffusion of oxygen and nutrients but is able to effectively prevent immune cells and antibodies from reaching the graft, thus delaying rejection. Encapsulation can be employed as a platform to deliver localized immunosuppression at the transplant site, thus avoiding the adverse effects of chronic systemic immunosuppression and can also be used to deliver nutrients and biological agents that will enhance islet survival and function after transplantation into patients [2]. While it is clear that encapsulation has several advantages over conventional islet transplantation, several roadblocks currently prevent translation of results from small animal and primate trials to human trials. This review addresses a few key issues that the authors believe can improve transplant outcomes and allow for expedited translation of this promising technology to clinical trials.

2 Key Factors Impacting Transplant Outcomes in Alginate Encapsulated Porcine Islet Xenotransplantation

2.1 Alginate Purity

Commercially available ultrapure alginates used in the fabrication of immunoisolation devices to encapsulate porcine islets have been found to contain pathogen-associated molecular patterns (PAMPS) such as peptidoglycan, lipotechoic acid, and flagellin, among other proteins, endotoxins, and polyphenols [3]. These impurities, either inherent in the alginate or incorporated into the substance during industrial extraction techniques, compromise the immunoprotective properties of alginate microcapsules, primarily by triggering recognition by the innate immune system (Fig. 1). These sensors stimulate a cascade of signal transduction pathways, starting with toll-like receptors (TLRs) and pattern-recognition receptors (PRR) [3], ultimately resulting in pericapsular fibrotic overgrowth (PFO) [4] (Fig. 2). PFO severely hinders graft survival due to obstruction of essential diffusive gradients necessary to prevent islet necrosis. However, direct adhesion of PAMP activated fibroblasts, macrophages, and other immune-cells onto capsules is not the only platform through which islet necrosis is triggered. It has been discovered

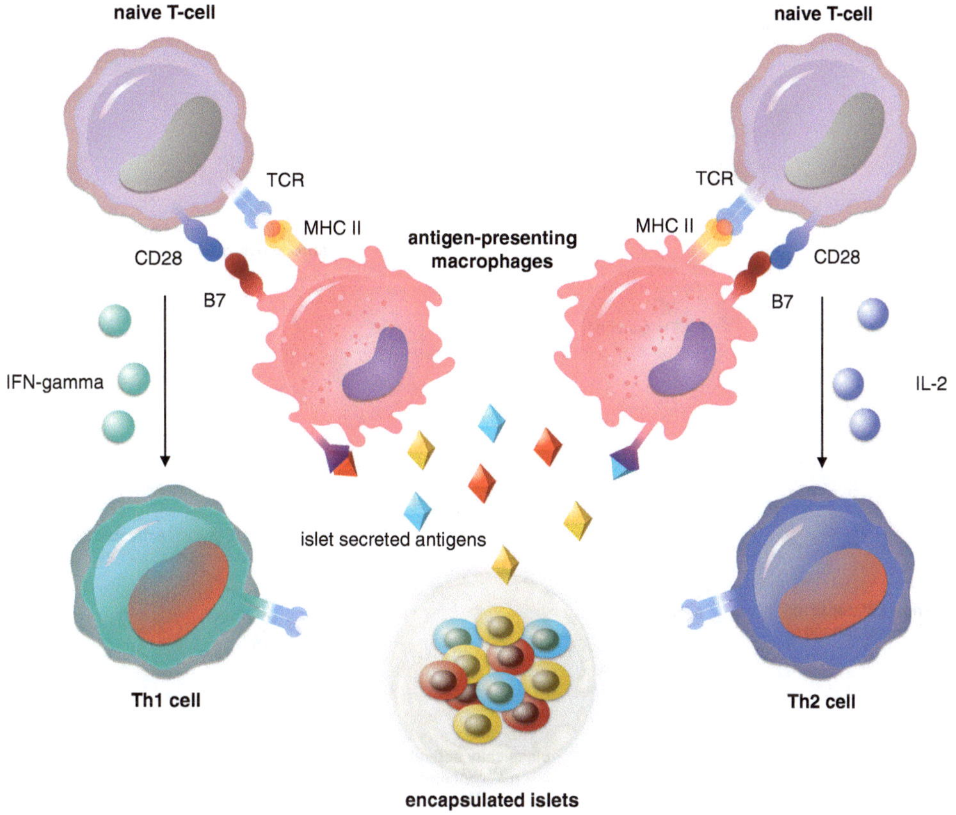

Fig. 1 A pathway by which the function of encapsulated islets are compromised. Capsule permeable antigens are secreted by islets, potentially as DAMPs due to hypoxic conditions generated by pericapsular fibrotic overgrowth. Antigenic small molecules are detected by antigen presenting cells, which then internalize, modify, and present the antigen to naiive T-cells. These naiive T-cells differentiate into either Th1 or Th2 cells

that PAMP-activated macrophages, granulocytes, and basophils can cause deleterious effects through biologically active small molecules such as chemokines and cytokines secreted in the graft milieu [5]. Should the morphological properties of the device surfaces lend itself to cellular adhesion, graft viability may be lowered by PFO, but barring such properties, the same outcome may be provoked via secreted molecules capable of diffusing through the pores of the alginate matrices [5].

PAMPS present in alginate microcapsules are mainly detected via membrane-associated receptors of innate immune cells, known as Toll-like Receptors (TLRs) [6]. These sensory receptors are one of many PRRs responsible for initiation of the innate immune response [7]. TLRs, upon recognition and binding of PAMPS to the receptor surface, initiate an intracellular signaling cascade ultimately resulting in the secretion of a host of inflammatory cytokines attributed to translocation of the Nf-kB (nuclear factor kappa-light-chain-enhancer of activated B cells) into the nucleus [8].

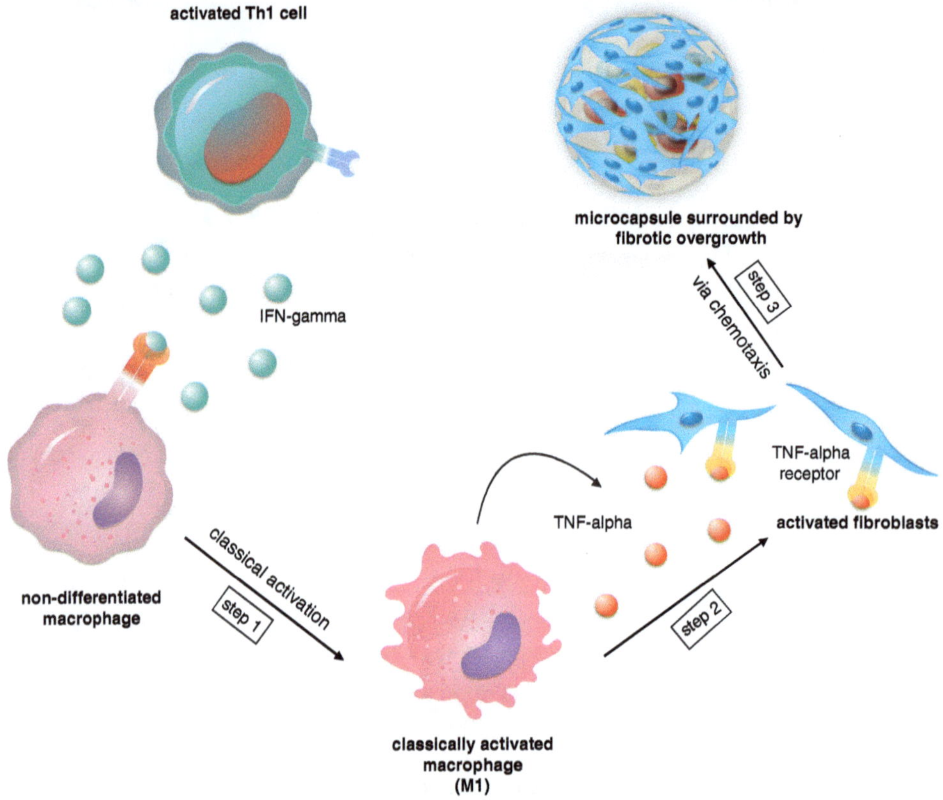

Fig. 2 Once activated, Th1 cells are responsible for activating non-differentiated macrophages into classically activated, TNF-alpha producing M1 macrophages, via IFN-gamma secretion. The M1 macrophages go on to activate and recruit fibroblasts, which eventually form an overgrowth layer on the capsule surface. Deprived of nutrients, oxygen and unable to secrete metabolic wastes, the encapsulated islets undergo necrosis. Subsequently, they secrete antigenic DAMPS, and lose functionality via overgrowth and both the cellular, and humoral immune systems

In 1997, De Vos et al. reported that purification consisting of filtration, precipitation and protein extraction of alginate-polylysine (PLL) led to retrieval of capsules after 12 months of which 10% or less had been affected by PFO, as compared to nearly 100% of crude, unpurified alginate [9]. However, the subsequent development of multiple, albeit unstandardized, purification techniques have warranted further comparative investigations. In an evaluation of the effectiveness of three in-house alginate purification procedures designed by Prokop and Wang, de Vos et al., and Klock et al. as opposed to using crude, unpurified, and commercially available purified alginate, all three groups reported varying results [10]. The three methods were specifically aimed at lowering concentrations of polyphenols, endotoxins and proteins in crude alginate, of which they were all successful, albeit to varying degrees. However, the most convincing results were obtained by Klock et al. Investigators achieved a decrease in polyphenol content from 13.83 AFU (Arbitrary Fluorescence Units) to 0.47 AFU, endotoxin content from 4000 EU/g (Endotoxin Units/g) to 8 EU/g, and protein content from 5.88 to 2.94 mg/g [10]. Upon further modification to the technique devised by Klock et al. incorporating size exclusion preparative chromatography and further dialysis, subsequent endeavors resulted in protein concentrations that were reduced by up to 90%. In vivo effects of these purification techniques in relation to PFO were quantitatively measured via cell adhesion scores upon retrieval from implantation sites [10].

Subsequent to the treatment of alginate by in-house purification protocols, many involving chemical extraction procedures [11], the properties of the polymer were also susceptible to change. Increased hydrophilicity of the capsule surface, along with increased solution viscosity contributed to lower immunogenicity [12]. The lack of a standardized protocol for crude alginate purification, inconsistencies in the quality of the starting material, and the substantial amount of potential contaminants present in commercially purified alginate [10] justifies a need for consistent and regulated guidelines for acceptable concentrations of immunogenic components in alginates prepared for use in clinical transplantation trials.

Studies delving into the specific mechanisms by which these contaminants bring about decreases in graft viability have reported an increase in the levels of circulating pro-inflammatory cytokines. Elevated levels of IL-Iβ, TNF-α, and IL-6 were detected subsequent to exposure of alginate microcapsules to human or murine macrophages [13–15] either in vitro or post-transplantation. Another study reported that TLRs 2 and 5, receptors sensitive to PAMPs derived from peptidoglycan, lipotechoic acid, and flagellin, were activated after exposure to alginate microcapsules [16].

Activation of TLRs 8 and 9, for which the conventional agonists are single-stranded RNA and unmethylated DNA, also occurred, but with the lack of elucidation of the PAMP responsible [16]. It becomes a point of interest to note that due to the transcriptional factor utilized in this pathway of the innate immune system, these PAMPs are thought to be released from the alginate matrices and enact their effects by binding to receptors as a secreted molecule, rather than through cell-capsule surface adhesion. Upon activation, these TLRs signal via an MyD88-mediated pathway to stimulate transcriptional factor NF-kB (Fig. 3). Concentrations of pro-inflammatory cytokines and chemokines characteristic of the innate immune response are upregulated, resulting in recruitment of basophils and other granulocytes, and macrophages [5, 17]. These immune cells, along with fibroblasts, are mainly responsible for the generation of the PFO surrounding microcapsules excluded from processes aimed at removing the previously mentioned PAMPs (Fig. 4).

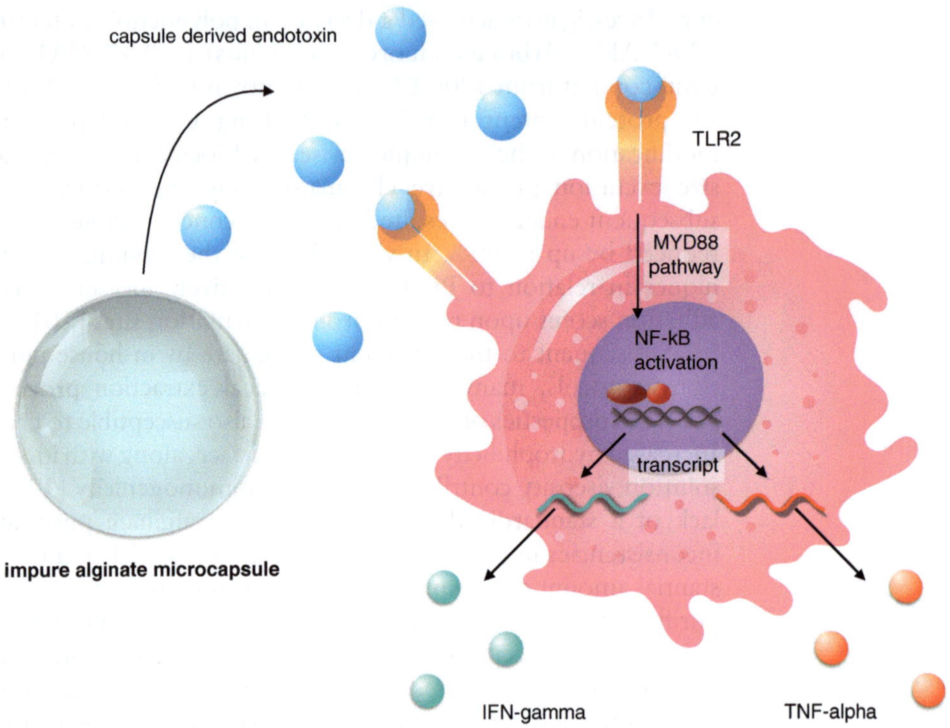

Fig. 3 An impure, blank alginate microcapsule is shown with soluble impurities. Endotoxins released from the alginate are detected by Toll-like receptors (TLR), specifically TLR2, on the surface of circulating macrophages. TLR2 pairing with its ligands triggers an intracellular signaling cascade, specifically, the MyD88 pathway. This pathway activates a transcription regulating protein complex, Nf-kB. Nf-kB upregulates the transcription of genes involved in production of cytokines and chemokines such as TNF-alpha and IFN-gamma

2.2 Polymer Composition and Morphology

Alginate, a polymer of 1–4 linked β-D-mannuronic acid (M) and 1-4 linked α-L-guluronic acid (G) can be extracted from several algal species. The polysaccharide can contain varying concentrations of M and G carbohydrates, thus conferring unique characteristics in terms of permeability, molecular weight, stability, and immunogenicity. Alginate can be classified, by the relative amount of M or G carbohydrates in their matrices, as high-G, intermediate-G, or low-G alginate. The cation responsible for cross-linking these molecules together, such as Ca^{2+}, Ba^{2+}, or Sr^{2+}, also confers distinctive features to the cross-linked alginate [3]. Furthermore, a spectrum of immune responses varying in severity can be attributed to certain surface modifications, commonly in the form of poly amino acids grafted onto the surface of the alginate microcapsules. Conventional modifications include the application of poly-D-lysine (PDL), polyethylene glycol (PEG), poly-L-ornithine (PLO), or poly-L-lysine (PLL) to confer additional stability to the membrane surface [18].

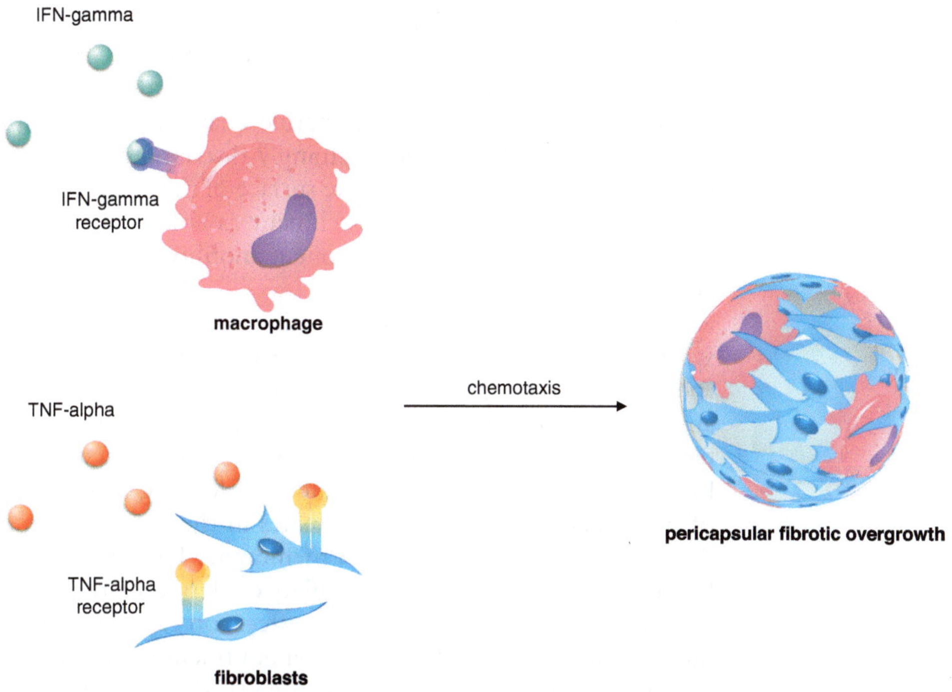

Fig. 4 Antigen activated macrophages secrete soluble, pro-inflammatory cytokines and chemokines such as TNF-alpha and IFN-gamma. These molecules activate and recruit neighboring macrophages and fibroblasts. These cells adhere to the surface of the microcapsule, forming a pericapsular fibrotic overgrowth layer and inhibiting the diffusive capacities of the capsule

Upon comparison of the different immunogenic properties attributable to the proportion of β-D-mannuronic acid (M) to α-L-guluronic acid (G) comprising the alginate matrices, historically, the tendency of mannuronic acid to provoke both an innate, and antibody-mediated adaptive immune response was often suggested as evidence to support the preference of high-G over Low-G (High-M) alginates. Therefore high-M alginate was often avoided when immunosuppression was the desired outcome [19, 20]. These results were independent of the type of cation used for cross-linking (Ca^{2+} or Ba^{2+}). High-M alginates have been shown to prompt macrophages to secrete various pro-inflammatory cytokines including IL-1, IL-6, and TNF-α through interactions with the monocyte CD14 receptor [20].

However, more recent studies seem to contradict these earlier findings, reporting less cell-adhesion on high-M comparison to high-G alginate microcapsules [21, 22]. Subsequent investigations have also suggested no clear relationship between M/G content of alginate and its immunogenicity upon thorough adherence to contaminant purification protocols [23]. Overall alginate immunogenicity was proposed to be mainly dependent on the purity and geometric associations of the polymer, parameters often influenced by the relative concentrations of M and G [24]. As to the underlying rationale behind the increased biocompatibility of intermediate-G capsules compared to high-G capsules, the decreased osmotic pressure, and increased polysaccharide chain stability intrinsic to the intermediate-G capsules seemingly discouraged cell-adhesion [25]. Protein adsorption onto the surface of these capsules, often a prerequisite to immune-cell attachment, is unfavorable due to the hydrophilicity and matrix component motility of intermediate-G alginates [26].

Another characteristic modification often manipulated in the synthesis of distinct types of alginate is the type of poly amino acid (polycation) to be grafted on the capsule surface for the purpose of augmentation of the immune-suppressive capabilities of the complex. This is achieved mainly by the polycations' function of stabilizing the negatively charged alginate, and allowing for manipulation of molecular weight, which culminates in the modification of the diffusive properties of the capsule membrane. The improvements or lack thereof to overall biocompatibility conferred by the grafting of these polycations is derived from its effects on capsule morphology, hinging on both the stability of the engraftment within the alginate as well as the type of gelling cation used (Ca^{2+}/Ba^{2+}) (Fig. 5).

PLL was one of the first poly amino acids to be used in alginate modification to confer membrane strength and allow for the manipulation of capsule membrane permeability. However, this polycation was reported to be associated with an undesirable propensity to induce an immune response rather than evade it, possibly due to inadequate binding of PLL to the capsule surface. In the unbound form, PLL has been known to stimulate an immune response characterized by complement activation, as well as IL-1, and TNF-α release, indicative of macrophage activation and fibroblast recruitment [5, 27, 28]. High-G alginate capsules tended to bind PLL less adequately than did Intermediate-G alginate capsules, consequently stimulating a stronger host immune response [29].

PLO has been characterized by its ability to prevent penetration of alginate membranes by large molecules as a direct result of its contribution to increase both swelling resistance and structural integrity of alginate capsules [18, 30]. However, according to Ponce et al. PLO was found to be inferior to PLL as a poly amino acid due to greater protein adsorption on PLO microcapsules [18], in combination with recruitment of significant numbers of lymphocytes, granulocytes, macrophages, and basophils in the vicinity of both PLO-grafted complexes.

On the other hand, PEG, reduced the tendency for protein adsorption, a precursor to cell-adhesion, to occur on alginate microcapsule surfaces [31]. Moreover, studies have reported that the polymer in its primitive form is non-immunogenic relative to PLL and PLO [32]. Owing to these two crucial traits, and the necessity of incorporating PLL to reduce the pore size of the

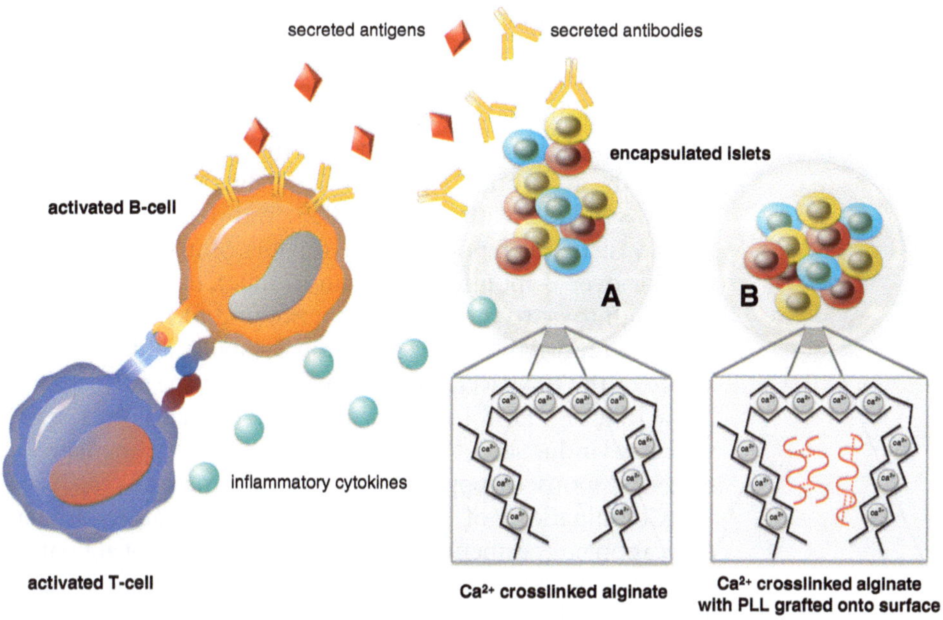

Fig. 5 Porcine islet xenografts protruding through a calcium cross-linked alginate microcapsule, lacking structural stability (**a**). The inadequate encapsulation allows for porcine antigens to be detected by B-cells. B-cells process and present antigens on MHC II surface molecules to naiive T-cells, which are activated. T-cells secrete inflammatory cytokines which ultimately lead to xenograft destruction via recruitment and activation of pro-inflammatory cells, such as plasma B-cells. Plasma B-cells secrete antibodies can directly contact these protruding islets, activating the complement system to compromise transplant viability. Porcine islet xenografts with an immuno-protective calcium cross-linked alginate barrier grafted with poly-L-lysine (PLL) (**b**). The PLL layer reduces the alginate pore size so large molecules are unable to traverse the membrane, and also confers structural stability, and mechanical strength to the surface itself so as to provide a sufficient barrier between the porcine tissue and foreign extra-capsular environment

alginate membrane, Spasojevic et al. explored the feasibility of applying a layer of PEG over a membrane already incorporated with PLL—as a PEG-b-PLL diblock copolymer association [17]. Although other polycations such as chitosan, PLO, and PDL have been attempted to improve the immunoprotective properties of the capsule membrane, they were all found to induce a strong host response in in vivo experiments [33]. Reduction in host immune response in terms of less macrophage and fibroblast capsular overgrowth detection was observed in in vivo studies utilizing a PEG-b-PLL grafted copolymer layer on top of the PLL layer in intermediate-G alginate microcapsules [17].

Cross-linking of divalent cations, mainly Ca^{2+} and Ba^{2+}, within the mannuronic and guluronic-acid matrices of alginate has been commonly utilized to enhance immunoprotection through greater microcapsule stability. Safley et al. compared biocompatibilities of alginate cross-linked with Ba^{2+} or Ca^{2+}, either with or without a PLL layer, and suggested Ba^{2+} alginate lacking a poly amino acid surface was the superior candidate [34]. Ba^{2+} capsules lacking PLL were then used to encapsulate adult porcine islets and were transplanted into the peritoneal cavity of diabetic NOD mice. Utilizing microscopic imaging, these capsules were shown to be free of pericapsular fibrotic overgrowth to a significantly greater extent than Ca^{2+} cross-linked capsules with PLL modifications [34]. In a separate study, a "hybrid" type of high-G alginate microcapsule made by cross-linking with 50 mM Ca^{2+} and 1 mM Ba^{2+} containing human islets was reported to be capable of inducing normoglycemia for over 3 months in immunocompetent diabetic mice [35]. Although toxic to the host in the unbound form, addition of Ba^{2+}, has been reported to confer additional stability and reduced permeability of the polysaccharide network comprising the capsule membrane [36, 37]. Histological examination upon eventual graft failure suggested induction of a nonspecific, macrophage-mediated inflammatory response as opposed to a humoral immune response [35].

Combinations of several modifications, namely in terms of polycation engraftment and cation cross-linkage, of alginate microcapsules have been evaluated for their effects on biocompatibility. Tam SK et al. reported that microcapsules made with intermediate-G alginate cross-linked with Ca^{2+} and coated with PLL resulted in a mean cell adhesion score of 4.5, while replacing PLL with PLO gave a score of 2.8 and replacing Ca^{2+} or intermediate-G alginate with Ba^{2+} or high-G alginate gave a score of 8.3 [21]. The latter result could be attributed to the dissociation of PLL from the capsule membrane which would be highly immunogenic in the unbound form, due to an increased affinity of the alginate to Ba^{2+} or to each other [9, 18, 37, 38]. The lower immunogenicity of Ca^{2+}-PLO compared to Ca^{2+}-PLL microcapsules could be attributed to the greater hydrophilicity of PLO, which would inhibit protein and cellular surface adhesion [39]. As implicated above, addition of a poly amino acid coating increased the extent to which PFO occurred as opposed to alginate cross-linked only with divalent cations. However, this phenomenon could be attributed to a lack of adequate binding of polycations such as PLL, which have been proven to be highly immunogenic unless stable bonds are formed at the alginate surface [3].

These physiochemical modifications, constituting different strategies aimed at decreasing the host response, result in capsular morphological changes as well. Although PLL/PLO is held more tightly by Ca^{2+} cross-linked intermediate-G alginate capsules, a decrease in resistance to osmotic gradients compromised the integrity of the shape of the complex. Smooth, and round topological features were associated with less fibrotic overgrowth, as opposed to collapsed, asymmetrical microcapsules. Given a geometrically favorable binding surface, as well as increased likelihood of encapsulated islet protrusion into the host environment, immune cells are more likely to cluster in the vicinity. However, utilization of Ba^{2+} to strengthen core mechanical stability and resistance to osmotic pressure may increase immunogenicity by less rigid binding of polyaminoacid layers [25]. Potential trade-offs are involved in any choice or combination of choices for alginate microcapsule modification.

3 Antigen Shedding and Danger Associated Molecular Patterns

The duration of normoglycemia achieved from transplantation of encapsulated islets, as well as the underlying mechanism posited for xenograft rejection fluctuate depending on the type of islet used (neonatal, fetal, or adult, or genetic knock-out], the type of encapsulation material high-g, int-g, Ca^{2+}, Ba^{2+}), and the animal model utilized for implantation (mouse, rat, nonhuman primate, and human). This review focuses on characterization of rejection mechanisms following porcine islet transplantation mainly in nonobese diabetic (NOD) or STZ-induced diabetic, immune-compromised, and wild type (C57BL/6) mice, to a minor extent, rats, and nonhuman primates are also discussed. Although individual results vary, overarching patterns in the type of immune response elicited suggest common macroscopic phenomena hypothesized to induce xenogeneic graft failure. In particular, pericapsular fibrotic overgrowth, capsule-permeable damage-associated molecular pattern (DAMP) secretion, and inadequate encapsulation of islets resulting in direct exposure to circulating sentinel cells are key catalytic events in the initiation of the foreign body response.

In many instances, microcapsule degradation, large islet size, and imprecise encapsulation techniques have led to deficient protection of the xenograft from the host environment [40]. The concept of donor islet protrusion from defective microcapsules leading to detection and rejection by circulating host leukocytes has been widely characterized in the literature. However, the DAMP secretion paradigm through which the viability of transplants encapsulated in mechanically stable devices is compromised is not as straightforward. Nevertheless, a primary hypothesis has been formed based on post-transplant analysis of cellular infiltration found on explanted capsules, and in the implant site itself. Investigators have consistently recovered encapsulated islets covered with a pericapsular overgrowth composed of macrophages as well as fibrinogen subsequent to transplant [41]. PFO has been shown to accelerate islet necrosis due to obstruction of diffusive gradients, and therefore justifies compromise of islet functionality. However, the layer itself fails to account for the presence of host anti-porcine antibodies, activated T-cells and associated cytokines [41]. These findings suggest capsule permeable porcine antigens must act as a key player in compromising the immune-compatibility of the graft. Furthermore, an increase in intragraft expression of porcine islet DAMPs [42], similarly capable of extrusion into the extra-capsular environment to potentially interact with host sentinel cells, forms the basis for a possible link between PFO, islet necrosis, and secretion of antigenic DAMPs [3].

The main sources of porcine islets come from fetal pig islet-like clusters (FP ICC), neonatal porcine pancreatic cell clusters (NPCC) and adult porcine islets. Each of these sources is associated with distinct advantages and disadvantages in terms of biocompatibility in foreign systems. Sources assert that FP ICCs and NPCCs express the porcine "Gal" epitope, an immunogenic carbohydrate chain incompatible with transplantation in humans, to a greater extent than adult porcine islets [43]. Should a fragment of a transplanted islet be incompletely encapsulated and protruding into the "extra-capsular" environment, preformed human anti-Gal IgM and IgG antibodies immediately recognize this epitope, trigger the complement cascade, and compromise porcine islet survival [44]. However, through genetic modifications, specifically, alpha 1,3-galactosyltransferase gene knockout (GTKO), porcine islets may be modulated to lack expression of this epitope. Nonetheless, non-encapsulated GTKO porcine islets and wild type islets were both rejected subsequent to nonhuman primate transplantation, suggesting alternative antigens are likely involved in rejection of tissue lacking complete enclosure [45]. Interestingly, a drawback of utilizing adult porcine islets with a dormant Gal epitope was that certain inflammatory cytokines detrimental to adult beta-cell survival were found to be less hostile to FP ICC and NPCC survival [46]. On the other hand, the principal advantage of using adult porcine islets is immediate functionality following transplantation, as opposed to the significant delay associated with the FP ICCs and NPCCs, which require weeks to months to differentiate into fully functional insulin-secreting beta cells. However, FP ICCs and NPCCs are more resistant to hypoxia, can be efficiently purified while lasting longer in culture and the isolation procedure is simple and scalable [46, 47]. We begin this section by an in-depth analysis of the antigenicity, and performance of encapsulated NPCCs, the most commonly utilized of the three, followed by an evaluation of literature concerning FP ICCs and adult porcine islets.

3.1 Porcine Neonatal Pancreatic Cell Clusters [NPCCs]

A brief discussion on incompletely encapsulated NPCCs is warranted as surface imperfections of microcapsules are commonly detected, and problems during the unstandardized encapsulation process resulting in inadequate protection of the xenogeneic tissue have also been reported [48]. The antigenicity of carbohydrate chains present on the surface of neonatal porcine islets has been well characterized. These include N-linked sugar chains such as gal-α-gal, and sialic acid antigens such as the Hanganutziu-Deicher (H-D) glycoconjugate. Expression of the gal-α-gal epitope on islets in an in vitro FACS analysis triggered the deposition of human antibodies IgG and IgM, while the H-D antigen was recognized by a specific anti-H-D antibody [49]. Actual transplantation of naked neonatal porcine islets [simulating islet exposure] resulted in a severe immune response mediated by indirectly activated CD4+ T-cells in response to surface antigen recognition [50]. Host APCs detect epitopes ubiquitous on the surface of neonatal porcine islets, and go on to activate CD4 T-cells via MHC class II peptide interaction with the T-cell and co-stimulatory receptors [50, 51]. This indirect cellular mechanism of neonatal xenograft rejection, particularly within the porcine donor to murine recipient framework, is validated by the acceptance of transplanted naked islets in MHC class II deficient recipients.

While detection of these exposed antigens may account for the xenogeneic graft rejection of incompletely encapsulated NPCCs, it becomes a point of interest to explore immune-response provocation in transplants where the morphology of the capsule has maintained its integrity. The hypothesis of soluble antigen secretion has emerged to explain why intact alginate microcapsules, verifiably providing complete coverage of the islets via histological/microscopic examination [52], often failed to fully protect the xenografts from recognition by the adaptive immune system.

Upon further analysis of the immune mechanisms associated with rejection of encapsulated NPCCs, it was found that a CD4+ T-helper cell-dependent pathway, similar to the response activated against non-encapsulated islets, was largely responsible [53]. Transplantation of encapsulated neonatal porcine islets in diabetic C57Bl/6 mice with competent immune systems resulted in no effect on hyperglycemia, indicative of host rejection [54]. However, administration of a combination therapy consisting of anti-CD154 and anti-LFA-1 monoclonal antibodies, which bind specifically to co-stimulatory surface proteins expressed on activated T-cells, significantly prolonged the survival of the xenografts. The immunomodulatory effect of CD154 ligand inhibition, a molecule essential to promoting differentiation of B-cells and macrophages, implies their involvement in catalyzing graft failure. Furthermore, the longer periods of normoglycemia achieved by administration of these antibodies justify the pivotal role T-cells are purported to play in triggering rejection [54, 55]. The involvement of CD4+ T-helper cells in the host immune response was additionally validated, not only by detection of IFN-γ produced by Th1 cells, and IL-10 secreted by Th2 cells, but also by the fact that administration of CD4+ T-helper cells in immunocompromised mice resulted in graft failure, while reconstitution with CD8+ T-cells did not [53].

Interestingly, a small number of B-cells, in addition to CD4+ T-cells and macrophages, were detected on the surface of NPCC microcapsules, while murine IgG antibodies were observed to have infiltrated into membrane [53, 56]. Opsonization by these antibodies onto the islets, which experiments have shown are permeable to certain alginate microcapsules with unconventionally large pores, can activate complement proteins such as C3 and C4, which may produce cytotoxic effects on the encapsulated islets [53, 56]. As the above mechanisms of rejection pertain solely to NPCCs, antigens other than Gal α- [1, 3] gal, a nonspecific antigen in most mammalian species, are likely responsible for the activation of the adaptive immune response seen in immunocompetent mouse recipients. Unfortunately, humans are the only mammalian species lacking expression of gal α- [1, 3] gal, and therefore, the only species prone to mount a response upon detection of this epitope that is ubiquitous in porcine islets.

Although the concept of antigen-shedding opens up the possibility of graft destruction via complement activation, CD4+ T-cell recruitment and anti-porcine IgM/IgG antibody deposition, a large portion of literature on the subject indicates the prevalence of macrophages as the predominant immune cell detected on the surface of microcapsules containing porcine islets. All branches of the immune system recruit, activate and utilize macrophages as effector cells, and their role in xenogeneic tissue rejection has become increasingly apparent. Increased expression of the MHC class II receptor on macrophages recovered subsequent to contact with transplanted encapsulated islets suggest an inclination for them to function as antigen-presenting cells [57]. IFN-γ secreted by CD4+ helper T-cells, as well as an upregulation detection of chemokines such as MIP-1alpha/beta, and MCP-1 has been shown to result in both macrophage recruitment and fetal porcine islet rejection [58]. Macrophage depletion has also been correlated with prolonged normoglycemia in diabetic rodents transplanted with NPCC xenografts [59]. Additionally, although relatively large human anti-islet antibodies may be unable to cross most alginate microcapsule membranes, especially those with improved diffusive selectivity due to grafted poly-L-lysine, they may bind to extruded porcine antigens, thus forming a complex that can trigger macrophages to release cytotoxic chemokines via binding with the F_c receptor. Some of these damaging macrophage products include IL-1 and TNF-α, which may directly induce islet destruction, or promote indirect damage by triggering inflammatory pathways [60].

3.2 Fetal Porcine Islet-Like Cell Clusters

Transitioning to an overview of FP ICCs, fetal islets share many of the same advantages and disadvantages as neonatal islets compared to adult islets. FP ICCs do not differentiate into fully functional insulin secreting β-cells until approximately 3 weeks post transplantation, as detectable by fluctuations in porcine C-peptide levels in the host serum—a principal disadvantage of FP ICCs [42]. Levels of porcine C-peptide are directly indicative of the levels of insulin originating from porcine β-cells. Although less differentiated than neonatal islets, thereby yielding a lower proportion of β-cells, following transplantation, a majority of progenitor pancreatic cells differentiate into insulin producing cells. The capability of

FP ICCs to reverse hyperglycemia by stabilizing glycemic indices in immunosuppressed mice has been reported [61]. In regard to the antigenicity of FP ICCs, "naked" forms of these xenografts lacking encapsulation are rejected via a cellular pathway similar to that of neonatal islets, with minor differences. Evidence indicates increased intragraft mRNA transcript expression of chemokines such as monocyte chemotactic protein 1 [MCP1], and MIP-1alpha bind to chemokine receptor 2 [CCR2] subsequent to islet transplantation, prompting the recruitment of macrophages and CD4 T-cells to the transplant site [62]. Similar to neonatal islets, fetal porcine islet transplantation into C57BL6 mice primarily resulted in a CD4+ helper T-cell mediated indirect pathway of rejection [63]. However, CD4+ helper T-cell-mediated direct response, as well as a weak CD8+ cytotoxic T-cell, NK cell and eosinophil response was also detected in mice lacking CD4+ T-cells [63]. More specifically, analysis of mRNA expression of cytokines at graft sites via RT-PCR revealed an initial Th1 response, characterized by activated macrophage intrusion into the graft site and aggregation of T-cells peripheral to the foreign tissue. The Th1 mechanism, responsible for a majority of the graft destruction, was then followed by a Th2 response resulting in elevated anti-FP ICC IgG levels along with eosinophil infiltration [64].

Encapsulation of FP ICCs within alginate microcapsules theoretically reduces the extent of the immune response by allowing for the evasion of detection by immune cells. While protection from the host environment afforded by cation cross-linked alginate membranes augmented graft survival rates and prolonged normoglycemia in immunodeficient mouse recipients, the results have proven difficult to replicate in immunocompetent rodents and nonhuman primate models [65]. Again, similar to encapsulated neonatal islets, differences in the host-rejection mechanism of "naked" vs. encapsulated FP ICCs stem from an immune response initiated indirectly via antigenic DAMP release as opposed to direct contact between host cells and epitopes present on the foreign tissue surface [53]. Previous studies had established that an immune response is not elicited ensuing transplantation of empty barium alginate capsules [66, 67]. These findings further validate the hypothesis that antigenic DAMP shedding may play a role in mediating graft failure after encapsulated FP ICC transplantation given that the alginate used was pure, capsules were generated as per lab-specific protocol, and the islets depicted in microscopy and histologic cross-sections were adequately enclosed [42]. Furthermore, Evaluation of alginate microcapsule pore size by Vaithilingam et al. revealed that molecules smaller than 250 kDa were still able to diffuse through the alginate-cation cross-linked membrane [66]. Within the graft, an increase in the mRNA expression of porcine antigenic transcripts such as MIP-1 α, IL-8, HMGB1 and HSP90 substantiated claims that release of DAMPs originating from islets small enough to traverse the microcapsule membrane is a likely phenomenon in the post-transplantation setting [42]. The prevalence of macrophages comprising of the capsular overgrowth, the upregulation of chemotactic molecules specific for the recruitment of monocytes such as MIP-1 α, and the detection of peaks in the pro-inflammatory cytokine TNF-α corresponding with peaks in mRNA expression of HSP90 [42] all suggest a delayed hypersensitivity response composed of macrophages and NK-cells targeting FP ICCs [68, 69]. The ability of HSP proteins to activate and prompt inflammatory cytokine secretion upon contact with macrophages has been described in the literature [70, 71]. Furthermore, increases in TNF-α levels stemming from recruited macrophages justify the onset of the PCO surrounding the microcapsules. The propensity of these cytokines to induce myofibroblast proliferation, thereby stimulating deposition of extracellular matrix proteins (i.e., collagen, proteoglycans, glycoproteins) has also been previously characterized, upholding the assertion that macrophages and associated pro-inflammatory mediators are primarily responsible for graft failure due to PCO obstructing nutrient/metabolite diffusion. The necrosis of encapsulated FP ICCs has been commonly attributed to the formation of a nutrient/O_2 barrier [72]. It is interesting to note that for FP ICCs encapsulated within barium alginate microcapsules, PCO composed of macrophages and CD4+ T cells was detected much later than with calcium alginate microcapsules. Graft tissue viability was also higher with the former [42].

3.3 Adult Porcine Islets [APIs]

The therapeutic benefits, along with the distinctive challenges corresponding to the utilization of adult porcine islets as opposed to FP ICCs and neonatal porcine islets warrant further analysis. Although APIs are composed of fully differentiated β-cells and result in detection of an immediate rise in porcine C-peptide levels in host serum post-transplantation, they tolerate in vitro culture poorly, and are prone to hypoxic-injury [73]. Therefore, they are economically inefficient to harvest, isolate and culture compared to FP ICCs and neonatal islets. Like neonatal islets, adult porcine islets express gal alpha [1, 3] gal antigens, albeit to a significantly lesser extent [43]. Consequently, anti-gal antibodies present in human sera are responsible for the lysis of exposed neonatal islet cells, but have a diminished role in the eradication of incompletely encapsulated adult islets [74]. Much of the immunogenicity of adult porcine islets stems from N-linked sugars such as sialic acids (i.e., Hanganutziu-Deicher (HD) antigens) [73]. Although terminal the H-D antigen may function as an epitope recognized by human anti-HD antibodies, alternative sialic acid epitopes, such as those present in N-linked sugars, were found to be integral to recognition by circulating APCs [73].

The adaptive response against capsule-compromised APIs subsequent to transplantation in STZ-treated mice was found to be similar to that against FP ICCs and NPCCs [75–77]. An indirectly activated CD4+ helper T-cell response was detected in which APCs were crucial components [51]. Genetically modified CD4-Knockout mouse strains were able to maintain normoglycemia, while CD8-Knockout and C57BL/6 wild-type strains completely rejected the implanted xenogeneic tissue. Interestingly, the CD8-Knockout strain rejected the adult islets at an earlier point in time as compared to the wild-type, implying that a lack of CD8+ regulatory T-cells in CD8-KO mice resulted in earlier xenograft destruction [51]. As for the humoral arm of the host immune system, circulating antibodies specific to porcine islet epitopes were detected in mouse serum subsequent to transplantation. Moreover, antibody mediated complement pathways resulting in xenograft destruction has been described [78, 79].

Experiments investigating outcomes of alginate-encapsulated APIs transplanted into diabetic NOD mice suggest that the mechanisms by which immune-mediated graft failure occurs are similar to those reported with encapsulated FP ICCs and NPCCs, with a few key differences. According to Cui et al. subsequent to graft retrieval and confirmation of microcapsule structural integrity, histological evaluation demonstrated significant PFO and islet necrosis in APIs. Upon further evaluation, eosinophils, macrophages, and to a lesser extent, CD4+ T-cells, CD8+ T-cells, B-cells and neutrophils were detected in the peritoneal fluid [80]. Serum examined from these immune-competent mice had increased levels of circulating anti-porcine IgG relative to baseline values observed in the control group. Since the microcapsules were demonstrated to be impermeable to 150 kDa dextrans, the antibodies were thought to have an indirect effect on graft health and survival. The hypothesis that rejection was initiated despite no sign of microcapsule degradation hinges on antigenic small-molecule secretion by APIs. Islet antigens and other DAMPs can be detected by antibodies in the extracapsular space and activate peritoneal macrophages via F_c receptor interactions [80]. In addition to capsule-permeable toxic mediators secreted by macrophages, in immunosuppressed mice, xenograft destruction was also reported to correlate with eosinophil recruitment, and a primarily humoral response [81]. Interestingly, the Ca^{2+}-cross-linked capsules with a PLL layer performed significantly worse at ensuring islet survival than Ba^{2+} cross-linked capsules without a PLL layer, reinforcing the importance of biomaterial device composition and design and the resultant capsule characteristics in maintaining adequate xenograft protection. PFO-induced necrosis was noted with islets encapsulated within Ca^{2+}-PLL microcapsules, while Ba^{2+} microcapsules lacked substantial host-cell adherence, indicative of barium encapsulated xenograft destruction chiefly mediated by a membrane permeable cytotoxic factors [i.e., cytokines, reactive-oxygen species] [34].

4 Understanding Xenograft Rejection in Nonhuman Primate Recipients

The adaptive arm of the host immune system plays a role in inadequately encapsulated porcine graft rejection in the nonhuman primate model. There is evidence of both humoral and cellular involvement in the graft rejection process. IgM antibodies specific to both anti-Gal and other epitopes on porcine islets were detected, and shown to contribute to complement activation and neutrophil invasion [73]. Furthermore, subsequent to the B-cell response, the cellular response was found to consist of mainly of CD8+ cytotoxic T-cells, along with CD4+ T-cells, neutrophils, and macrophages, which had been detected in a sequential order along the periphery of the implanted islets [82, 83]. The involvement of T-cells in destruction of porcine islets in nonhuman primates have been verified by the increase in graft survival times following administration of antibodies aimed at blocking the CD40/CD154 co-stimulation pathway between APCs and host T-cells [84]. A review of the literature shows that unencapsulated porcine xenograft islets have only been able to maintain viability for substantial periods of time in the presence of intense immunosuppressive regimens [85–87].

Key differences exist which warrant caution when applying conclusions drawn from primate outcomes in the clinical trial setting. In addition to having a distinct adaptive response, investigators have found that a much greater transplant mass is necessary to achieve normoglycemia in diabetic nonhuman primates as opposed to humans due to variations in glucose metabolism, with nonhuman primates, particularly macaques, reportedly requiring greater insulin levels in the plasma to achieve normoglycemia [88]. Although humans and primates utilize immune systems comprised of pathways that are in some ways heterogeneous, the physiological environment of NHPs are overall similar to that found in humans relative to other animal models. Thus, preclinical outcomes in primates are a necessary prerequisite to the translation of encapsulated porcine islet transplantation to human clinical trials.

In the earliest trial of encapsulated islet transplantation in a primate model to correct for hyperglycemia, alginate-PLL-alginate cross-linked with Ca^{2+} was utilized to protect APIs. Seven of the nine spontaneously diabetic monkeys became euglycemic and the period of insulin independence ranged from 120 to 804 days [89]. While these results proved difficult to replicate, a few primate trials were subsequently initiated, with varying degrees of success. In 2005, Elliot et al. demonstrated that upon explant of microencapsulated neonatal porcine islets from the intraperitoneal compartment of cynomolgus monkeys, they stained positive for insulin and were free of PFO at 8 weeks post transplantation [40]. In conjunction with this outcome, the same research group discovered that the same graft implanted into diabetic cynomolgus primates resulted in a steady decrease in exogenous insulin requirements for up to 24 weeks post transplantation [40]. A separate study examined microencapsulated adult porcine islet viability in nondiabetic cynomolgus monkeys and concluded that despite a marked increase in circulating IgM and IgG following implantation as compared with the control group [blank microcapsules], a majority of the grafts were intact, responsive to stimulation by glucose, and free of PFO for up to 6 months post-transplant [52]. Since anti-porcine antibody levels were higher in the treatment group, it was concluded that some islets were inadequately encapsulated and/or antigenic secretion was occurring. Finally, the most recent of these studies showed that both microencapsulation and macroencapsulation devices could protect adult porcine islets from humoral rejection and T-cell mediated cytotoxicity, implanted in either the kidney capsule or peritoneal compartment of primates. The macroencapsulation device also corrected hyperglycemia in the NHP models for up to 6 months without immunosuppression [90]. Lymphocytes and monocytes were nevertheless detected adjacent to the transplant site, and anti-porcine IgG levels were again strongly increased, suggesting encapsulation was unable to fully immunoisolate the xenograft. Furthermore, the findings reinforced the similarity of immune-mediated rejection of encapsulate islets in both the primate and mouse/rat models [90].

In the primate framework of preclinical trials, as in the mouse model, the rat model, the outcome in terms of graft functionality and mechanism of rejection are dependent of a variety of factors such as transplantation site, type of porcine islet utilized, encapsulation parameters, and specific investigator protocol. Pros and cons of each combination of elements have been characterized but not to the extent that a standard has been created, warranting further investigation into these variable issues.

5 Role of Immunomodulation in Preventing Xenograft Rejection

In regard to the immunological challenges facing translation of encapsulated islet transplantation to human clinical trials, substantial progress has been made in optimizing capsule parameters, and multiple pathways involved in graft rejection have been successfully characterized. However, major flaws remain in current approaches to adequately immunoisolate islet xenografts. Examples of which include, but aren't limited to, persistent contamination of alginates with immunogenic impurities, use of biomaterial device constructs with suboptimal biocompatibility profiles, an incomplete understanding of islet DAMP release and lack of strategies to reduce the resulting deleterious effects. The aforementioned shortcomings act as obstacles to standardization of encapsulation protocols, which only further compounds problems. These deficiencies ultimately result in immune-mediated rejection of islet transplants, albeit with different groups achieving highly variable durations of transiently induced normoglycemia. To address these issues, novel strategies are being evaluated to supplement conventional microencapsulation techniques with immunomodulatory therapeutic agents. The aim is to improve the biocompatibility of encapsulated islet xenografts. Notably, certain researchers have proposed concurrently transplanting immunomodulatory mesenchymal stem cells (MSCs) with encapsulated islets, in addition to various alternative microcapsule modifications and even genetic alterations, in an attempt at evading immune rejection and preserving islet function. This section focuses primarily on MSCs, with a minor review of other ongoing endeavors applying an alternative novel framework to achieve complete xenograft compatibility.

5.1 Immunomodulation and Tolerance Induction Using Mesenchymal Stem Cells

MSCs, in addition to their ability to differentiate into a diverse array of cell lines, have been of particular interest to transplant researchers due to their ability to suppress both the innate and adaptive immune response [91]. It was discovered that they are able to achieve this outcome both by secreting soluble signaling factors detected by circulating effectors of the immune system, and by direct contact with T-cells and dendritic cells [92].

MSCs have been shown to be capable of functioning as APCs when exposed to appropriate environmental cues. Interestingly, when stimulated by high extracellular concentrations of the cytokine IFN-γ, MSCs have been shown to secrete Indoleamine 2, 3-dioxygenase (IDO) [93]. IDO functions to downregulate effector T-cell proliferation [94], polarize macrophage differentiation to the M2 (anti-inflammatory) phenotype [95], and induce tolerogenic dendritic cell (DC) propagation [94]. MSCs have also been shown to secrete prostaglandin E-2 (PGE-2), IL-10, IL-6, and nitrogen oxide (NO), all of which play a role in dampening T-cell mediated inflammation by modulation of cell-line differentiation, and activating apoptotic pathways. PGE-2 secretion has been associated with induction of tolerogenic DCs and macrophages [96], IL-6 has been suggested to stunt DC maturation [97], and NO, purported to have a role in T-cell apoptosis [98].

MSCs are also able to initiate immunosuppressive pathways via direct contact with active players in the acute and adaptive immune response. MSCs as functional APCs lack essential co-stimulatory molecules such as CD80, CD86, and CD40. These ligands are responsible for binding to and providing the proper intracellular signal to prompt the progression of naïve T-cells to pro-inflammatory effector T-cells in conjunction with antigen presentation via the MHC II complex [99]. Although the MHC II complex, and therefore the antigen-presenting capabilities of MSCs, is not constitutively expressed, exposure to the appropriate microenvironment transcriptionally activates the MHC II gene [100].

Under high IFN-γ concentrations, MSCs have been found to express the MHC II complex. This phenomenon allows for MSC interaction with undifferentiated T-cells via both the MHC II complex and programmed death-1 (PD-1) surface protein to induce T-cell anergy or polarization favoring T-regulatory cell proliferation. Effector T-cell activation and proliferation is circumvented as a result of the absence of CD80/86 and CD40 ligand–receptor interaction, transcripts for which MSCs lack expression [101]. Proliferation of effector T-cells, particularly CD4+ T-cells has been shown to be fundamental to immune rejection of encapsulated porcine islet xenografts [53], thus the blockade of this pathway could potentially promote prolonged transplant survival and function. Additionally, it has been hypothesized that through contact-dependent notch, and jagged-2 signaling, MSCs may downregulate DC maturation and co-stimulatory markers to inhibit effector T, and B-cell proliferation through induction of anergy upon contact with naïve cells [102].

Although a majority of these findings were elucidated through in vitro studies, the potential for MSCs to preserve encapsulated islet xenograft function via immunosuppressive mechanisms have been investigated in vivo in both murine and nonhuman primate models. When rat islets, bone-marrow derived MSCs, and fibroblasts were transplanted into immunocompetent diabetic mice, normoglycemia was achieved for up to four times as long as mice receiving islets or islets with only fibroblasts [103]. Furthermore, when human bone-marrow derived MSCs were transplanted into the intraperitoneal compartment of humanized diabetic mice along with human islets, they stabilized recipient blood glucose levels in a dose-dependent fashion, with increased concentration of MSC injection resulting in longer periods of normoglycemia [104]. Unfortunately, aside from these few investigations, there is currently a dearth in the number of experiments exploring the effect of xenogeneic MSC co-transplantation with islets, but an abundance of studies in allogenic transplant models. The results of studies utilizing autologous MSC transplantation also support the immunosuppressive capabilities of MSCs when co-transplanted with allogenic islets as evidenced by extension of the duration of normoglycemia in the recipient [105–107]. However, in an attempt restoration of standard insulin secretion in response to plasma glucose levels for type 1 diabetics via xenogeneic transplants (islets co-transplanted with MSCs), current investigators have begun to focus on genetic modification of porcine MSCs in hopes of achieving cross-species compatibility, and functionality [108].

5.2 Localized Immunosuppression Using Chemotherapeutic Agents

A ubiquitous component of xenogeneic graft failure has been PFO on the surface of microcapsules, which not only induces necrosis due to inhibition of nutrient/gas exchange, but also may facilitate islet antigenic DAMP excretion. Targeting this phenomenon, the emphasis of current immune-modulatory efforts other than MSC utilization has been surface modifications of microcapsules in an attempt to mitigate fibrosis and the provocation of an immune response, which consistently follows. Park et al. has shown that incorporation of rapamycin [an FDA approved immunosuppressive drug] into the PEG layer of alginate microcapsules significantly decreased macrophage adherence and proliferation on the capsule surface compared to the microcapsules lacking rapamycin [109]. Similarly, Vaithilingam et al. reported that macromolecular heparin conjugates could improve transplant biocompatibility when coated onto the surface of highly purified, int-G, Ba $^{2+}$ microcapsules. Heparin had no detectable effect on capsule permeability, and size, but did have a positive effect on tensile strength, and most importantly, reduced PFO in allogeneic rat transplantation models [110]. Although the permeability was not directly affected, it was believed that improved intra-capsular islet survival secondary to an uninhibited diffusion gradient probably downregulated the rate of antigenic DAMP secretion. Some groups have proposed more extensive methods of improving islet preservation in microcapsules such the core–shell model, which simultaneously encapsulates islets in the core alginate (Ba^{2+} cross-linked) solution while a shell solution adds a protective buffer layer to center the islets and prevent protrusion [111]. To address the issue of diffusion gradient interference, and increased transplant volume load inherent in utilizing microcapsules with surface modifications, investigators have produced capsules with a focus on reducing thickness of the capsule. Coating of islets with a micrometer thin layer of alginate would promote nutrient and oxygen transport, insulin secretion, and lower total graft volume [112]. The caveat of this functionality would be an increased risk of islet exposure to the extra-capsular environment.

5.3 The Emerging Role of Biological Agents as Immunomodulators

Co-encapsulation of islets with bioactive small molecules offers yet another approach to delay xenograft rejection. Researchers observed that curcumin, a synthetic compound with anti-inflammatory properties, when incorporated into high-G, Ba^{2+} cross-linked alginate capsules and utilized to encapsulated rat islets, resulted in improved outcomes in terms of glycemic control and biocompatibility [2]. Upon qPCR analysis of the explanted capsule surface and surrounding tissue, curcumin reduced immunological markers indicative of the presence of macrophages, dendritic cells, B-cells, inflammatory cytokines, collagen, and actin, while fluorescent staining revealed less detectable fibrosis as compared with the control group [2]. Chen et al. reported that incorporation of stromal derived factor 1α (SDF-1α, also known as CXCL12) into ultrapure, low-viscosity alginate cross-linked with Ca^{2+} improved islet function and viability. The mechanism by which the treatment group experienced superior results over the control involved selective recruitment of regulatory T-cells and repulsion of effector cells [113]. Modification of encapsulation parameters does not have to be limited to material composition and structure, but can also be extended to the graft genome. Park et al. reported that transduction of the exendin-4 gene into rat islets using lentiviral vectors improved both the functionality and survival times of the PEG/PLL grafted encapsulated transplants in vivo [114]. Although substantial progress has been made in translating microencapsulated xenogeneic islets for diabetic patients to clinical trials, biocompatibility challenges remain. Multiple innovative perspectives offer potential solutions to the problem of graft functionality loss attributable to the distressingly transient immune-suppressive capacity of microcapsules.

6 Conclusion

Utilization of encapsulated porcine islets has produced improved results in terms of graft functionality and survival time. However, there remain a variety of parameters to be optimized such as optimization of alginate purification processes, composition, use of engraftable poly-amino acids to achieve modifications to microcapsule surface and configuration, identification of the optimal transplantation site, and identification of the best porcine donor strain and age before successful translation to human clinical trials can be achieved. Studies elucidating the possible mechanisms of immunological rejection of encapsulated islets have proven tremendously useful, as they have provided specific targets in innate and adaptive pathways fundamental to transplant failure. Co-transplantation of encapsulated islets with immunomodulatory MSCs providing yet another layer of protection from rejection processes represents one of many potentially viable approaches for successful translation of the encapsulated islet transplant procedure to human clinical trials, but further research into these approaches must be initiated before we can move closer to this goal.

Acknowledgements

The authors gratefully acknowledge Esther Yu-Tin Chen from the Department of Biomedical Engineering, University of California, Irvine for providing illustrations for this chapter.

References

1. Brennan DC, Kopetskie HA, Sayre PH, Alejandro R, Cagliero E, Shapiro AM, Goldstein JS, DesMarais MR, Booher S, Bianchine PJ (2015) Long-term follow-up of the Edmonton protocol of islet transplantation in the United States. Am J Transplant. doi:10.1111/ajt.13458
2. Dang TT, Thai AV, Cohen J, Slosberg JE, Siniakowicz K, Doloff JC, Ma M, Hollister-Lock J, Tang KM, Gu Z, Cheng H, Weir GC,

Langer R, Anderson DG (2013) Enhanced function of immuno-isolated islets in diabetes therapy by co-encapsulation with an anti-inflammatory drug. Biomaterials 34(23):5792–5801. doi:10.1016/j.biomaterials.2013.04.016

3. Paredes Juarez GA, Spasojevic M, Faas MM, de Vos P (2014) Immunological and technical considerations in application of alginate-based microencapsulation systems. Frontiers Bioeng Biotechnol 2:26. doi:10.3389/fbioe.2014.00026
4. Mallett AG, Korbutt GS (2009) Alginate modification improves long-term survival and function of transplanted encapsulated islets. Tissue Eng Part A 15(6):1301–1309. doi:10.1089/ten.tea.2008.0118
5. de Vos P, van Hoogmoed CG, de Haan BJ, Busscher HJ (2002) Tissue responses against immunoisolating alginate-PLL capsules in the immediate posttransplant period. J Biomed Mater Res 62(3):430–437. doi:10.1002/jbm.10345
6. Kumar S, Ingle H, Prasad DV, Kumar H (2013) Recognition of bacterial infection by innate immune sensors. Crit Rev Microbiol 39(3):229–246. doi:10.3109/1040841X.2012.706249
7. Kawai T, Akira S (2010) The role of pattern-recognition receptors in innate immunity: update on Toll-like receptors. Nat Immunol 11(5):373–384. doi:10.1038/ni.1863
8. Pearl JI, Ma T, Irani AR, Huang Z, Robinson WH, Smith RL, Goodman SB (2011) Role of the Toll-like receptor pathway in the recognition of orthopedic implant wear-debris particles. Biomaterials 32(24):5535–5542. doi:10.1016/j.biomaterials.2011.04.046
9. De Vos P, De Haan B, Van Schilfgaarde R (1997) Effect of the alginate composition on the biocompatibility of alginate-polylysine microcapsules. Biomaterials 18(3):273–278
10. Dusseault J, Tam SK, Menard M, Polizu S, Jourdan G, Yahia L, Halle JP (2006) Evaluation of alginate purification methods: effect on polyphenol, endotoxin, and protein contamination. J Biomed Mater Res A 76(2):243–251. doi:10.1002/jbm.a.30541
11. Zimmermann U, Thurmer F, Jork A, Weber M, Mimietz S, Hillgartner M, Brunnenmeier F, Zimmermann H, Westphal I, Fuhr G, Noth U, Haase A, Steinert A, Hendrich C (2001) A novel class of amitogenic alginate microcapsules for long-term immunoisolated transplantation. Ann N Y Acad Sci 944:199–215
12. Tam SK, Dusseault J, Polizu S, Menard M, Halle JP, Yahia L (2006) Impact of residual contamination on the biofunctional properties of purified alginates used for cell encapsulation. Biomaterials 27(8):1296–1305. doi:10.1016/j.biomaterials.2005.08.027
13. de Haan BJ, Rossi A, Faas MM, Smelt MJ, Sonvico F, Colombo P, de Vos P (2011) Structural surface changes and inflammatory responses against alginate-based microcapsules after exposure to human peritoneal fluid. J Biomed Mater Res A 98(3):394–403. doi:10.1002/jbm.a.33123
14. de Vos P, van Hoogmoed CG, van Zanten J, Netter S, Strubbe JH, Busscher HJ (2003) Long-term biocompatibility, chemistry, and function of microencapsulated pancreatic islets. Biomaterials 24(2):305–312
15. Halle JP, Bourassa S, Leblond FA, Chevalier S, Beaudry M, Chapdelaine A, Cousineau S, Saintonge J, Yale JF (1993) Protection of islets of Langerhans from antibodies by microencapsulation with alginate-poly-L-lysine membranes. Transplantation 55(2):350–354
16. Paredes-Juarez GA, de Haan BJ, Faas MM, de Vos P (2013) The role of pathogen-associated molecular patterns in inflammatory responses against alginate based microcapsules. J Control Release 172(3):983–992. doi:10.1016/j.jconrel.2013.09.009
17. de Vos P, Spasojevic M, de Haan BJ, Faas MM (2012) The association between in vivo physicochemical changes and inflammatory responses against alginate based microcapsules. Biomaterials 33(22):5552–5559. doi:10.1016/j.biomaterials.2012.04.039
18. Ponce S, Orive G, Hernandez R, Gascon AR, Pedraz JL, de Haan BJ, Faas MM, Mathieu HJ, de Vos P (2006) Chemistry and the biological response against immunoisolating alginate-polycation capsules of different composition. Biomaterials 27(28):4831–4839. doi:10.1016/j.biomaterials.2006.05.014
19. Espevik T, Otterlei M, Skjak-Braek G, Ryan L, Wright SD, Sundan A (1993) The involvement of CD14 in stimulation of cytokine production by uronic acid polymers. Eur J Immunol 23(1):255–261. doi:10.1002/eji.1830230140
20. Otterlei M, Ostgaard K, Skjak-Braek G, Smidsrod O, Soon-Shiong P, Espevik T (1991) Induction of cytokine production from human monocytes stimulated with alginate. J Immunother 10(4):286–291
21. Tam SK, Bilodeau S, Dusseault J, Langlois G, Halle JP, Yahia LH (2011) Biocompatibility and physicochemical characteristics of alginate-polycation microcapsules. Acta Biomater 7(4):1683–1692. doi:10.1016/j.actbio.2010.12.006
22. Duvivier-Kali VF, Omer A, Parent RJ, O'Neil JJ, Weir GC (2001) Complete protection of islets against allorejection and autoimmunity by a simple barium-alginate membrane. Diabetes 50(8):1698–1705

23. Omer A, Duvivier-Kali V, Fernandes J, Tchipashvili V, Colton CK, Weir GC (2005) Long-term normoglycemia in rats receiving transplants with encapsulated islets. Transplantation 79(1):52–58
24. Orive G, Hernandez RM, Rodriguez Gascon A, Calafiore R, Chang TM, de Vos P, Hortelano G, Hunkeler D, Lacik I, Pedraz JL (2004) History, challenges and perspectives of cell microencapsulation. Trends Biotechnol 22(2):87–92
25. Tam SK, Dusseault J, Bilodeau S, Langlois G, Halle JP, Yahia L (2011) Factors influencing alginate gel biocompatibility. J Biomed Mater Res A 98(1):40–52. doi:10.1002/jbm.a.33047
26. Park H, Park K (1996) Biocompatibility issues of implantable drug delivery systems. Pharm Res 13(12):1770–1776
27. Juste S, Lessard M, Henley N, Menard M, Halle JP (2005) Effect of poly-L-lysine coating on macrophage activation by alginate-based microcapsules: assessment using a new in vitro method. J Biomed Mater Res A 72(4):389–398. doi:10.1002/jbm.a.30254
28. Vandenbossche GM, Bracke ME, Cuvelier CA, Bortier HE, Mareel MM, Remon JP (1993) Host reaction against empty alginate-polylysine microcapsules. Influence of preparation procedure. J Pharm Pharmacol 45(2):115–120
29. Thu B, Bruheim P, Espevik T, Smidsrod O, Soon-Shiong P, Skjak-Braek G (1996) Alginate polycation microcapsules. I. Interaction between alginate and polycation. Biomaterials 17(10):1031–1040
30. Darrabie MD, Kendall WF Jr, Opara EC (2005) Characteristics of poly-L-ornithine-coated alginate microcapsules. Biomaterials 26(34):6846–6852. doi:10.1016/j.biomaterials.2005.05.009
31. Gon S, Fang B, Santore MM (2011) Interaction of cationic proteins and polypeptides with biocompatible cationically-anchored PEG brushes. Macromolecules 44(20)
32. Veronese FM, Mero A (2008) The impact of PEGylation on biological therapies. BioDrugs 22(5):315–329
33. van Schilfgaarde R, de Vos P (1999) Factors influencing the properties and performance of microcapsules for immunoprotection of pancreatic islets. J Mol Med 77(1):199–205
34. Safley SA, Cui H, Cauffiel S, Tucker-Burden C, Weber CJ (2008) Biocompatibility and immune acceptance of adult porcine islets transplanted intraperitoneally in diabetic NOD mice in calcium alginate poly-L-lysine microcapsules versus barium alginate microcapsules without poly-L-lysine. J Diabetes Sci Technol 2(5):760–767
35. Qi M, Morch Y, Lacik I, Formo K, Marchese E, Wang Y, Danielson KK, Kinzer K, Wang S, Barbaro B, Kollarikova G, Chorvat D Jr, Hunkeler D, Skjak-Braek G, Oberholzer J, Strand BL (2012) Survival of human islets in microbeads containing high guluronic acid alginate crosslinked with Ca2+ and Ba2+. Xenotransplantation 19(6):355–364. doi:10.1111/xen.12009
36. Morch YA, Qi M, Gundersen PO, Formo K, Lacik I, Skjak-Braek G, Oberholzer J, Strand BL (2012) Binding and leakage of barium in alginate microbeads. J Biomed Mater Res A 100(11):2939–2947. doi:10.1002/jbm.a.34237
37. Morch YA, Donati I, Strand BL, Skjak-Braek G (2006) Effect of Ca2+, Ba2+, and Sr2+ on alginate microbeads. Biomacromolecules 7(5):1471–1480. doi:10.1021/bm060010d
38. Li HB, Jiang H, Wang CY, Duan CM, Ye Y, Su XP, Kong QX, Wu JF, Guo XM (2006) Comparison of two types of alginate microcapsules on stability and biocompatibility in vitro and in vivo. Biomed Mater 1(1):42–47. doi:10.1088/1748-6041/1/1/007
39. Elbert DL, Herbert CB, Hubbell JA (1999) Thin polymer layers formed by polyelectrolyte multilayer techniques on biological surfaces. Langmuir 15(16)
40. Elliott RB, Escobar L, Tan PL, Garkavenko O, Calafiore R, Basta P, Vasconcellos AV, Emerich DF, Thanos C, Bambra C (2005) Intraperitoneal alginate-encapsulated neonatal porcine islets in a placebo-controlled study with 16 diabetic cynomolgus primates. Transplant Proc 37(8):3505–3508. doi:10.1016/j.transproceed.2005.09.038
41. Tuch BE, Keogh GW, Williams LJ, Wu W, Foster JL, Vaithilingam V, Philips R (2009) Safety and viability of microencapsulated human islets transplanted into diabetic humans. Diabetes Care 32(10):1887–1889. doi:10.2337/dc09-0744
42. Vaithilingam V, Fung C, Ratnapala S, Foster J, Vaghjiani V, Manuelpillai U, Tuch BE (2013) Characterisation of the xenogeneic immune response to microencapsulated fetal pig islet-like cell clusters transplanted into immunocompetent C57BL/6 mice. PLoS One 8(3):e59120. doi:10.1371/journal.pone.0059120
43. Bennet W, Bjorkland A, Sundberg B, Davies H, Liu J, Holgersson J, Korsgren O (2000) A comparison of fetal and adult porcine islets with regard to Gal alpha (1,3)Gal expression and the role of human immunoglobulins and complement in islet cell cytotoxicity. Transplantation 69(8):1711–1717

44. Marigliano M, Bertera S, Grupillo M, Trucco M, Bottino R (2011) Pig-to-nonhuman primates pancreatic islet xenotransplantation: an overview. Curr Diab Rep 11(5):402–412. doi:10.1007/s11892-011-0213-z
45. Rayat GR, Rajotte RV, Hering BJ, Binette TM, Korbutt GS (2003) In vitro and in vivo expression of Galalpha-(1,3)Gal on porcine islet cells is age dependent. J Endocrinol 177(1):127–135
46. Bai L, Tuch BE, Hering B, Simpson AM (2002) Fetal pig beta cells are resistant to the toxic effects of human cytokines. Transplantation 73(5):714–722
47. Nagaraju S, Bottino R, Wijkstrom M, Trucco M, Cooper DK (2015) Islet xenotransplantation: what is the optimal age of the islet-source pig? Xenotransplantation 22(1):7–19. doi:10.1111/xen.12130
48. de Vos P, Wolters GH, van Schilfgaarde R (1994) Possible relationship between fibrotic overgrowth of alginate-polylysine-alginate microencapsulated pancreatic islets and the microcapsule integrity. Transplant Proc 26(2):782–783
49. Omori T, Nishida T, Komoda H, Fumimoto Y, Ito T, Sawa Y, Gao C, Nakatsu S, Shirakura R, Miyagawa S (2006) A study of the xenoantigenicity of neonatal porcine islet-like cell clusters (NPCC) and the efficiency of adenovirus-mediated DAF (CD55) expression. Xenotransplantation 13(5):455–464. doi:10.1111/j.1399-3089.2006.00335.x
50. Rayat GR, Johnson ZA, Beilke JN, Korbutt GS, Rajotte RV, Gill RG (2003) The degree of phylogenetic disparity of islet grafts dictates the reliance on indirect CD4 T-cell antigen recognition for rejection. Diabetes 52(6): 1433–1440
51. Chitilian HV, Laufer TM, Stenger K, Shea S, Auchincloss H Jr (1998) The strength of cell-mediated xenograft rejection in the mouse is due to the CD4+ indirect response. Xenotransplantation 5(1):93–98
52. Dufrane D, Goebbels RM, Saliez A, Guiot Y, Gianello P (2006) Six-month survival of microencapsulated pig islets and alginate biocompatibility in primates: proof of concept. Transplantation 81(9):1345–1353. doi:10.1097/01.tp.0000208610.75997.20
53. Kobayashi T, Harb G, Rajotte RV, Korbutt GS, Mallett AG, Arefanian H, Mok D, Rayat GR (2006) Immune mechanisms associated with the rejection of encapsulated neonatal porcine islet xenografts. Xenotransplantation 13(6):547–559. doi:10.1111/j.1399-3089.2006.00349.x
54. Kobayashi T, Harb G, Rayat GR (2005) Prolonged survival of microencapsulated neonatal porcine islets in mice treated with a combination of anti-CD154 and anti-LFA-1 monoclonal antibodies. Transplantation 80(6): 821–827
55. Rayat GR, Gill RG (2005) Indefinite survival of neonatal porcine islet xenografts by simultaneous targeting of LFA-1 and CD154 or CD45RB. Diabetes 54(2):443–451
56. Rayat GR, Rajotte RV, Ao Z, Korbutt GS (2000) Microencapsulation of neonatal porcine islets: protection from human antibody/complement-mediated cytolysis in vitro and long-term reversal of diabetes in nude mice. Transplantation 69(6):1084–1090
57. Siebers U, Horcher A, Brandhorst H, Brandhorst D, Hering B, Federlin K, Bretzel RG, Zekorn T (1999) Analysis of the cellular reaction towards microencapsulated xenogeneic islets after intraperitoneal transplantation. J Mol Med 77(1):215–218
58. Yi S, Hawthorne WJ, Lehnert AM, Ha H, Wong JK, van Rooijen N, Davey K, Patel AT, Walters SN, Chandra A, O'Connell PJ (2003) T cell-activated macrophages are capable of both recognition and rejection of pancreatic islet xenografts. J Immunol 170(5):2750–2758
59. Omer A, Keegan M, Czismadia E, De Vos P, Van Rooijen N, Bonner-Weir S, Weir GC (2003) Macrophage depletion improves survival of porcine neonatal pancreatic cell clusters contained in alginate macrocapsules transplanted into rats. Xenotransplantation 10(3):240–251
60. Andersen HU, Jorgensen KH, Egeberg J, Mandrup-Poulsen T, Nerup J (1994) Nicotinamide prevents interleukin-1 effects on accumulated insulin release and nitric oxide production in rat islets of Langerhans. Diabetes 43(6):770–777
61. Tu J, Khoury P, Williams L, Tuch BE (2004) Comparison of fetal porcine aggregates of purified beta-cells versus islet-like cell clusters as a treatment of diabetes. Cell Transplant 13(5):525–534
62. Solomon MF, Kuziel WA, Mann DA, Simeonovic CJ (2003) The role of chemokines and their receptors in the rejection of pig islet tissue xenografts. Xenotransplantation 10(2):164–177
63. Koulmanda M, Laufer TM, Auchincloss H Jr, Smith RN (2004) Prolonged survival of fetal pig islet xenografts in mice lacking the capacity for an indirect response. Xenotransplantation11(6):525–530.doi:10.1111/j.1399-3089.2004.00174.x

64. Krook H, Hagberg A, Song Z, Landegren U, Wennberg L, Korsgren O (2002) A distinct Th1 immune response precedes the described Th2 response in islet xenograft rejection. Diabetes 51(1):79–86
65. Foster JL, Williams G, Williams LJ, Tuch BE (2007) Differentiation of transplanted microencapsulated fetal pancreatic cells. Transplantation 83(11):1440–1448. doi:10.1097/01.tp.0000264555.46417.7d
66. Vaithilingam V, Kollarikova G, Qi M, Lacik I, Oberholzer J, Guillemin GJ, Tuch BE (2011) Effect of prolonged gelling time on the intrinsic properties of barium alginate microcapsules and its biocompatibility. J Microencapsul 28(6):499–507. doi:10.3109/02652048.2011.586067
67. Vaithilingam V, Oberholzer J, Guillemin GJ, Tuch BE (2010) The humanized NOD/SCID mouse as a preclinical model to study the fate of encapsulated human islets. The review of diabetic studies. Rev Diabet Stud 7(1):62–73. doi:10.1900/RDS.2010.7.62
68. Candinas D, Belliveau S, Koyamada N, Miyatake T, Hechenleitner P, Mark W, Bach FH, Hancock WW (1996) T cell independence of macrophage and natural killer cell infiltration, cytokine production, and endothelial activation during delayed xenograft rejection. Transplantation 62(12):1920–1927
69. Lin Y, Vandeputte M, Waer M (1997) Natural killer cell- and macrophage-mediated rejection of concordant xenografts in the absence of T and B cell responses. J Immunol 158(12):5658–5667
70. Basu S, Binder RJ, Suto R, Anderson KM, Srivastava PK (2000) Necrotic but not apoptotic cell death releases heat shock proteins, which deliver a partial maturation signal to dendritic cells and activate the NF-kappa B pathway. Int Immunol 12(11):1539–1546
71. Byrd CA, Bornmann W, Erdjument-Bromage H, Tempst P, Pavletich N, Rosen N, Nathan CF, Ding A (1999) Heat shock protein 90 mediates macrophage activation by Taxol and bacterial lipopolysaccharide. Proc Natl Acad Sci U S A 96(10):5645–5650
72. Vilcek J, Palombella VJ, Henriksen-DeStefano D, Swenson C, Feinman R, Hirai M, Tsujimoto M (1986) Fibroblast growth enhancing activity of tumor necrosis factor and its relationship to other polypeptide growth factors. J Exp Med 163(3):632–643
73. Komoda H, Miyagawa S, Kubo T, Kitano E, Kitamura H, Omori T, Ito T, Matsuda H, Shirakura R (2004) A study of the xenoantigenicity of adult pig islets cells. Xenotransplantation11(3):237–246.doi:10.1111/j.1399-3089.2004.00121.x
74. Bennet W, Sundberg B, Lundgren T, Tibell A, Groth CG, Richards A, White DJ, Elgue G, Larsson R, Nilsson B, Korsgren O (2000) Damage to porcine islets of Langerhans after exposure to human blood in vitro, or after intraportal transplantation to cynomolgus monkeys: protective effects of sCR1 and heparin. Transplantation 69(5):711–719
75. Mandel TE, Dillon H, Koulmanda M (1995) The effect of a depleting anti-CD4 monoclonal antibody on T cells and fetal pig islet xenograft survival in various strains of mice. Transpl Immunol 3(3):265–272
76. Simeonovic CJ, Ceredig R, Wilson JD (1990) Effect of GK1.5 monoclonal antibody dosage on survival of pig proislet xenografts in CD4+ T cell-depleted mice. Transplantation 49(5):849–856
77. Wilson JD, Simeonovic CJ, Ting JH, Ceredig R (1989) Role of CD4+ T-lymphocytes in rejection by mice of fetal pig proislet xenografts. Diabetes 38(Suppl 1):217–219
78. Benda B, Karlsson-Parra A, Ridderstad A, Korsgren O (1996) Xenograft rejection of porcine islet-like cell clusters in immunoglobulin- or Fc-receptor gamma-deficient mice. Transplantation 62(9):1207–1211
79. Simeonovic CJ, McKenzie KU, Wilson JD, Zarb JC, Hodgkin PD (1998) Role of anti-donor antibody in the rejection of pig proislet xenografts in mice. Xenotransplantation 5(1):18–28
80. Cui H, Tucker-Burden C, Cauffiel SM, Barry AK, Iwakoshi NN, Weber CJ, Safley SA (2009) Long-term metabolic control of autoimmune diabetes in spontaneously diabetic nonobese diabetic mice by nonvascularized microencapsulated adult porcine islets. Transplantation 88(2):160–169. doi:10.1097/TP.0b013e3181abbfc1
81. Safley SA, Kapp LM, Tucker-Burden C, Hering B, Kapp JA, Weber CJ (2005) Inhibition of cellular immune responses to encapsulated porcine islet xenografts by simultaneous blockade of two different costimulatory pathways. Transplantation 79(4):409–418
82. Hering BJ, Wijkstrom M, Graham ML, Hardstedt M, Aasheim TC, Jie T, Ansite JD, Nakano M, Cheng J, Li W, Moran K, Christians U, Finnegan C, Mills CD, Sutherland DE, Bansal-Pakala P, Murtaugh MP, Kirchhof N, Schuurman HJ (2006) Prolonged diabetes reversal after intraportal xenotransplantation of wild-type porcine islets in immunosuppressed nonhuman primates. Nat Med 12(3):301–303. doi:10.1038/nm1369
83. Kirchhof N, Shibata S, Wijkstrom M, Kulick DM, Salerno CT, Clemmings SM, Heremans

Y, Galili U, Sutherland DE, Dalmasso AP, Hering BJ (2004) Reversal of diabetes in non-immunosuppressed rhesus macaques by intraportal porcine islet xenografts precedes acute cellular rejection. Xenotransplantation 11(5):396–407. doi:10.1111/j.1399-3089.2004.00157.x

84. Thompson P, Cardona K, Russell M, Badell IR, Shaffer V, Korbutt G, Rayat GR, Cano J, Song M, Jiang W, Strobert E, Rajotte R, Pearson T, Kirk AD, Larsen CP (2011) CD40-specific costimulation blockade enhances neonatal porcine islet survival in nonhuman primates. Am J Transplant 11(5):947–957. doi:10.1111/j.1600-6143.2011.03509.x

85. Cardona K, Korbutt GS, Milas Z, Lyon J, Cano J, Jiang W, Bello-Laborn H, Hacquoil B, Strobert E, Gangappa S, Weber CJ, Pearson TC, Rajotte RV, Larsen CP (2006) Long-term survival of neonatal porcine islets in nonhuman primates by targeting costimulation pathways. Nat Med 12(3):304–306. doi:10.1038/nm1375

86. Casu A, Bottino R, Balamurugan AN, Hara H, van der Windt DJ, Campanile N, Smetanka C, Cooper DK, Trucco M (2008) Metabolic aspects of pig-to-monkey (Macaca fascicularis) islet transplantation: implications for translation into clinical practice. Diabetologia 51(1):120–129. doi:10.1007/s00125-007-0844-4

87. van der Windt DJ, Bottino R, Casu A, Campanile N, Smetanka C, He J, Murase N, Hara H, Ball S, Loveland BE, Ayares D, Lakkis FG, Cooper DK, Trucco M (2009) Long-term controlled normoglycemia in diabetic non-human primates after transplantation with hCD46 transgenic porcine islets. Am J Transplant 9(12):2716–2726. doi:10.1111/j.1600-6143.2009.02850.x

88. Graham ML, Mutch LA, Rieke EF, Kittredge JA, Faig AW, DuFour TA, Munson JW, Zolondek EK, Hering BJ, Schuurman HJ (2011) Refining the high-dose streptozotocin-induced diabetic non-human primate model: an evaluation of risk factors and outcomes. Exp Biol Med 236(10):1218–1230. doi:10.1258/ebm.2011.011064

89. Sun Y, Ma X, Zhou D, Vacek I, Sun AM (1996) Normalization of diabetes in spontaneously diabetic cynomolgus monkeys by xenografts of microencapsulated porcine islets without immunosuppression. J Clin Invest 98(6):1417–1422. doi:10.1172/JCI118929

90. Dufrane D, Goebbels RM, Gianello P (2010) Alginate macroencapsulation of pig islets allows correction of streptozotocin-induced diabetes in primates up to 6 months without immunosuppression. Transplantation 90(10):1054–1062. doi:10.1097/TP.0b013e3181f6e267

91. Nauta AJ, Fibbe WE (2007) Immunomodulatory properties of mesenchymal stromal cells. Blood 110(10):3499–3506. doi:10.1182/blood-2007-02-069716

92. English K (2013) Mechanisms of mesenchymal stromal cell immunomodulation. Immunol Cell Biol 91(1):19–26. doi:10.1038/icb.2012.56

93. English K, Barry FP, Field-Corbett CP, Mahon BP (2007) IFN-gamma and TNF-alpha differentially regulate immunomodulation by murine mesenchymal stem cells. Immunol Lett 110(2):91–100. doi:10.1016/j.imlet.2007.04.001

94. Ge W, Jiang J, Arp J, Liu W, Garcia B, Wang H (2010) Regulatory T-cell generation and kidney allograft tolerance induced by mesenchymal stem cells associated with indoleamine 2,3-dioxygenase expression. Transplantation 90(12):1312–1320. doi:10.1097/TP.0b013e3181fed001

95. Francois M, Romieu-Mourez R, Li M, Galipeau J (2012) Human MSC suppression correlates with cytokine induction of indoleamine 2,3-dioxygenase and bystander M2 macrophage differentiation. Mol Ther 20(1):187–195. doi:10.1038/mt.2011.189

96. Nemeth K, Leelahavanichkul A, Yuen PS, Mayer B, Parmelee A, Doi K, Robey PG, Leelahavanichkul K, Koller BH, Brown JM, Hu X, Jelinek I, Star RA, Mezey E (2009) Bone marrow stromal cells attenuate sepsis via prostaglandin E(2)-dependent reprogramming of host macrophages to increase their interleukin-10 production. Nat Med 15(1):42–49. doi:10.1038/nm.1905

97. English K, Barry FP, Mahon BP (2008) Murine mesenchymal stem cells suppress dendritic cell migration, maturation and antigen presentation. Immunol Lett 115(1):50–58. doi:10.1016/j.imlet.2007.10.002

98. Ren G, Zhang L, Zhao X, Xu G, Zhang Y, Roberts AI, Zhao RC, Shi Y (2008) Mesenchymal stem cell-mediated immunosuppression occurs via concerted action of chemokines and nitric oxide. Cell Stem Cell 2(2):141–150. doi:10.1016/j.stem.2007.11.014

99. Klyushnenkova E, Mosca JD, Zernetkina V, Majumdar MK, Beggs KJ, Simonetti DW, Deans RJ, McIntosh KR (2005) T cell responses to allogeneic human mesenchymal stem cells: immunogenicity, tolerance, and suppression. J Biomed Sci 12(1):47–57. doi:10.1007/s11373-004-8183-7

100. Le Blanc K, Ringden O (2007) Immunomodulation by mesenchymal stem cells and clinical experience. J Intern Med 262(5):509–525. doi:10.1111/j.1365-2796.2007.01844.x
101. Augello A, Tasso R, Negrini SM, Amateis A, Indiveri F, Cancedda R, Pennesi G (2005) Bone marrow mesenchymal progenitor cells inhibit lymphocyte proliferation by activation of the programmed death 1 pathway. Eur J Immunol 35(5):1482–1490. doi:10.1002/eji.200425405
102. Li YP, Paczesny S, Lauret E, Poirault S, Bordigoni P, Mekhloufi F, Hequet O, Bertrand Y, Ou-Yang JP, Stoltz JF, Miossec P, Eljaafari A (2008) Human mesenchymal stem cells license adult CD34+ hemopoietic progenitor cells to differentiate into regulatory dendritic cells through activation of the Notch pathway. J Immunol 180(3):1598–1608
103. Perez-Basterrechea M, Obaya AJ, Meana A, Otero J, Esteban MM (2013) Cooperation by fibroblasts and bone marrow-mesenchymal stem cells to improve pancreatic rat-to-mouse islet xenotransplantation. PLoS One 8(8): e73526. doi:10.1371/journal.pone.0073526
104. Wu H, Wen D, Mahato RI (2013) Third-party mesenchymal stem cells improved human islet transplantation in a humanized diabetic mouse model. Mol Ther 21(9):1778–1786. doi:10.1038/mt.2013.147
105. Berman DM, Willman MA, Han D, Kleiner G, Kenyon NM, Cabrera O, Karl JA, Wiseman RW, O'Connor DH, Bartholomew AM, Kenyon NS (2010) Mesenchymal stem cells enhance allogeneic islet engraftment in non-human primates. Diabetes 59(10):2558–2568. doi:10.2337/db10-0136
106. Ding Y, Xu D, Feng G, Bushell A, Muschel RJ, Wood KJ (2009) Mesenchymal stem cells prevent the rejection of fully allogenic islet grafts by the immunosuppressive activity of matrix metalloproteinase-2 and -9. Diabetes 58(8):1797–1806. doi:10.2337/db09-0317
107. Longoni B, Szilagyi E, Quaranta P, Paoli GT, Tripodi S, Urbani S, Mazzanti B, Rossi B, Fanci R, Demontis GC, Marzola P, Saccardi R, Cintorino M, Mosca F (2010) Mesenchymal stem cells prevent acute rejection and prolong graft function in pancreatic islet transplantation. Diabetes Technol Ther 12(6):435–446. doi:10.1089/dia.2009.0154
108. Mundra V, Wu H, Mahato RI (2013) Genetically modified human bone marrow derived mesenchymal stem cells for improving the outcome of human islet transplantation. PLoS One 8(10):e77591. doi:10.1371/journal.pone.0077591
109. Park HS, Kim JW, Lee SH, Yang HK, Ham DS, Sun CL, Hong TH, Khang G, Park CG, Yoon KH (2015) Antifibrotic effect of rapamycin containing polyethylene glycol-coated alginate microcapsule in islet xenotransplantation. J Tissue Eng Regen Med. doi:10.1002/term.2029
110. Vaithilingam V, Kollarikova G, Qi M, Larsson R, Lacik I, Formo K, Marchese E, Oberholzer J, Guillemin GJ, Tuch BE (2014) Beneficial effects of coating alginate microcapsules with macromolecular heparin conjugates-in vitro and in vivo study. Tissue Eng Part A 20(1–2):324–334. doi:10.1089/ten.TEA.2013.0254
111. Ma M, Chiu A, Sahay G, Doloff JC, Dholakia N, Thakrar R, Cohen J, Vegas A, Chen D, Bratlie KM, Dang T, York RL, Hollister-Lock J, Weir GC, Anderson DG (2013) Core-shell hydrogel microcapsules for improved islets encapsulation. Adv Healthcare Mater 2(5):667–672. doi:10.1002/adhm.201200341
112. Tomei AA, Manzoli V, Fraker CA, Giraldo J, Velluto D, Najjar M, Pileggi A, Molano RD, Ricordi C, Stabler CL, Hubbell JA (2014) Device design and materials optimization of conformal coating for islets of Langerhans. Proc Natl Acad Sci U S A 111(29):10514–10519. doi:10.1073/pnas.1402216111
113. Chen T, Yuan J, Duncanson S, Hibert ML, Kodish BC, Mylavaganam G, Maker M, Li H, Sremac M, Santosuosso M, Forbes B, Kashiwagi S, Cao J, Lei J, Thomas M, Hartono C, Sachs D, Markmann J, Sambanis A, Poznansky MC (2015) Alginate encapsulant incorporating CXCL12 supports long-term allo- and xenoislet transplantation without systemic immune suppression. Am J Transplant 15(3):618–627. doi:10.1111/ajt.13049
114. Park JB, Jeong JH, Lee M, Lee DY, Byun Y (2013) Xenotransplantation of exendin-4 gene transduced pancreatic islets using multi-component (alginate, poly-L-lysine, and polyethylene glycol) microcapsules for the treatment of type 1 diabetes mellitus. J Biomater Sci Polym Ed 24(18):2045–2057. doi:10.1080/09205063.2013.823071

Chapter 25

Microencapsulation in Clinical Islet Xenotransplantation

Masayuki Shimoda and Shinichi Matsumoto

Abstract

Allogeneic islet transplantation has become a viable treatment for patients with unstable type 1 diabetes; however, donor shortage and the necessity for immunosuppressive drugs are the major drawbacks of this approach. Microencapsulated porcine islet xenotransplantation could solve these drawbacks. Clinical porcine islet xenotransplantation as well as microencapsulated islet transplantation has been conducted without significant side effects. However, these transplantations are not as efficacious as allogeneic naked islet transplantation. High quality porcine islets, biocompatible capsules, and appropriate implant sites should be the key factors for improving efficacy. With improved efficacy, microencapsulated islet xenotransplantation will solve the major drawbacks associated with current islet transplantation.

Key words Porcine islet, Encapsulation, Alginate, APA capsule, Xenotransplantation

1 Introduction

Type 1 diabetes is caused by the loss of beta cells and requires insulin therapy. Current insulin therapy is unable to prevent the development of acute and chronic diabetic complications, including diabetic retinopathy, nephropathy, and neuropathy, which can be alleviated by intensive insulin therapy. However, intensive insulin therapy increases the risk of hypoglycemia, which is one of its acute complications [1]. Especially, severe hypoglycemia, which is defined as the necessity of third party help, and hypoglycemia unawareness, that is, hypoglycemia without symptoms, are dangerous complications. Beta cell replacement therapies, including pancreas and islet transplantation, can normalize hemoglobin A1c (HbA1c) without increasing the risk of hypoglycemia [2]. Traditionally, pancreas transplantation has been more efficacious than islet transplantation, but it requires major surgery. Recent advances in islet transplantation have made this approach nearly as efficacious as pancreas transplantation; therefore, it has become the preferred treatment option [3].

Emmanuel C. Opara (ed.), *Cell Microencapsulation: Methods and Protocols*, Methods in Molecular Biology, vol. 1479,
DOI 10.1007/978-1-4939-6364-5_25, © Springer Science+Business Media New York 2017

The major drawbacks of islet transplantation are donor shortage and the necessity for immunosuppressive drugs. Donor shortage can be alleviated by using marginal [4] and living donors [5]; however, finding a donor other than a human donor is the ultimate solution to this problem. Xenotransplantation using porcine pancreas and islets generated from embryonic stem cells or induced pluripotent stem cells would resolve the shortage of human donors. Immunological tolerance and immune isolation are the solutions for avoiding the use of immunosuppressive drugs. The combination of these approaches, that is, islet transplantation using a nonhuman donor and the avoidance of immunosuppressive drugs, would resolve the major drawbacks of islet transplantation. Microencapsulated porcine islet xenotransplantation is one of the ultimate solutions for islet transplantation.

In this chapter, clinical trials of islet xenotransplantation and islet transplantation with microencapsulation and future directions are described.

2 Porcine Islet Xenotransplantation

Porcine islet xenotransplantation has been conducted in several countries (Table 1). Some of these trials used immunosuppressive drugs, but some were conducted without immunosuppression.

2.1 Initial Trial in Sweden

Historically, Groth et al. performed the first islet xenotransplantation using porcine islet-like cell clusters (ICCs) in 10 insulin-dependent diabetic kidney transplanted patients [6]. ICCs were transplanted into the portal vein in 8 cases and under the kidney capsule in 2 cases. All patients received immunosuppressive therapy, which included induction therapy using anti-thymocyte globulin or 15-deoxyspergualin, and maintenance therapy using standard immunosuppressive drugs. This trial demonstrated that porcine islets could survive in the human body.

2.2 Chinese Trial

Wang et al. performed neonatal porcine islet transplantation on 22 type 1 diabetic patients via the hepatic artery [7]. The first 14 patients received cyclosporine, mycophenolate mofetil (MMF), and prednisolone, two patients received cyclosporine and MMF, and the remaining six patients received OKT3, tacrolimus, sirolimus, and MMF. Insulin dose and HbA1c levels were reduced in the first 14 patients, and a therapeutic effect was not observed in two patients. The other six patients had normalized HbA1c levels for 3 months, reduced insulin dose, and were positive for porcine C-peptide, indicating that the porcine islets were functional. This study demonstrated that porcine islets can be used as a functional replacement for human islets.

Table 1
Clinical islet xenotransplantation

Location	Pig source	Antirejection	Efficacy	References
Sweden	Fetal	Anti-thymocyte globulin or 15-deoxyspergualin and standard immunosuppression	Positive porcine C-peptide but no metabolic improvement	[6]
China	Neonatal	Standard immunosuppression	Reduction in insulin dose and HbA1c with positive porcine C-peptide	[7]
Mexico	Neonatal	Sertoli cells	Transient insulin independence in 2 cases	[8, 9]
New Zealand	Fetal	Alginate–PLL–alginate capsules	Reduction in insulin dose and HbA1c with positive porcine C-peptide	[10]
Russia	Neonatal	Alginate–PLO–alginate capsules	Two patients achieved insulin independence and 6 patients improved HbA1c	[11]
New Zealand	Neonatal	Alginate–PLO–alginate capsules	Reduction in hypoglycemia unawareness episodes	[14]
Argentina	Neonatal	Alginate–PLO–alginate capsules	Reduction in HbA1c, insulin dose, and hypoglycemia unawareness episodes	[11]

2.3 Mexican Trial

Macroencapsulated neonatal porcine islets and Sertoli cells were transplanted into adolescent type 1 diabetic patients without immunosuppressive drugs [8]. Two devices were implanted subcutaneously in the upper anterior wall of the abdomen. The devices consisted of a surgical grade stainless steel mesh tube with a polytetrafluoroethylene (PTFE) rod in its interior. After the inserted PTFE rod was removed, 250,000 islets with 30–100 Sertoli cells were implanted into the devices. The insulin dose was decreased in six patients and increased in six patients. Of note, two patients achieved transient insulin independence.

The same group extended the clinical trial and analyzed a total of 21 type 1 diabetic patients [9]. HbA1c levels were reduced to less than 7% after the first transplantation, but increased to more than 8% after the second transplantation. Urinary porcine C-peptide was detected. This study demonstrated the possibility of using porcine islet xenotransplantation without immunosuppression.

3 Microencapsulation in Clinical Islet Xenotransplantation (New Zealand, Russian, and Argentinian Trials)

3.1 Initial Case

Microencapsulation in clinical islet xenotransplantation could solve two of the major issues of current islet transplantation, that is, donor shortage and the necessity for immunosuppression.

A clinical trial of alginate–poly-lysine–alginate microencapsulated fetal porcine islets in a type 1 diabetic patient was conducted in New Zealand [10]. The patient received a total of 1,305,000 islet equivalent (IEQ) islets into the peritoneal cavity. At 12 weeks after transplantation, their daily insulin dose was temporarily reduced by approximately 30%. Urinary porcine C-peptide was detected for up to 11 months after transplantation. HbA1c levels were reduced to 7.8% at 14 months from a pre-transplant level of 9.3%. Interestingly, at 9.5 years after transplantation, a biopsy sample was both insulin- and glucagon-positive by immunohistochemistry. The retrieved capsule produced a small amount of insulin when placed in high glucose concentrations in vitro.

3.2 Collaboration Between Living Cell Technologies Ltd. (LCT) and the University of Perugia

Following this initial case, LCT and a group at the University of Perugia started a collaboration in 1999 with an exchange of their respective products (neonatal porcine islets from LCT and alginate–poly-L-ornithine [PLO]–alginate [APA] encapsulation technology from the University of Perugia) [11]. The islet isolation and encapsulation procedures are described below.

Islet isolation procedure:

Neonatal porcine islet isolation was performed based on the Korbutt method [12] with some modifications [13]. Pancreata were procured from neonatal pigs under general anesthesia. The procured pancreata were cut into pieces and digested with 0.75 mg/mL Liberase MTF C/T (collagenase and thermolysin) (Roche) [13]. Digested pancreatic tissues were cultured in 500-mL disposable spinner flasks at a concentration of 40 mL/g pancreatic tissue in RPMI medium plus 0.3 g/L ciprofloxacin, 1% human serum albumin, and 0.12% wt/vol nicotinamide [13]. The flasks were stirred on a four-position slow speed magnetic stirrer at 80–90 rpm in a humidified 5% CO_2 incubator at 37 °C [13]. Culture medium was replaced daily (100%) following sedimentation of the cells for 15 min.

Capsulation procedure:

Approximately 1.8% (wt/vol) alginate with a molecular weight of 140–150 kDa was used for encapsulation and diluted 1/10 with saline for the outer coat [13]. PLO with a molecular weight 10–12 kDa (Sigma) was dissolved in saline to a final concentration of 0.1 and 0.05% (wt/vol) and all washing steps were performed with saline solution [13]. Porcine pancreatic islets were mixed with 1.8% (wt/vol) alginate. The mixed porcine islets and alginate solution was pumped through a needle supplied with a coaxial airflow. This system allows the creation of droplets that are then dropped into a 109 mM $CaCl_2$ solution. After 15 min, alginate and calcium ions create cross-links that result in the formation of encapsulated islets. The encapsulated islets were further coated with 0.1% (wt/vol) PLO for 10 min, followed by a second coat of 0.05% (wt/vol) PLO for 6 min [13]. Then, a final layer of 0.18% (wt/vol) alginate was added to the encapsulated islets to create APA encapsulated islets, which were cultured for an additional 3 weeks before transplantation.

3.3 Clinical Trials of APA Encapsulated Porcine Islet Xenotransplantation

Following the first case, a phase I/IIa clinical trial was conducted in Russia using neonatal porcine islets encapsulated with APA [14]. Seven type 1 diabetic patients received 1–3 implantations. Five patients demonstrated improved blood glucose levels and two patients achieved insulin independence for up to 32 weeks [14].

Following the previous clinical trials, islet xenotransplantation under a regulatory framework has been conducted in New Zealand for the first time [15]. This study was registered at ClinicalTrial.gov (NCT00940173). To date, no safety concerns have been reported in 14 recipients of encapsulated islets. This study consists of four different doses (5000 IEQ/kg $n=4$, 10,000 IEQ/kg $n=4$, 15,000 IEQ/kg $n=4$, and 20,000 IEQ/kg $n=2$) [15]. The data for up to 52 weeks post-transplantation have been analyzed. There were four serious adverse events that were resolved without residual effects. The number of hypoglycemia unawareness events was reduced at 1 year after transplantation compared with pre-transplantation in all dose groups. In terms of HbA1c, the levels were less than 7% before transplantation in 1 out of 14 patients and four patients reached less than 7% at 1 year after transplantation.

Following this trial, phase I/IIa and IIb clinical trials of immunoprotected porcine islets for xenotransplantation have been registered in Argentina [11]. Interim analysis revealed a significant reduction in HbA1c levels, insulin dose, and hypoglycemia unawareness episodes [11]. However, no patient achieved insulin independence; therefore, further improvement is needed [11].

The safety of microencapsulation in clinical islet xenotransplantation has been demonstrated by this series of clinical trials; however; an improvement of efficacy is necessary [11].

4 Microencapsulation in Clinical Islet Allotransplantation

There are three clinical trials of microencapsulation in allogeneic islet transplantation (Table 2). The first trial was performed in California (USA), followed by trials at the University of New South Wales (Australia) and the University of Perugia (Italy)

Table 2
Clinical trials of microencapsulated allogeneic islet transplantation

	Capsule	Efficacy	Reference
USA	Alginate	Insulin independence for 9 months	[16]
Australia	Ba-alginate	Positive human C-peptide for up to 2.5 years in 1 case with multiple transplantation	[17]
Italy	Alginate–PLO–alginate	Transient insulin independence in 1 case	[18]

4.1 First Microencapsulated Allogeneic Islet Transplantation in the USA

Alginate microencapsulated allogeneic human islets (10,000 IEQ/kg body weight) were transplanted intraperitoneally into a type 1 diabetic patient with low-dose immunosuppression [16]. This patient received a second transplant (5000 IEQ/kg body weight) at 6 months after the first transplant. The patient recovered from surgery without any complications, and achieved insulin independence with tight glycemic control for 9 months.

4.2 Australian Trial

A clinical trial of Ba-alginate encapsulated human islet transplantation was conducted at the University of New South Wales [17]. Four type 1 diabetic patients received an average of 178,200 IEQ islets into the intraperitoneal cavity. None of the patients were positive for C-peptide at 1–4 weeks after the initial transplantation. One patient received 4 islet infusions and the total amount of transplanted islets was 779,000 IEQ. C-peptide was detected at 6 weeks after the third infusion and remains detectable at 2.5 years; however, neither insulin dose nor glycemic control was affected. Capsules recovered at 16 months were surrounded by fibrous tissue and contained necrotic islets. This demonstrated that even though a sufficient islet mass was provided, the Ba-alginate encapsulated human islets were unable to improve glycemic control due to bio-incompatibility. The prevention of fibrosis seems to be the major factor that would improve the efficacy of Ba-alginate encapsulation.

4.3 Italian Trial

The University of Perugia performed a clinical trial of allogeneic islet transplantation using APA microcapsules [18]. Four type 1 diabetic patients received APA encapsulated human islets intraperitoneally. All patients showed a decline of exogenous daily insulin consumption and one patient was able to suspend insulin injections temporarily. C-peptide was detected in all cases and was present for 100–480 days. No anti-MHC class I–II, anti-GAD65, or islet cell antibodies were detected in any of the transplanted patients throughout the 5 years of post-transplantation follow-up. This suggested that APA microcapsules could prevent the detection of foreign antigens by the host even without immunosuppression.

The encapsulated islets were retrieved at 5 years after transplantation and the capsules appeared mostly intact, though the majority contained only necrotic debris.

5 Research Targets for Microencapsulated Porcine Islet Xenotransplantation

5.1 Lessons from the Clinical Trials

On the basis of the clinical trials of islet xenotransplantation and encapsulated islet allotransplantation, there were no serious adverse events that required termination of the trials. Under appropriate compliance with the regulations and monitoring for both donor source and patients, islet xenotransplantation can be justified. Encapsulation helps the acceptance of xenotransplantation because this technology can avoid the use of immunosuppression. The International Xenotransplantation Association published consensus guidelines for islet xenotransplantation, which help to justify the use of clinical islet xenotransplantation [19].

Improving the efficacy of this approach should be the next target since none of the clinical trials of islet xenotransplantation achieved insulin independence constantly. To improve efficacy, the quality of islets, biocompatibility of capsules, and implant sites are key factors.

5.2 Improving the Quality of Porcine Islets

Donor age, isolation method, and culture methods are major factors that determine the quality of isolated porcine islets.

Regarding donor age, all previous clinical trials used fetal or neonatal pigs. This is mainly due to the difficulty of maintaining a designated pathogen-free status for a long period of time. For practical reasons, it might be necessary to use fetal or neonatal pigs in clinical trials.

Islet isolation consists of pancreas procurement, pancreas preservation, pancreas digestion, and islet purification. Avoidance of warm ischemic injury is the most important factor during pancreas procurement. We have demonstrated pancreas preservation using the two-layer method [20, 21], and pancreatic ductal preservation improved the outcome of islet isolation [22]. These technologies are worth testing in the clinical setting. Neonatal pancreas digestion is conducted based on the Korbutt method [12]; however, the Ricordi method [23, 24] should be considered when large-scale islet isolation is necessary. Digested neonatal pancreatic tissues are cultured for purification and this process can be improved by modification of the culture medium.

5.3 Improving Capsules

Among the materials used to generate capsules, alginate is the most favored because of its superior biocompatibility, hydrophilicity, simple gelation process, stable architecture, abundant availability, relative ease of procurement, low manufacturing costs, and prolonged stability in vivo [25]. In addition, considering that alginate

Table 3
Variables for alginate capsules

Purity	Heavy metals, endotoxins, proteins
Size	Small, large (1.5 mm)
M:G ratio	Low (solid), high (soft)
Cations	Ca^{2+}, Ba^{2+}
Polycations	Poly-L-lysine, poly-L-ornithine
Shape	Core–shell, double core–shell, solid/liquid core

is the only clinically applied material, improving alginate capsules is a practical research target. Variables that affect the stability, permselectivity, and biocompatibility of capsules have been examined (Table 3). However, the gold standard has not yet been established. For the fabrication of microcapsules using highly purified material with improved stability, low levels of heavy metals, proteins, and endotoxin content are necessary [25].

Capsule size is another important factor that should be considered. Traditionally, smaller capsules were believed to improve efficacy because of the ease of exchange of materials through the capsule [11]. Recently, it was demonstrated that spherical materials that are ≥1.5 mm in diameter proved to be significantly more biocompatible than their smaller-sized or differently shaped counterparts [26]. An in vivo test demonstrated that 1.5-mm alginate encapsulated rat islets were able to restore blood glucose control in streptozotocin-induced diabetic C57BL/6 mice for up to 180 days. This suggests that biocompatibility seems to be more important than ease of exchange of materials with respect to efficacy.

Alginate consists of linear binary copolymers of beta-D-mannuronate and alpha-L-guluronate. The length and sequence of mannuronate and guluronate chains and the ratio of mannuronate to guluronate (M:G ratio) in alginate hydrogels determine the mechanical strength, elasticity, durability, permeability, and swelling characteristics of the alginate [25]. The use of multivalent cations (Ca^{2+}, Ba^{2+}) and polycations (poly-L-lysine or poly-L-ornithine) during alginate synthesis modifies its properties [25]. For example, alginate poly-L-ornithine alginate capsules provide a high degree of permselectivity, better stability, and improved mechanical strength, but induce pericapsular cellular overgrowth and macrophage activation, resulting in fibrotic overgrowth of the capsules. Multivalent cations such as barium can avoid such fibrotic overgrowth, but reduce permselectivity.

5.4 Implant Sites

The intraperitoneal cavity has been selected for the clinical trials of microencapsulated islet transplantation mainly due to its lower restriction on graft volume. However, this site has several

Table 4
Implant sites of encapsulated islets

	Close access to blood	Stable distribution	Avoidance of aggregation	Ease of retrieval
Intraperitoneal cavity	No	No	No	No
Renal subcapsule	Yes	Yes	No	No
Subcutaneous	No	Yes	Yes	Yes
Pre-vascularized subcutaneous	Yes	Yes	Yes	Yes
Omentum pouch	Yes	Yes	Yes	Yes

disadvantages, including lack of close contact to the blood stream, uncertain distribution of transplanted encapsulated islets, stacking of capsules in the pelvic cavity of bipeds resulting in aggregation, and difficulty of retrieval. Other implant sites have been examined to solve these issues (Table 4).

Use of the renal subcapsule has been commonly used in mouse models. Considering the large amount of encapsulated islets that are used for clinical transplantation, this site cannot prevent the aggregation of encapsulated islets. The subcutaneous space can be used for a large amount of encapsulated islets; however, this site is notorious for its poor blood access. The pre-vascularized subcutaneous space seems promising for either device or device-less approaches [27]; however, this area requires two operations for pre-vascularization and actual transplantation.

The omentum pouch can also be used without the need for two operations; therefore, this might be one of the ideal sites for the implantation of encapsulated islets. In fact, the long-term function of encapsulated islets in the omentum pouch was shown using an immune-competent diabetic rat model [28].

6 Conclusions

Both porcine xeno islets and microencapsulated islets have been applied clinically without significant side effects. On the other hand, the efficacy of this approach is still not ideal compared with allogeneic naked islet transplantation. Research to improve the quality of islets, to enhance the biocompatibility of capsules, and to identify a suitable implant site is key to the practical development of this therapy. Further studies should make this approach as efficacious as allogeneic naked islet transplantation. This will represent a real breakthrough to overcome donor shortage and avoid the side effects associated with the use of immunosuppressive drugs.

References

1. The Diabetes Control and Complications Trial Research Group (1997) Hypoglycemia in the Diabetes Control and Complications Trial. Diabetes 46:271–286
2. Matsumoto S (2010) Islet cell transplantation for type 1 diabetes. J Diabetes 2:16–22
3. Shapiro AMJ (2012) Islet transplantation in type 1 diabetes: ongoing challenges, refined procedures, and long-term outcome. Rev Diabet Stud 9:385–406
4. Matsumoto S, Okitsu T, Iwanaga Y et al (2006) Successful islet transplantation from nonheart-beating donor pancreata using modified Ricordi islet isolation method. Transplantation 82:460–465
5. Matsumoto S, Okitsu T, Iwanaga Y et al (2005) Insulin independence after living-donor distal pancreatectomy and islet allotransplantation. Lancet 365:1642–1644
6. Groth CG, Korsgren O, Tibell A et al (1994) Transplantation of porcine fetal pancreatic to diabetic patients. Lancet 344:1402–1404
7. Wang W, Mo Z, Ye B et al (2011) A clinical trial of xenotransplantation of neonatal pig islets for diabetic patients. Zhong Nan Da Xue Xue Bao Yi Xue Ban 36:1134–1140
8. Valdes-Gonzalez RA, Dornates LM, Garibay GN et al (2005) Xenotransplantation of porcine neonatal islets of Langerhans and Sertoli cells: a 4-year study. Eur J Endocrinol 153: 419–427
9. Valdes-Gonzalez R-VAL, White DJ et al (2010) Long-term follow-up of patients with type 1 diabetes transplanted with neonatal pig islets. Clin Exp Immunol 162:537–542
10. Elliott RB, Escobar L, Tan PLJ et al (2007) Live encapsulated porcine islets from a type 1 diabetic patient 9.5 yr after xenotransplantation. Xenotransplantation 14:157–161
11. Calafiore R, Basta G (2014) Clinical application of microencapsulated islets: actual prospective on progress and challenges. Adv Drug Deliv Rev 67–68:84–92
12. Korbutt GS, Elliott JF, Ao Z et al (1996) Large scale isolation, growth and function of porcine neonatal islet cells. J Clin Invest 97:270–280
13. Hillberg A, Kathirgamanathan K, Lam JBB et al (2013) Improving alginate-poly-L-ornithine-alginate capsule biocompatibility through genipin crosslinking. J Biomed Mater Res B Appl Biomater 101:258–268
14. Zhu HT, Lu L, Liu XY et al (2015) Treatment of diabetes with encapsulated pig islets: an update on current developments. J Zhejiang Univ Sci B 16:329–343
15. Matsumoto S, Tan P, Baker J et al (2014) Clinical porcine islet xenotransplantation under comprehensive regulation. Transplant Proc 46:1992–1995
16. Soon-Shiong P, Heintz RE, Merideth N et al (1994) Insulin independence in a type 1 diabetic patient after encapsulated islet transplantation. Lancet 343:950–951
17. Tuch BE, Keogh GW, Williams LJ et al (2009) Safety and viability of microencapsulated human islets transplanted into diabetic humans. Diabetes Care 32:1887–1889
18. Basta G, Montanucci P, Luca G et al (2011) Long-term metabolic and immunological follow-up of non-immunosuppressed patients with type 1 diabetes treated with microencapsulated islet allografts: four cases. Diabetes Care 34:2406–2409
19. Hering BJ, Cooper DKC, Cozzi E et al (2009) The international xenotransplantation association consensus statement on conditions for undertaking clinical trials of porcine islet products in type 1 diabetes- Executive summary. Xenotransplantation 16:196–202
20. Noguchi H, Levy MF, Kobayashi N, Matsumoto S (2009) Pancreas preservation by the two-layer method: does it have a beneficial effect compared with simple preservation in University of Wisconsin solution? Cell Transplant 18: 497–503
21. Qin H, Matsumoto S, Klintmalm GB et al (2011) A meta-analysis for comparison of the two-layer and university of Wisconsin pancreas preservation methods in islet isolation. Cell Transplant 20:1127–1137
22. Matsumoto S, Noguchi H, Shimoda M et al (2010) Seven consecutive successful clinical islet isolation with pancreatic ductal injection. Cell Transplant 19:291–297
23. Ricordi C, Lacy PE, Finke EH et al (1988) Automated method for islet isolation of human pancreatic islets. Diabetes 37:413–420
24. Ricordi C, Socci C, Davalli AM et al (1990) Isolation of the elusive pig islet. Surgery 107:688–694
25. Krishnan R, Alexander M, Robles L et al (2014) Islet and stem cell encapsulation for clinical transplantation. Rev Diabet Stud 11:84–101
26. Veiseh O, Doloff JC, Ma M et al (2015) Size- and shape-dependent foreign body immune

response to materials implanted in rodents and non-human primates. Nat Mater 14:643–651

27. Pepper AR, Gala-Lopez B, Pawlick R et al (2015) A prevascularized subcutaneous device-less site for islet and cellular transplantation. Nat Biotechnol 33:518–523

28. Pareta R, McQuilling JP, Sittadjody S et al (2014) Long-term function of islets encapsulated in a re-designed alginate microcapsule construct in omentum pouch of immune-competent diabetic rats. Pancreas 43: 605–613

Chapter 26

Methods for Microencapsulated Porcine Islet Production

Masayuki Shimoda and Shinichi Matsumoto

Abstract

Allogeneic islet transplantation has become a viable option for the treatment of unstable type 1 diabetes; however, donor shortage and the necessity for immunosuppressive drugs are two major drawbacks of this treatment. Microencapsulated porcine islets can solve both of these drawbacks, and clinical trials have been conducted. Previous clinical trials demonstrated the safety and marginal efficacy of this treatment; therefore, it is necessary to improve its efficacy. The production of microencapsulated porcine islets consists of pancreas procurement, islet isolation, and microencapsulation. In this chapter, we describe our current methods for porcine pancreas procurement, islet isolation, and microencapsulation.

Key words Pancreas procurement, Porcine islets, Microencapsulation, Islet transplantation, Donor shortage, Type 1 diabetes

1 Introduction

Unstable type 1 diabetes caused by the complete loss of beta cells in the pancreas is a devastating disease due to its acute and chronic complications. Especially, hypoglycemic attacks are a serious and life-threatening acute complication. Severe hypoglycemia, which requires help from a third party to manage the condition, and hypoglycemia unawareness, which results in sympathetic nerve damage, are two major dangerous conditions. Beta cell replacement therapies, which consist of allogeneic pancreas and islet transplantation, are the best approaches to manage hypoglycemia [1]. Traditionally, pancreas transplantation has been more efficacious than islet transplantation; however, recent improvements have resulted in islet transplantation becoming as efficacious as pancreas transplantation [2]. Since islet transplantation removes the necessity for major surgery, this treatment has become the preferred option for unstable type 1 diabetes. Allogeneic islet transplantation still has two major drawbacks: donor shortage and the necessity for immunosuppressive drugs [3]. Microencapsulated porcine islet transplantation can solve both of these drawbacks. Microencapsulated

Emmanuel C. Opara (ed.), *Cell Microencapsulation: Methods and Protocols*, Methods in Molecular Biology, vol. 1479, DOI 10.1007/978-1-4939-6364-5_26, © Springer Science+Business Media New York 2017

islet transplantation requires two techniques: porcine islet isolation and microencapsulation. The first clinical trial of microencapsulated porcine islets was conducted by a group in New Zealand who used fetal porcine islets and alginate poly-lysine alginate capsules [4]. After the first trial, this group adapted the Korbutt method for islet isolation [5] and the Perugia method for microencapsulation [6]. The Korbutt method uses neonatal porcine pancreata, which provide more islets than fetal pancreata. Recently, it has been revealed that young (14–21 days) porcine pancreata can provide more islets than neonatal porcine pancreata (1–5 days), and the most recent clinical trial conducted by the New Zealand group used young porcine pancreata [7].

Conversely, we have developed Baylor's islet isolation method [8], which includes pancreatic ductal preservation [9, 10], two-layer pancreas preservation [11], and Ricordi pancreas digestion. Since young porcine islets can be isolated using this protocol, we developed an islet isolation method for young porcine pancreata, known as the NCGM method.

In terms of microencapsulation, the Perugia method has become one of the standard methods [12], and this method has been used with minor modifications [13, 14].

In this chapter we describe the NCGM method for young porcine islet isolation and microencapsulation based on the Perugia method.

2 Materials

2.1 Porcine Pancreas Procurement and Preservation

Pigs: LWD pigs (14–21 days)

1. Perfluorodecalin (PFC; Wako Chemical, Ltd.)
2. Medical grade oxygen.
3. Sterile wide-mouth plastic jar with cover.
4. ETK solution (1 L; Otsuka Pharm Factory, Inc.)
5. Ulinastatin (5 mL; Mochida Pharmaceutical Co., Ltd.)
6. Cooler box and slush.
7. Sterile stainless steel metal plate (Umihira).
8. Sterile needle for pancreatic duct cannulation (22G).
9. Oxygen filtration system (syringe filter, silicon tube, and sterile extension tube).
10. Sterile surgical instruments (forceps, scissors, hemostats, and electronic knife).
11. Sterile isolation bags (×2).
12. Sterile slush (saline).
13. 10 mL syringes and 18G needles.

14. 14G needles for aortic cannulation.
15. Extension tubes.
16. Sterile basins (×2).
17. Scale.
18. Bottles for blood collection.

2.2 Porcine Islet Isolation and Evaluation: NCGM Method

Reagents and Media

1. Hank's balanced salt solution (HBSS; Mediatech, Inc.)
2. Fetal bovine serum (FBS).
3. Collagenase and Thermolysin (Liberase MTF C/T GMP Grade; Roche Life Science).
4. Culture media (M-199; Thermo Fisher Scientific, Inc.)
5. Nicotinamide (Sigma-Aldrich Co., LLC.)
6. Antibiotic solution.
7. Betadine (Meiji Seika Pharma Co., Ltd.)
8. Heparin (Mochida Pharmaceutical Co., Ltd.)
9. UW solution (Viaspan; Bristol-Myers Squibb Pharmaceutical, Ltd.)
10. Dilution solution (Mediatech, Inc.)
11. Dithizone (Sigma-Aldrich Co., LLC.)
12. Washing solution (Mediatech, Inc.)
13. Priming solution (Mediatech, Inc.)
14. Fluorescein diacetate (FDA) and propidium iodide (PI) (Sigma-Aldrich Co., LLC.)

Disposables

1. 250 and 50 mL conical tubes.
2. 22G needles, 20 mL syringes.
3. Pipettes.
4. 1 L beakers.
5. Master flex tubing.
6. Vacuum filter flasks.
7. Counting dishes (35 mm).
8. 24-well plates.
9. Spinner flasks (500 mL).

Equipment

1. Sterile surgical instruments (forceps and scissors).
2. Sterile transport pan.
3. Scale.

4. Centrifuge.
5. Peristaltic pump.
6. Thermometer.
7. Chiller units.
8. Heated water bath.
9. Pipette aid.
10. Biological safety cabinet (BSC).
11. Ricordi chamber w/marbles (Biorep Technologies, Inc.)
12. Ricordi chamber screen mesh (Biorep Technologies, Inc.)

2.3 Microencapsulation

1. 1.8 % sodium alginate solution.
   Keltone LVCR Lot. 67432A, stored at 2–8 °C light protected
   Molecular weight 120,000–190,000 kDa
   Mannuronic–guluronic acid ratio: 61 %/39 %
   Viscosity: 1.8 % 100–300 cps
   pH: 7.2–7.6 (temperature 20 °C)
   Microbiology (bacteria, yeast, and fungi) sterile
   Endotoxin (EU/mL) < 0.5 EU/mL
   Protein content 2–2.5 μg/mL
2. Sodium chloride: 0.9 %.
3. Poly-L-ornithine (Sigma).
4. 109 mM $CaCl_2$ solution.
5. 55 mM sodium citrate.

Equipment (Fig. 1)

1. Sterile beakers.
2. Peristaltic pump.
3. Pump tube.
4. Coaxial droplet generator.
5. Scale.

3 Methods

3.1 Porcine Pancreas Procurement and Preservation

1. Filter approximately 300 mL PFC and keep cold on ice at 4 °C.
2. In a sterile basin on a sterile table, pour the PFC into a wide-mouth plastic jar (lower layer).
3. Mix ETK solution and ulinastatin (M-ETK).
4. Pour 400 mL M-ETK into the jar (upper layer).
5. Keep the jar cold with sterile saline slush in the basin.

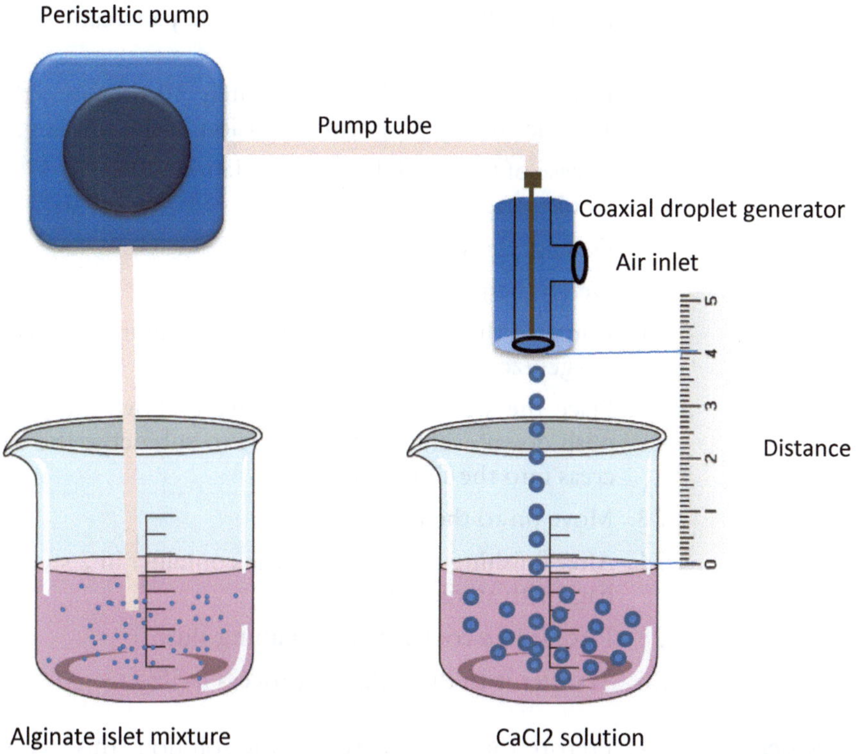

Fig. 1 The islet microencapsulation system consists of an alginate islet mixture solution, peristaltic pump, pump tube, coaxial droplet generator, and $CaCl_2$ solution. The alginate islet mixture solution is pumped using the peristaltic pump and encapsulated islets are generated by the coaxial droplet generator. Alginate beads including islets are cross-linked in a 109 mM $CaCl_2$ solution

6. Cool 10 mL M-ETK in a 10 mL syringe in sterile slush.
7. Bubble oxygen into the PFC layer, using a filter system (syringe filter, silicon tube, and sterile extension tube).
8. Hold in place and bubble with oxygen (approximately 2 L/min) for at least 30 min.
9. Start surgical operation under general anesthesia.
10. Open abdomen by midline incision.
11. Identify the pancreas and gently detach adhesions around the pancreas.
12. Identify the aorta near the bifurcation and gently detach the connective tissues.
13. Insert a 14G needle or cannula into the aorta and connect it to the extension tubes.
14. Start to remove the blood. The blood is collected into bottles. Quickly, insert slush into the abdominal cavity and cool the pancreas.
15. After blood outflow stops, cut the vena cava in the chest.

16. Insert a 22G needle or tube into the pancreatic duct near the duodenum.
17. Infuse cold M-ETK (approximately 10 mL) into the pancreatic duct slowly. Confirm expansion of the whole pancreas.
18. In case of failure of ductal cannulation, infuse M-ETK into the whole pancreas directly.
19. Procure the whole pancreas.
20. Weigh the pancreas.
21. Immerse the pancreas in the jar containing the M-ETK and oxygenized PFC.
22. Place the sterile stainless steel plate above the pancreas and push down evenly and vertically to submerge the whole pancreas into the PFC layer.
23. Move on to the next pig.
24. After finishing all pigs, place the container in 2 sterile isolation bags with sterile slush.
25. Place it in a cooler box containing slush.
26. Transfer it to the isolation process facility.

3.2 Porcine Islet Isolation and Evaluation: NCGM Method

1. Prepare the collagenase enzyme solution (this step should be started 30 min before isolation). Dissolve Liberase (1 vial) and Thermolysin (1 vial) in 700 mL HBSS.
2. Prepare the digestion process circuit. Connect the Ricordi chamber to 2 master flex tubes. Attach the peristaltic pump to one of the tubes. At the end of both tubes, connect the candy cane-shaped stainless tubes (one for inflow and the other for outflow). Hold both stainless tubes in a single 250 mL conical tube. The Ricordi chamber contains the metal marbles and mesh screen. Insert a heating metal coil into the circuit and immerse it in a 40 °C water bath.
3. Receive the pig pancreas at the islet isolation laboratory.
4. Open the shipping container and take out the organ container. Keep it sterile.
5. Put it into a BSC and open the cover.
6. For decontamination, place the pancreas quickly into cold Betadine solution. Next, immerse it into cold antibiotic solution for 5 min. After that, dip and rinse the pancreas in cold HBSS three times. Place the pancreas in a sterile pan.
7. In the pan, perform a direct infusion of the collagenase solution into the pancreas using a 27G needle (5 mL per pancreas). Confirm the expansion of the pancreas.
8. Cut the pancreas into small pieces. Every piece should be approximately 1 cm in width.

9. Place the dissected pancreas into the Ricordi chamber.
10. Pour the rest of the collagenase solution and fill the digestion circuit.
11. Measure the temperature of the enzyme solution using a sterile thermo-sensor probe connected to a thermal readout.
12. Remove all air bubbles from the circuit.
13. Start the pancreas digestion step as follows.
14. Start the peristaltic pump at a rate of 300 mL/min. Start timer.
15. Begin hand-shaking the Ricordi chamber at approximately 100 rpm.
16. Once 34 °C is reached, withdraw the heating coil from the water bath and slow the pump rate to 150 mL/min.
17. Monitor temperature. Record it every 2 min. The temperature should be maintained between 28.0 and 32.0 °C. Regulate the temperature by withdrawing and submerging the coil in the water bath.
18. After 2 min, start taking approximately 1 mL solution sample with a syringe through the sample port of the circuit tube and place into a well of a 24-well plate containing dithizone solution and observe under a microscope.
19. Additional samples should be taken every 2 min until the end of digestion and dilution.
20. Evaluate the 1 mL samples for pancreatic tissue. Most of the tissue is not stained with dithizone.
21. The digestion step is complete when most of the pancreas in the Ricordi chamber is softened, but the tissue in the sample is not granulated. The size of the cell clusters should be in the range of 50–400 μm in diameter. Prevent over-digestion.
22. Switch to the dilution and collection step.
23. The candy cane-shaped stainless tube for outflow is placed into a 250 mL conical tube containing 15 mL cold FBS. The collagenase solution with pancreas tissues in the digestion circuit is being collected. The collection tubes are cooled quickly at 4 °C using ice or cooling blocks.
24. Continue to shake the chamber.
25. The other candy cane-shaped stainless tube (for inflow) is placed into the dilution solution to supply the dilution solution continuously into the Ricordi chamber.
26. Adjust pump speed to 300 mL/min.
27. Centrifuge the collected solution in conical tubes at 170 × *g* for 1 min.
28. Remove the supernatant. Add UW solution and resuspend the pellet.

29. Continue to take and check tissue samples every 2 min.
30. Continue supplying fresh dilution solution to the chamber until the dilution process has ended. The process is complete when the tissue in the chamber is no longer being digested. This can be ascertained as when no pancreatic tissue is detected in the tissue sample.
31. After dilution is complete, remove undigested tissue from the chamber.
32. Weigh the undigested tissue using a scale.
33. Collect all pellets, pooling them together into a single 250 mL conical tube. Keep the tubes at 4 °C.
34. Wash the pooled tissue twice with dilution solution.
35. Take a 100 μL sample from 200 mL solution for counting.
36. Determine the cell cluster count and percentage fragmented tissue.
37. Check viability using FDA and PI.
38. Move to the culture step as follows.
39. Prepare the culture medium (500 mL M-199 with 10% FBS, antibiotics, and 10 mM nicotinamide).
40. The pancreatic cell clusters derived from 2 pig pancreata are resuspended in 350 mL culture medium in a 500 mL spinner flask.
41. Start to culture the cells at 37 °C, 5% CO_2 and rotate the stirrer wings at 72 rpm.
42. Replace the culture medium every 12 h.

3.2.1 Evaluation

1. Cell cluster count based on islet equivalent [15].
2. Viability (FDA/PI)

   Inspect 50 cell clusters and determine visually their individual viability, calculate their average viability.
3. Gross morphology scoring (Table 1) [15].

Table 1
Gross morphology scoring

	0	1	2
Shape	Flat	In between 0 and 2	Spherical
Border	Irregular	In between 0 and 2	Well rounded
Integrity	Fragmented	In between 0 and 2	Solid-compact
Uniformity of stain	Not uniform	In between 0 and 2	Uniform
Diameter	All <100 μm	In between 0 and 2	More than 10% >200 μm

3.3 Microencapsulation

Pre-run

1. Pour 1.8 % (wt/vol) alginate solution into a beaker.
2. Adjust pump flow, air flow, and distance between the droplet generator and $CaCl_2$ solution.
3. Run the pump to create an empty capsule.
4. Check the size of the empty capsule (approximately 500 μm).
5. If the size is not appropriate, adjust pump flow and/or air flow.

Run (Fig. 1)

1. Mix the porcine islets with 1.8 % (wt/vol) alginate.
2. Pump the homogenous mixture through a needle supplied with a coaxial airflow.
3. Cross-link the alginate beads with 109 mM $CaCl_2$ for 15 min.
4. Coat the alginate beads with 0.1 % (wt/vol) PLO for 6 min.
5. Coat the PLO-coated alginate capsules with a final layer of 0.18 % (wt/vol) alginate for 10 min.
6. Partially liquefy the microcapsules by incubation in a 55 mM sodium citrate solution.

4 Notes

4.1 Porcine Pancreas Procurement

1. Operation time of pancreas procurement is usually less than 10 min for 1 piglet.
2. The pancreas should be in contact with the oxygenated PFC for two-layer preservation.

4.2 Porcine Islet Isolation

1. Digestion time is usually less than 20 min.
2. Collection time is usually more than 30 min.

4.3 Microencapsulation

1. The distance between the droplet generator and $CaCl_2$ solution is usually approximately 4 cm.
2. Pump flow and air flow are changed for each droplet generator.

References

1. Bartlett ST, Markmann JF, Johnson P, et al (2015) Report from IPITA-TTS opinion leaders meeting on the future of beta cell replacement. Transplantation (in press)
2. Bruni A, Gala-Lopez B, Pepper AR et al (2014) Islet cell transplantation for the treatment of type I diabetes: recent advances and future challenges. Diabetes Metab Syndr Obes 7:211–223
3. Matsumoto S (2010) Islet cell transplantation for type 1 diabetes. J Diabetes 2:16–22
4. Elliott RB, Escobar L, Tan PLJ et al (2007) Live encapsulated porcine islets from a type 1

diabetic patient 9.5 yr after xenotransplantation. Xenotransplantation 14:157–161

5. Korbutt GS, Elliott JF, Ao Z et al (1996) Large scale isolation, growth and function of porcine neonatal islet cells. J Clin Invest 97:270–280

6. Calafiore R, Basta G (2014) Clinical application of microencapsulated islets: actual prospective on progress and challenges. Adv Drug Deliv Rev 67–68:84–92

7. Matsumoto S, Tan P, Baker J et al (2014) Clinical porcine islet xenotransplantation under comprehensive regulation. Transplant Proc 46:1992–1995

8. Matsumoto S, Takita M, Chaussabel D et al (2011) Improving efficacy of clinical islet transplantation with iodixanol-based islet purification, thymoglobulin induction, and blockage of IL-1beta and TNF-alpha. Cell Transplant 20:1641–1647

9. Matsumoto S, Noguchi H, Shimoda M et al (2010) Seven consecutive successful clinical islet isolations with pancreatic ductal injection. Cell Transplant 19:291–297

10. Shimoda M, Noguchi H, Fujita Y et al (2012) Improvement of collagenase distribution with the ductal preservation for human islet isolation. Islets 4:130–137

11. Qin H, Matsumoto S, Klintmalm GB et al (2011) A meta-analysis for comparison of the two-layer and university of Wisconsin pancreas preservation methods in islet isolation. Cell Transplant 20:1127–1137

12. Calafiore R, Basta G, Luca G et al (2006) Standard technical procedures for microencapsulation of human islets for graft into non-immunosuppressed patients with type 1 diabetes mellitus. Transplant Proc 38:1156–1157

13. Hillberg A, Kathirgamanathan K, Lam JBB et al (2013) Improving alginate-poly-L-ornithine-alginate capsule biocompatibility through genipin crosslinking. J Biomed Mater Res B Appl Biomater 101:258–268

14. Kitzmann JP, Law L, Shome A et al (2012) Real-time assessment of encapsulated neonatal porcine islets prior to clinical xenotransplantation. Xenotransplantation 19:333–336

15. Matsumoto S, Qualley SA, Goel S et al (2002) Effect of the two-layer (University of Wisconsin solution-perfluorochemical plus O2) method of pancreas preservation on human islet isolation, as assessed by the Edmonton isolation protocol. Transplantation 2002(74):1414–1419

Chapter 27

Microencapsulation of Parathyroid Cells for the Treatment of Hypoparathyroidism

Patricio Cabané Toledo, Ricardo L.Rossi, and Pablo Caviedes

Abstract

Cell encapsulation is an alternative to avoid rejection of grafted tissue, thus bringing an interesting alternative in cell therapy. It is particularly relevant in ailments where only the implant of small quantities of tissues is warranted. In such circumstances, the use of immunosuppressive therapy in patients implanted with tissues from donors is debatable, yet unavoidable at present in order to prevent rejection and/or sensitization of the host to the tissue, in turn jeopardizing the success of successive implants. Hence, a new line of thought, which aims to provide an immunoprivileged site for the grafted tissue, while at the same time insure its nutrition, as well as its survival and continued function, appears as a most attractive possibility. To achieve these goals, cells or tissues harvested for transplant could be encapsulated in biologically compatible matrices. Among the matrices currently in existence, sodium alginate is the most widely used polymer for tissue encapsulation.

In the present chapter, we present a technique used to encapsulate parathyroid tissue, for use as cell transplant therapy in patients with secondary hypoparathyroidism. With this procedure, implanted tissue survives and remains functional for up to 18 months.

Key words Cell encapsulation, Alginate, Microencapsulation, Parathyroid tissue, Hypoparathyroidism

1 Introduction

The most important complication in thyroid surgery is hypocalcemia. When it becomes permanent (1 %), [1] the medical treatment based on the oral administration of calcium and vitamin D generally yields poor clinical results, as it lacks the principle of control of a regulated variable. One therapeutic alternative appeared with the development of recombinant PTH for the treatment of these patients, yet this alternative is hindered by high costs and the lack of knowledge on the correct daily dose [2]. In an attempt to develop a definitive, curative treatment for hypoparathyroidism, some experimental studies have been carried out using allotransplantation of parathyroid tissue and cells in animal models and humans, with

Emmanuel C. Opara (ed.), *Cell Microencapsulation: Methods and Protocols*, Methods in Molecular Biology, vol. 1479,
DOI 10.1007/978-1-4939-6364-5_27, © Springer Science+Business Media New York 2017

or without immunosuppressant treatment [3–5]. At present, there are some groups working with stromal parathyroid stem cells, testing several differentiating protocols to improve secretion of PTH [6, 7]. Yet immunogenic rejection is still a problem [8].

Cell encapsulation (micro or macro encapsulation) allows isolated cells and tissue to be used as transplant therapy. Using such techniques, a critical goal undertaken is to avoid rejection and the consequent graft failure. To create these microspheres, sodium alginate, derived from sea algae, is widely used. Several gelation protocols for encapsulation do exist, that control different parameters, such as size, mechanical strength, permeability, and inflammatory reaction of the host towards the microspheres. These parameters can be manipulated with the composition of alginate, namely the use of high or low ratio of guluronic to mannuronic acid residues depending on the kind of original algae harvested, the cation used for gelation (Ca^{+2}, Ba^{+2}), and other supplements like polylysine or albumin in order to improve some mechanical properties. Each type of encapsulation technique may induce different cell reactions (differentiation, increased function, etc.). Hence, this must be specifically determined for any particular cell or tissue of interest. [9] For parathyroid cells, the most frequent gelation technique in cell therapy uses Barium Chloride to induce the polymerization of alginate. With this protocol, the spheres are formed with the necessary resistance to avoid the entrance of lymphocytes and complement to the graft, and a sufficient permeability to allow cell metabolites, nutrients and smaller molecules to freely flow between the encapsulated tissue and the external milieu. [5]

The aim of this chapter is to present a cryopreservation technique for parathyroid cells and tissue that ensures maximal survival post thawing, and an encapsulation technique used in the successful transplant of parathyroid tissue in a patient with severe hypoparathyroidism.

2 Materials

2.1 Cell Culture

1. 0.2% collagenase II.
2. Culture medium (DMEM/F12).

2.2 Tissue Cryopreservation

1. *Parathyroid tissue*: Obtained with the proper informed consent from patients with secondary hyperparathyroidism due to chronic kidney disease. This protocol has been used for the last 10 years in our institution, with approval of the Ethics Committee.
2. *Cryotubes (1.8 ml).*
3. *99.9% dimethyl sulfoxide (DMSO).*
4. *Autologous serum*: obtained by clotting and centrifugation at 1500 rpm for 15 min at 4 °C.

5. *99.7% isopropyl alcohol.*
6. *Cryopreservation medium for tissue*: Add 90% autologous patient serum and 10% dimethyl sulfoxide.
7. *Cryopreservation medium for cell suspension*: Add DMEM/F12, 20% fetal bovine serum and 10% dimethylsulfoxide.

2.3 Encapsulation

1. *2% w/v medium viscosity alginate (Sigma®)*: 2 g of alginate are weighed and diluted in 100 ml of distilled water. Later, it is mixed with a heated magnetic stirrer until the solution is homogeneous. The solution is later autoclaved for 30 m at 121 °C and 1 kg/cm^2, and later filtered by passage through 0.2 μm filters (*see* **Note 1**).
2. *20 mM barium chloride (BaCl2), Sodium Chloride (NaCl) 0.9% (pH 7.0–290 mOsm)*: (MW $BaCl_2$: 208.25) (MW NaCl: 58.44). Add 0.20825 g of $BaCl_2$ and 0.366 g of NaCl to 50 ml of distilled water and bring to pH 7.0 using 5 mM histidine solution.
3. *5 mM histidine*: (MW: 209.63). Add 0.524075 g of histidine to 50 ml of distilled water (5 mM). Sterilize by filtering through Millipore 0.22 or 0.45 μm.
4. *0.9% sodium chloride (NaCl)*: Weigh 0.9 g per 100 ml of distilled water. Sterilize by filtering through Millipore 0.22 or 0.45 μm.
5. *5 mM sodium sulfate (Na2SO4)*: (MW: 142.05). Weigh 0.213 g of Na_2SO_4 to 250 ml of distilled water (pH. 7.0). Sterilize by filtering through Millipore 0.22 or 0.45 μm.

Solutions 2–5 must be adjusted to 290 mOsm with 115 mM NaCl, and pH 7.0 with 5 mM histidine.

3 Methods

3.1 Primary Culture: Cell Suspension

1. Identify and reserve parathyroid parenchyma and mince to 1 × 1 mm pieces with scissors.
2. Then the pieces of tissues are enzymatically digested with 0.2% collagenase II and incubated for 30 min at 37 °C.
3. Subsequently the digested tissue is disintegrated by pipetting and then centrifuged at 1500 rpm for 10 min to remove the digestion medium.
4. The cells are then resuspended in culture medium and seeded onto plastic dishes.

3.2 Cryopreservation

3.2.1 Tissue Cryopreservation Procedure

1. In the operating room, cut the tissue into small pieces of 1 × 1 mm, using two scalpel blades.
2. Then place the tissue into 1.8 ml cryotubes (5–8 pieces per cryotube).

3. Transport the tissue to the laboratory, and fill each cryotube with 1.2 ml of a cryopreservation media.
4. Place the cryotubes in a cryopreservation container with 99.7% isopropyl alcohol.
5. Transfer the container into a –85 °C ultrafreezer for 2 h.
6. Finally transfer the cryotubes to liquid nitrogen (–198 °C).

3.2.2 Cell Cryopreservation Procedure

For experimental purposes or future cell transplants, we now describe the technique of cell cryopreservation.

1. Prepare a cell suspension as we describe before in Subheading 3.1.
2. Wash and centrifuge the cell suspension for 10 min at 500 rpm.
3. Resuspend the pellet in cryopreservation medium.
4. Transfer cell suspension into 1.8 ml cryotubes.
5. Place the cryotubes in a cryopreservation container with 99.7% isopropyl alcohol.
6. Transfer the container into a –85 °C ultrafreezer for 2 h.
7. Finally transfer the cryotubes to liquid nitrogen (–198 °C).

3.3 Encapsulation Technique

Encapsulation was carried out according to Schneider et al. [10], with modifications.

3.3.1 Cell Encapsulation

1. Prepare fresh or previously cryopreserved cell suspension.
2. Thaw the cell suspension in 1 min in a warm bath set at 37 °C.
3. Wash vigorously the cell suspension or tissue in PBS.
4. Wash and centrifuge the cell suspension for 10 min at 500 rpm.
5. Resuspend the pellet in 2% medium viscosity alginate at 4 °C (*see* **Note 2**). The number of cells depends on the cell type used (5×10^5–1×10^6 cells/ml alginate).
6. Place the alginate-cell suspension into a syringe (23G needle).
7. Drop the alginate-cell suspension into 30 ml of a 20 mM $BaCl_2$ solution placed in 45 ml tubes kept on ice (4 ° C) (*see* **Note 3**).
8. Alginate spheres of 2–3 mm are rapidly formed upon contact with the barium solution.
9. Incubate the spheres for 3 min at 37 °C in 6 mM Na_2SO_4.
10. Wash three times with a 0.9% NaCl solution.
11. Incubate in complete medium at 37 °C in culture flasks.

3.3.2 Tissue Encapsulation

This protocol was developed to carry out an allogeneic transplant of parathyroid tissue in a patient with severe postsurgical hypoparathyroidism, fully dependent on a continuous IV infusion of calcium to survive. After the 3rd catheter-derived sepsis, the surgical team decided to perform a microencapsulated parathyroid tissue allograft. [11]

1. Prepare fresh or previously cryopreserved parathyroid tissue.
2. Thaw the tissue in 1 min in a warm bath set at 37 °C.
3. Wash vigorously the cell suspension or tissue in PBS.
4. Wash and centrifuge the cell suspension for 10 min at 500 rpm.
5. Resuspend the pellet in 2 % medium viscosity alginate at 4 °C (*see* **Note 2**).
6. Place the alginate–tissue suspension into a syringe (14G needle).
7. Drop the alginate–tissue suspension into 30 ml of a 20 mM $BaCl_2$ solution placed in 45 ml tubes kept on ice (4 °C) (*see* **Note 3**).
8. Alginate spheres of 3–4 mm are rapidly formed upon contact with the Barium solution (Fig. 1).
9. Wash three times with a 0.9 % NaCl solution.
10. Incubate in complete medium at 37 °C in culture flasks until transplant.

After 3 months of incubation of 5 spheres reserved for this purpose, intact PTH levels were measured in the culture medium. The levels detected exceeded the maximum detection limit of the kit (>2500 pg/ml).

3.4 Transplant

The surgical procedure is performed in an operating room under local anesthesia (1 % lidocaine).

1. A 10-cm incision is made in the right forearm, exposing the underlying muscle and creating pockets of approximately 4 mm in depth with atraumatic avulsion for each sphere.

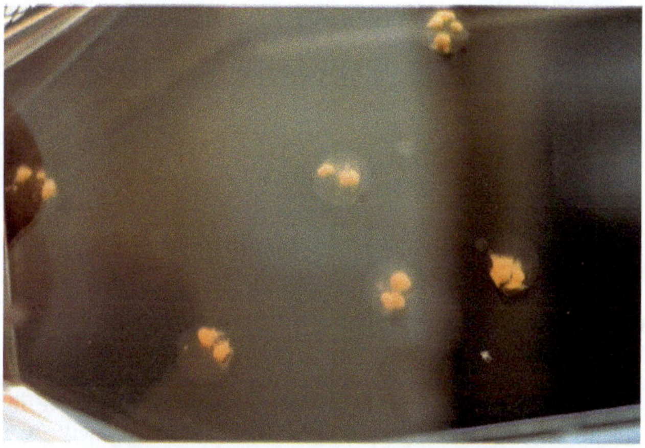

Fig. 1 Alginate spheres with encapsulated parathyroid tissue in cell culture flask. After 3 months of incubation of 5 spheres reserved for this purpose, intact PTH levels were measured in the culture medium. The levels detected exceeded the maximum detection limit of the kit (>2500 pg/ml) (Immulite 2000 intact PTH Siemens)

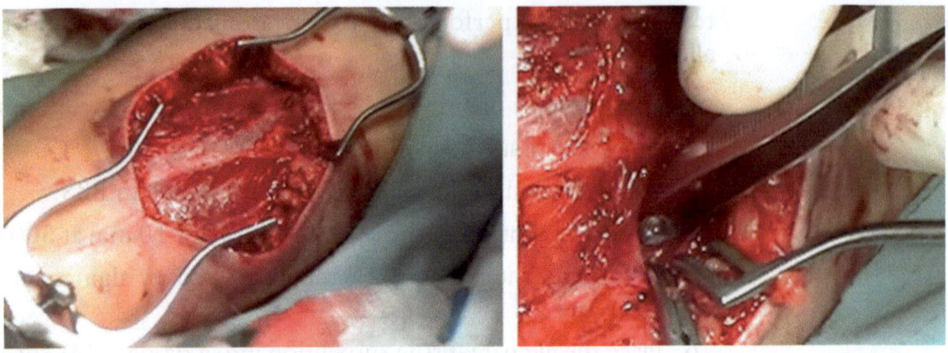

Fig. 2 Exhibition of forearm muscles (*left*) and microspheres with parathyroid tissue in pockets of forearm muscle (*right*). A demonstrative video can be seen in http://www.youtube.com/watch?v=JobLDGUcOt4

2. Deposit gently each sphere in a pocket.
3. The pockets were then closed with Biosyn 3-0.
4. Transplant between 20 and 30 microspheres, corresponding to approximately three normal parathyroid glands.
5. Finally, the skin was sutured by layers (Fig. 2).

4 Notes

1. The process of filtering alginate trough 0.22 Millipore filters may be quite difficult. A vacuum pump system may prove very useful to facilitate this procedure.
2. It is very important to work in 4 °C with all material and solutions. The container with cationic solution must be kept in ice all the time.
3. The ideal height to the alginate solution and tissue into the cationic solution, should be sufficient so that the droplet submerges completely, and not remain on the surface. It is also important to maintain a slow rocking movement in the cationic solution; otherwise the droplets with the alginate will not agglomerate. If there are remains of alginate on the surface, they must be removed.

References

1. Gac P, Cabané P, Amat J, Huidobro F, Rossi R, Rodríguez F, Ferrada C et al (2007) Incidencia de hipocalcemia pos tiroidectomía total. Rev Med Chil 135(1):26–30
2. Cusano NE, Rubin MR, McMahon DJ, Zhang C, Ives R, Tulley A, Bilezikian JP (2012) Therapy of hypoparathyroidism with PTH (1–84): a prospective four-year investigation of efficacy and safety. J Clin Endocrinol Metab 98(1):137–144
3. Hasse C, Klöck G, Schlosser A et al (1997) Parathyroid allo-transplantation without immunosuppression. Lancet 350:1296–1297

4. Tolloczko T, Wozniewicz B, Sawicki A, Gòrski A (1997) Allotransplantation of cultured human parathyroid cells: present status and perspectives. Transplant Proc 29:998–1000
5. Nawrot I, Wozniewicz B, Tolloczko T, Sawicki A, Gorski A, Chudzinski W et al (2007) Alotransplantation of cultured parathyroid progenitor cells without immunosuppression: clinical results. Transplantation 83(6):734–740
6. Bingham EL, Cheng SP, Woods Ignatoski KM, Doherty GM (2009) Differentiation of human embryonic stem cells to a parathyroid-like phenotype. Stem Cells Dev 18(7):1071–1080
7. Karaöz E, Arıkan S, Aksoy A, Ünal ZS, Adaş M, Adaş G, Duruksu G (2013) Stromal stem cells from parathyroid glands of patients with secondary hyperparathyroidism demonstrate higher telomerase activity and osteogenic differentiation ability than normal bone marrow derived stromal stem cells. Br J Medicine Medical Rese 3(3):654–680
8. Flechner SM, Berber E, Askar M, Stephany B, Agarwal A, Milas M (2010) Allotransplantation of Cryopreserved Parathyroid Tissue for Severe Hypocalcemia in a Renal Transplant Recipient. Am J Transplant 10:2061–2065
9. Wilson JL, Najia MA, Saeed R, McDevitt TC (2014) Alginate encapsulation parameters influence the differentiation of microencapsulated embryonic stem cell aggregates. Biotechnol Bioeng 111(3):618–631
10. Schneider S, Feilen P, Cramer H, Hillgartner M, Brunnenmeier F, Zimmermann H et al (2003) Beneficial effects of human serum albumin on stability and functionality of alginate microcapsules fabricated in different ways. J Microencapsul 20:627–636
11. Cabané P, Gac P, Amat J, Pineda P, Rossi R, Caviedes R, Caviedes P (2009) Allotransplant of microencapsulated parathyroid tissue in severe postsurgical hypoparathyroidism: a case report. Transplant Proc 41(9):3879–3883

Index

Emmanuel C. Opara (ed.), *Cell Microencapsulation: Methods and Protocols*, Methods in Molecular Biology, vol. 1479, DOI 10.1007/978-1-4939-6364-5, © Springer Science+Business Media New York 2017

MIX
Papier aus verantwortungsvollen Quellen
Paper from responsible sources
FSC® C105338

If you have any concerns about our products,
you can contact us on
ProductSafety@springernature.com

In case Publisher is established outside the EU,
the EU authorized representative is:
Springer Nature Customer Service Center GmbH
Europaplatz 3, 69115 Heidelberg, Germany

Printed by Libri Plureos GmbH
in Hamburg, Germany